Statistical
Thermodynamics
of Alloys

Statistical Thermodynamics of Alloys

N. A. Gokcen

Albany Research Center
Bureau of Mines
U.S. Department of the Interior
Albany, Oregon

Plenum Press • New York and London

Library of Congress Cataloging in Publication Data

Gokcen, N. A.
  Statistical thermodynamics of alloys.

  Includes bibliographies and index.
  1. Alloys. 2. Statistical thermodynamics. I. Title.
TN690.G673   1986   669′.94   86-6628
ISBN-13: 978-1-4684-5055-2   e-ISBN-13: 978-1-4684-5053-8
DOI: 10.1007/978-1-4684-5053-8

© 1986 Plenum Press, New York
Softcover reprint of the hardcover 1st edition 1986
A Division of Plenum Publishing Corporation
233 Spring Street, New York, N.Y. 10013

TO

Altan Gokcen, Pianist

Selma Gokcen, Cellist

with devotion and admiration

Preface

This book is intended for scientists, researchers, and graduate students interested in solutions in general, and solutions of metals in particular. Readers are assumed to have a good background in thermodynamics, presented in such books as those cited at the end of Chapter 1, "Thermodynamic Background." The contents of the book are limited to the solutions of metals + metals, and metals + metalloids, but the results are also applicable to numerous other types of solutions encountered by metallurgists, materials scientists, geologists, ceramists, and chemists. Attempts have been made to cover each topic in depth with numerical examples whenever necessary.

Chapter 2 presents phase equilibria and phase diagrams as related to the thermodynamics of solutions. The emphasis is on the binary diagrams since the ternary diagrams can be understood in terms of the binary diagrams coupled with the phase rule, and the Gibbs energies of mixing. The calculation of thermodynamic properties from the phase diagrams is not emphasized because such a procedure generally yields mediocre results. Nevertheless, the reader can readily obtain thermodynamic data from phase diagrams by reversing the detailed process of calculation of phase diagrams from thermodynamic data. Empirical rules on phase stability are given in this chapter for a brief and clear understanding of the physical and atomistic factors underlying the alloy phase formation.

Chapter 3 is a condensed and thorough summary of statistical thermodynamic methods necessary for pursuing the remaining chapters. Quantum mechanical postulates evoked in deriving a number of useful equations should not discourage the reader once he accepts the proposition that the end justifies the means for such postulates. The methods of quantum mechanics are therefore not always inductive and deductive, but often postulative. Several excellent texts are cited for the reader interested in delving deeper into statistical and quantum mechanics.

Theories of solutions are presented in Chapter 4 in their general and simpler forms that are believed to be free of incorrect earlier theories based on unrealistic assumptions and methods. Additional theories, their expansion, modification, and application, are given in the remaining chapters. Abstract models such as cell theories have been left alone, as perhaps they should be. Permutations of single bonds and half bonds are left to their demise because they yield results inconsistent with the actual numbers of configurations, which are enumerable for two-dimensional crystals. Additional results for the actual numbers of configurations in highly ordered structures are necessary to obtain a firmer basis for the order–disorder phenomena that are discussed in Chapter 5. However, the existing results are already based on realistic numbers of configurations, as discussed in detail in Chapter 5.

Interstitial solutions play important roles in theories and applications of phase equilibria. Therefore, Chapter 6, the longest chapter in the book, is devoted to this topic, and also contains computer calculations of the Fe–C phase diagram. The temptation was great here to delve into hydrogen storage in metals and alloys because in less than a quarter of a century a considerable portion of our energy will be based on hydrogen generated by solar energy. The Wagner theory on ternary interstitial solutions is presented here rather than in Chapter 4. Unfortunately, it is applicable only to interstitial or near-interstitial solutes dissolved in ideal binary metal solutions. Progress in this field is necessary to account for large deviations from symmetrical behavior in binary and dilute multicomponent solutions.

Chapter 7 deals with semiconductors within the area of expertise of the author. A new interpretation of the thermodynamic behavior of electrons and holes is given in this chapter. It was enticing to delve into p–n junctions in general, and solar cells in particular. A long section on tandem solar cells is presented to show that nearly perfect conversion of solar energy into electrical energy is possible within the limitations imposed by the second law of thermodynamics. The discussion of solar energy conversion here and that of hydrogen storage in Chapter 6 complement each other in the current atmosphere of intense research effort devoted to renewable sources of energy.

Appendixes A and B contain recent and evolving theories of alloy phase formation, and methods of estimation and correlation of thermodynamic properties. Appendix C deals with the correlation of thermodynamic properties in dilute solutions. The remaining appendixes contain selected properties of the elements, selected phase diagrams and their thermodynamic properties, a comprehensive list of symbols, and the periodic table.

The effort to write this book has been long and arduous. I am grateful for the patience, tolerance, and endurance of my family during the long hours that I spent reading numerous publications, writing, and correcting the manuscript. I am indebted to several U.S. Bureau of Mines employees, in particular Mrs. Sharon L. Brittain for her careful preparation of the manuscript, Mr. George E. Daut for his meticulous proofreading, and Mr. J. L. Wilderman, who kindly traced most of the figures as well as the periodic chart at the end of the book.

Professor Marvin C. Y. Lee, formerly of the Beijing Institute of Aeronautics and Astronautics, People's Republic of China, and currently a member of the research staff of the U.S. Bureau of Mines, kindly read the entire manuscript and made valuable suggestions. He also read the proofs and made the text as error-free as humanly possible. Very useful comments and criticisms were offered on Chapter 6 by Professor Taiji Nishizawa of Tohoku University, and on Chapter 7 by Professor Melvin Cutler of Oregon State University.

Expert and invaluable help on my summary of the Engel–Brewer Theories in Appendix A was very generously offered by Professor L. Brewer of the University of California. Dr. A. R. Miedema of Philips Research Laboratories in Holland kindly read and corrected Appendix B on the estimation of the enthalpy of alloy formation, a technique that was developed by his diligent group in Eindhoven. Finally, I am grateful to the staff of Plenum Press for their outstanding effort in publishing this book..

N. A. Gokcen

Albany, Oregon

Contents

Statistical
Thermodynamics
of Alloys

1

Thermodynamic Background

Introduction

The areas of thermodynamics with which this book is concerned are those dealing with the interrelationships of energy and the changes in properties of matter upon mixing various components to form various phases. A component in this book is usually a pure element, but occasionally it may be an intermetallic or a metal–metalloid compound of well-defined stoichiometry. A single phase is a substance that has uniform physical and compositional properties. At a phase boundary these properties change abruptly.

The concepts of thermodynamics are based on empirical observations of the macroscopic properties of matter and the results of these observations are expressed in mathematical functions. Statistical thermodynamics of solutions strives to obtain thermodynamic relationships based on the molecular behavior of matter. The mathematical background needed for our purposes will be developed as we proceed; a detailed background may be found elsewhere.[1,2] A basic knowledge of thermodynamics, as presented in Refs. 1, 3, or 4, is essential in pursuing the subject matter in this book.

We shall define a *term* as representing a clear and concise phenomenon, property, or concept, and designate it by a symbol. All the symbols in this book are defined as they occur in the text, and are compiled and redefined in Appendix F. A thermodynamic system, or briefly a system, is the substance under investigation, separated from the surroundings by rigid or movable boundaries that may or may not exchange energy or matter or both with the surroundings. The composition of a system is usually expressed in mole fractions, x_i, but the molality, i.e., moles per kilogram solvent, is used for solutions of electrolytes, and particles per cubic centimeter, for semiconductors. An extensive property $\mathscr{G}$ is a function of two of the three variables consisting of pressure P, volume V, and temperature T, as well as the number of moles n_i for an open system; thus,

$$\mathscr{G} = \mathscr{G}(P, T, n_1, n_2, \ldots, n_i) \tag{1.1}$$

where there are i chemically non-compound-forming species. For brevity we shall occasionally write n for all the components. The relationship correlatng P, V, and T is called an equation of state. An intensive property

Table 1.1. Physical and Chemical Constants[a]

Name	Symbol	Value and units
Fundamental constants:		
Ice + water + vapor point		
(triple point of water = 0.0100°C)		273.1600 K
Molar volume of perfect gas		
(0°C, 1 atm)	$V°$	22,413.83 cm^3 mole^{-1},
Avogadro number	N_0	6.022045×10^{23} mole^{-1},
		(molecules mole^{-1})
Gas constant	R	8.31441 J mole^{-1} K^{-1},
		1.98719 cal mole^{-1} K^{-1},
		82.05684 cm^3 atm mole^{-1} K^{-1},
Boltzmann constant	$k = R/N_0$	1.380662×10^{-23} J K^{-1}
Faraday constant	F	96484.6 C g-equivalent^{-1},
		23,060.4 cal
		V^{-1} g-equivalent^{-1}
Velocity of light in vacuum	ς	2.997925×10^8 m sec^{-1}
Planck constant	h	6.626189×10^{-34} J sec
Proton charge (electrtonic charge)	e^+	1.60219×10^{-19} C
Permittivity of vacuum	$e_0 = 10^7/(4\pi\varsigma^2)$	8.85419×10^{-12} C
		m^{-1} V^{-1}
Defined constants, and conversion factors:		
Thermochemical calorie	cal	4.1840 J
Standard gravity	$g°$	9.80665 m sec^{-2}
Atmosphere of pressure	atm	101,325 $\mathscr{N}$ m^{-2} (Pascals)
Torricelli of pressure	Tor (or Torr)	1/760 atm
Newton	$\mathscr{N}$	10^5 dynes
Joule	J	10^7 ergs
Milliliter	ml	1.000028 cm^3
Electron volt	eV	23,060.4 cal mole^{-1}
		= 96,484.6 J mole^{-1}
Kelvins (formerly degrees Kelvin)	K	273.1500 + °C

[a]From D. H. Whiffen, *Manual of Symbols and Terminology for Physiochemical Quantities and Units,* 1979 Ed., IUPAC, Pergamon Press (1979). See also *Dimensions/NBS,* January 1974, a pamphlet by NBS.

G may be expressed by

$$G = G(P, T, x_1, x_2, \ldots, x_{i-1}) \tag{1.2}$$

where there are $i - 1$ composition variables because $\sum x_i = 1$, and one of the composition variables is not independent. We shall write $\sum$ without its limits for brevity when the subscript i, as in x_i, is used as the running index for components. A single phase is defined to be homogeneous, in which P, T, and x_i are uniform throughout. A process or a change in a thermodynamic property is denoted by Δ, which, for any property such as $\varnothing$, signifies that $\Delta\varnothing = \varnothing(\text{final}) - \varnothing(\text{initial}) = \varnothing_2 - \varnothing_1$. The symbol Δ will always refer to the final property minus the initial property. Equations (1.1) and (1.2) are for any thermodynamic property in this section. The physical and chemical constants useful in expressing various thermodynamic properties are listed in Table 1.1.

Consequences of Laws of Thermodynamics

The first law of thermodynamics is a special consequence of the law of conservation of energy in physics. We enunciate it by stating that there is a single-valued function E, called the energy of a system, $E = E(V, T, n)$, so that when E changes from $E_1 = E(V_1, T_1, n)$ to $E_2 = E(V_2, T_2, n)$, then $\Delta E = E_2 - E_1$ is always the same irrespective of the process used in going from (V_1, T_1, n) to (V_2, T_2, n). For a closed homogeneous system in which n_i are not variables, we write

$$dE = dq - P\,dV + dW + dW' \cdots \tag{1.3}$$

where q is the thermal energy, $-P\,dV$ is the work of compression or expansion, and $W, W', \ldots$ are other types of work. At constant volume, if other types of work are absent for a closed system, $dE_v = dq_v$, and the heat capacity at constant volume is $(\partial E/\partial T)_v = C_v$, and therefore

$$dE = C_v\,dT - P\,dV; \quad [C_v = (\partial E/\partial T)_v] \tag{1.4}$$

Thus, E is a function of T and V, hence $E = E(T, V)$, because T and V are present as dT and dV in equation (1.4). We add PV to both sides of equation (1.3) without $W, W', \ldots$ to write

$$d(E + PV) \equiv dH = dq + V\,dp; \quad (H \equiv E + PV) \tag{1.5}$$

where $H \equiv E + PV$ is the enthalpy by definition. At constant pressure, $dH_p = dq_p$, and the heat capacity at constant pressure, C_p, is defined by $C_p \equiv (\partial H/\partial T)_p = (\partial q/\partial T)_p$. Substitution of C_p in equation (1.5) gives

$$dH = C_p \, dT + V \, dP; \qquad [C_p = (\partial H/\partial T)_p] \qquad (1.6)$$

where $H = H(T, P)$; therefore, H is a function of the more convenient variables T and P than the variables T and V for E in equation (1.4).

The existence of entropy is an important consequence of the second law of thermodynamics. The entropy S is given by

$$dS = \frac{dq(\text{reversible})}{T} = \frac{dE + P \, dV}{T} \qquad (1.7)$$

Consequently, the equations for dE and dH are

$$dE = T \, dS - P \, dV \qquad (1.8)$$

$$dH = d(E + PV) = T \, dS + V \, dP \qquad (1.9)$$

Comparison of equations (1.6) and (1.9) shows that for pure condensed substances in their stable states at $P = 1$ atm $= 101{,}325$ Pa, the standard entropy $S°$ is calculated from

$$S° - S°(T = 0) = \int_0^T \frac{C_p°}{T} \, dT \qquad (1.10)$$

where $S°(T = 0)$ is taken to be zero for a pure stable single component in a stable crystalline form on the basis of the third law convention. The third law requires that $S°$ approach zero about as fast or faster than T approaches zero. Some noncrystalline but annealable pure components also obey the third law convention. If there are phase transitions at T_1 and T_2 with $\Delta H_1°$ and $\Delta H_2°$, equation (1.10) must be modified into

$$S° = \int_0^{T_1} \frac{C_p°}{T} \, dT + \left(\frac{\Delta H_1°}{T_1}\right) + \int_{T_1}^{T_2} \frac{C_p'°}{T} \, dT + \left(\frac{\Delta H_2°}{T_2}\right) + \int_{T_2}^{T} \frac{C_p''°}{T} \, dT \qquad (1.11)$$

where $C_p°$, $C_p'°$, and $C_p''°$ are the heat capacities in the temperature ranges indicated by the limits of integration.

Helmholtz and Gibbs Energies

Equation (1.8) can be modified by subtracting $d(TS)$ from dE; thus,

$$d(E - TS) \equiv dA = -S\,dT - P\,dV; \qquad (A \equiv E - TS) \qquad (1.12)$$

where $A \equiv E - TS$ is the Helmholtz energy, formerly called the work function. Likewise, subtraction of $d(TS)$ from dH in equation (1.9) gives

$$d(H - TS) \equiv dG = V\,dP - S\,dT; \qquad (G \equiv H - TS) \qquad (1.13)$$

where $G \equiv H - TS$ is the Gibbs energy. For an ideal gas at a fixed temperature,

$$dG = V\,dP = RT d \ln P \qquad (1.14)$$

from which, by integration,

$$G = G°(P = 1) + RT \ln P \qquad (1.15)$$

where $G°$ is the standard Gibbs energy at standard pressure so that $G°$ is a function of temperature only, i.e., $G° = G°(T)$.

If a gas is not ideal, then a corrected pressure, called the fugacity, F, is used to obtain the correct values of G:

$$dG = RT d \ln F; \qquad \text{or} \quad G = G° + RT \ln F \qquad (1.16)$$

This is the definition of fugacity wherein $G°$ refers to the unit fugacity, and again $G°$ is a function of T only. For example, if the equation of state is $PV = ZRT$ where Z is the compressibility factor, and if for moderate pressures $Z = 1 + \alpha P$, with α as a function of T, i.e., $\alpha = \alpha(T)$, then the substitution of $V = ZRT/P = (1 + \alpha P)RT/P$ in $dG = V\,dP$ yields $dG = RT(d \ln P + \alpha\,dP)$; hence, with $dG = RT d \ln F$, we obtain $d \ln(F/P) = \alpha\,dP$, and since for $P \to 0$, $F/P = 1$, the integration of this equation from $P \to 0$ to higher values of P yields

$$\ln(F/P) = \alpha P = Z - 1 \qquad (1.17)$$

For example, at 300 K and 100 atm, the correct value of F for hydrogen is 106.3 and equation (1.17), with $Z = 1.061$ at 100 atm, also yields 106.3. At much greater pressures of hydrogen, e.g., hydrogen released from metals and alloys into confined cracks or crevices, it is necessary to use Z as a power series in P and carry out the foregoing procedure in detail to obtain the values of F.

Partial Molar Gibbs Energy

If a function $\mathscr{G} = \mathscr{G}(P, T, n_1, n_2, n_3)$ is homogeneous with respect to its variables n_1, n_2, n_3, then by the definition of homogeneous functions,

$$\mathscr{G}(P, T, kn_1, kn_2, kn_3) = k^\alpha \mathscr{G}(P, T, n_1, n_2, n_3) \tag{1.18}$$

where α is the degree of homogeneity of $\mathscr{G}$ in n_1, n_2, n_3. For example, $\mathscr{G} = n_1^2 + n_2^2$ is a second-degree homogeneous function, but $\mathscr{G} = n_1 + n_2^2$ is not homogeneous. For extensive thermodynamic properties, α is one, but $\mathscr{G}$ divided by n_i or $(n_1 + n_2 + n_3)$ is an intensive property so that $G = \mathscr{G}/(n_1 + n_2 + n_3)$ is a molar property and α is zero for G; hence,

$$G(P, T, kn_1, kn_2, kn_3) = G = G(P, T, x_1, x_2) \tag{1.19}$$

G is then a function of P, T, and two of the three independent composition variables such as x_1 and x_2. For example, $\mathscr{G} = An_1 + Bn_2$, with A and B as constants, is a homogeneous function of the first degree in n_1 and n_2 but $G = \mathscr{G}/(n_1 + n_2) = A[n_1/(n_1 + n_2)] + B[n_2/(n_1 + n_2)]$ is a homogeneous function of the zeroth degree. The coefficients of A and B are x_1 and x_2, respectively, and since $x_1 + x_2 = 1$, it is evident that G is a function of either x_1 or x_2, e.g., $G = Ax_1 + B(1 - x_1)$. An interesting property of an extensive thermodynamic function $\mathscr{G}$ is that

$$n_1 \frac{\partial \mathscr{G}}{\partial n_1} + n_2 \frac{\partial \mathscr{G}}{\partial n_2} + n_3 \frac{\partial \mathscr{G}}{\partial n_3} = \alpha \mathscr{G} = \mathscr{G} \tag{1.20}$$

where $\alpha = 1$, since the degree of homogeneity of $\mathscr{G}$ is unity for an extensive thermodynamic property. Equation (1.20) is the formal statement of the Euler theorem on homogeneous functions as can be verified by using an example such as $\mathscr{G} = An_1^2 + Bn_2^2$ (for proof see Ref. 1). The total differential of the extensive Gibbs energy $\mathscr{G} = \mathscr{G}(P, T, n_1, n_2, n_3)$ at constant P and T is

$$d\mathscr{G} = \frac{\mathscr{G}}{\partial n_1} dn_1 + \frac{\mathscr{G}}{\partial n_2} dn_2 + \frac{\partial \mathscr{G}}{\partial n_3} dn_3 \tag{1.21}$$

If, purely for brevity, we write

$$\frac{\partial \mathscr{G}}{\partial n_1} = \bar{G}_1, \frac{\partial G}{\partial n_2} = \bar{G}_2, \ldots \tag{1.22}$$

and call $\bar{G}_i$ the partial molar Gibbs energy of i, then equations (1.20) and (1.21) respectively become

$$n_1\bar{G}_1 + n_2\bar{G}_2 + n_3\bar{G}_3 = \varnothing \qquad (1.23)$$

$$d\varnothing = \bar{G}_1\,dn_1 + \bar{G}_2\,dn_2 + \bar{G}_3\,dn_3 \qquad (1.24)$$

It is sufficient to write $\bar{G}_i$ as G_i, but the use of a bar in $\bar{G}_i$ is an established practice. Differentiation of equation (1.23) yields $d\varnothing = \sum n_i\,d\bar{G}_i + \sum \bar{G}_i dn_i$, and comparison with equation (1.24) yields the Gibbs–Duhem relation, i.e.,

$$n_1\,d\bar{G}_1 + n_2\,d\bar{G}_2 + n_3\,d\bar{G}_3 = 0 \qquad (1.25)$$

Division of equations (1.23) and (1.25) by $\sum n_i$ yields two important and convenient equations:

$$x_1\bar{G}_1 + x_2\bar{G}_2 + x_3\bar{G}_3 = G \qquad (1.26)$$

$$x_1\,d\bar{G}_1 + x_2\,d\bar{G}_2 + x_3\,d\bar{G}_3 = 0 \qquad (1.27)$$

Equations (1.25) and (1.27), first derived by Gibbs (1875) and later independently by Duhem (1886), are called the Gibbs–Duhem relations. It should be noted that n_i ($i = 1, 2, \ldots$) are the variables of state when a phase is open.

It is important to bear in mind that equations (1.18) to (1.27) are also applicable to other thermodynamic properties such as E, H, S, and C_p.

Activity

For a pure liquid or solid (crystalline) phase, the Gibbs energy $[G_i(\text{l, or c})]$ is equal to the Gibbs energy of its vapor $G_i(\text{g})$ at equilibrium; hence, from equation (1.16),

$$G_i^\circ(\text{l, or c}) = G_i(\text{g}) = G_i^\circ(\text{g}) + RT \ln F_i^* \qquad (1.28)$$

where F_i^* refers to the fugacity over the pure condensed phase, and $G_i^\circ(\text{g})$ is a function of temperature only, i.e., $G_i^\circ(\text{g}) = G_i^\circ(T)$ and $G_i^\circ(\text{l, or c}) = G_i^\circ(P, T)$. However, $G_i^\circ(\text{l, or c})$ is so weakly dependent on P that for most purposes this dependence for condensed phases will be generally ignored at ordinary pressures. When the liquid or the solid is a solution containing component i as one of the components, then

$$\bar{G}_i(\text{l, or c}) = G_i^\circ(\text{g}) + RT \ln F_i \qquad (1.29)$$

where F_i without the superscript * is the fugacity over the solution. Solving for $G_i°(g)$ from equation (1.28), i.e., $G_i°(g) = G_i°(l, \text{ or c}) - RT \ln F_i^*$ and eliminating $G_i°(g)$ in equation (1.29) leads to

$$\bar{G}_i(l, \text{ or c}) = G_i°(l, \text{ or c}) + RT \ln F_i/F_i^* \qquad (1.30)$$

The ratio F_i/F_i^*, representing the fugacity of component i over a solution divided by the fugacity over pure i, is called the activity of i in solution a_i; therefore,

$$\bar{G}_i(l, \text{ or c}) = G_i°(l, \text{ or c}) + RT \ln a_i; \qquad (a_i = F_i/F_i^*) \qquad (1.31)$$

The term $G_i°(l, \text{ or c})$ refers to the standard state, which is the pure stable condensed phase at 1 atm. Therefore, $G_i°$ for brevity without the parenthetic notations is a function of T only. For the standard state, the activity is always unity.

The dimensionless quantity that relates the activity to the mole fraction is denoted by γ_i and called the activity coefficient; it is defined by

$$a_i \equiv \gamma_i x_i \approx P_i/P_i^* \qquad (1.32)$$

where the approximate equality is due to the close equality of F_i/F_i^* and P_i/P_i^* at low pressures.

A component in a solution obeys Raoult's law (1887) when its activity and mole fraction are equal, i.e., $a_i = x_i$, or $\gamma_i = 1$. It has been observed experimentally that Raoult's law is obeyed in the limiting case when x_i approaches unity, i.e., $a_i \rightarrow x_i$ or $\gamma_i \rightarrow 1$ when $x_i \rightarrow 1$. The usefulness of this limiting law is that γ_i is unity in a small but finite range of composition approaching $x_i \rightarrow 1$ because otherwise $\gamma_i = 1$ is the definition of the standard state. The activity scale based on the pure component is called the Raoultian activity.

The activity of component i becomes also proportional to its mole fraction when x_i approaches zero, i.e., $a_i = \gamma_i° x_i$ for $x_i \rightarrow 0$ with $\gamma_i°$ becoming a constant, hence the superscript (°). Combination of this equation with the definition of activity yields

$$a_i \equiv F_i/F_i^* = \gamma_i° x_i; \qquad F_i = F_i^* \gamma_i° x_i; \qquad P_i \approx P_i^* \gamma_i° x_i = k' x_i \qquad (1.33)$$

where the last equality is the classical statement of Henry's law (1800), i.e., the pressure of i over a dilute solution of i is proportional to its mole fraction if $F_i \approx P_i$ and $F_i^* \approx P_i^*$, with k' as the proportionality constant. Substitution of $a_i = \gamma_i° x_i$ in equation (1.31), with $G_i° \equiv G_i°(l, \text{ or c})$, gives

$$\bar{G}_i = G_i° + RT \ln \gamma° + RT \ln x_i \equiv G_i^\bullet + RT \ln x_i; \qquad (x_i \rightarrow 0) \qquad (1.34)$$

where $G_i^{\bullet}$ is given by

$$G_i^{\bullet} = G_i^{\circ} + RT \ln \gamma_i^{\circ} \qquad (1.35)$$

As x_i increases, the deviation from the Henrian behavior of equation (1.34) is accounted for by using the activity coefficient f_i defined by

$$\bar{G}_i = G_i^{\bullet} + RT \ln f_i x_i; \qquad \mathring{a}_i = f_i x_i \qquad (1.36)$$

where $\mathring{a}_i$ is called the Henrian activity. Since $\bar{G}_i$ is independent of the choice of standard states, $\bar{G}_i$ in equations (1.31) and (1.36) are equal at any concentration; therefore,

$$G_i^{\circ} - G_i^{\bullet} = -RT \ln \frac{a_i}{\mathring{a}_i} = -RT \ln \frac{\gamma_i}{f_i} \qquad (1.37)$$

The left-hand side is a constant at a given temperature; therefore, the ratio of activities or activity coefficients of i based on these two different scales is also a constant independent of composition. The left-hand side, from equation (1.34), is $-RT \ln \gamma_i^{\circ}$; hence,

$$\gamma_i^{\circ} = a_i / \mathring{a}_i = \gamma_i / f_i \qquad (1.38)$$

Therefore,

$$\mathring{a}_i = a_i / \gamma_i^{\circ} = F_i / F_i^* \gamma_i^{\circ} \qquad (1.39)$$

Thus, the activity on one scale differs from that on another scale by a constant factor at a given temperature when the concentrations are in mole fractions. The activities and the activity coefficients based on Raoult's law and Henry's law are summarized as follows:

Basis	Definition	Standard state	Reference state
Raoult's law	$F_i / F_i^* = a_i = \gamma_i x_i$	Pure i, i.e., $a_i = 1$ for $x_i = 1$	Unnecessary
Henry's law	$\mathring{a}_i = F_i / F_i^* \gamma_i^{\circ} = f_i x_i$	$\mathring{a}_i = 1$ for $x_i = 1$ hypothetical (physically nonexisting)	$x_i \to 0$ (for which $f_i = 1$)

It should be noted that the standard state for both activity bases is the state of unit activity. The state of unit activity for the Henrian scale is physically

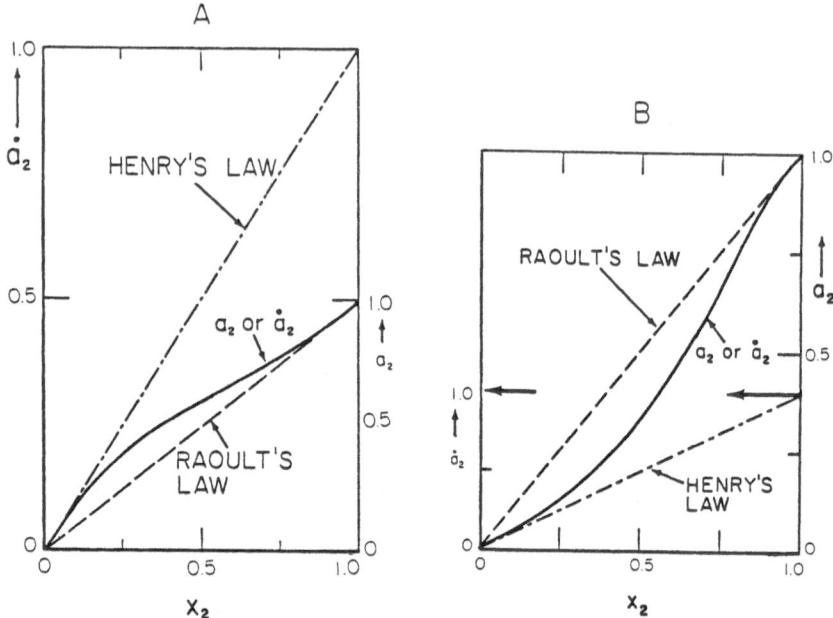

Figure 1.1. Activity of component 2 in hypothetical binary solutions. (A) Deviation from Raoult's law is positive; from Henry's law, negative. (B) Deviation from Raoult's law is negative; from Henry's law, positive. Vertical right scale is for a_2 with standard state as pure component 2, and vertical left scale for $\mathring{a}_2$ with reference state $x_2 \to 0$. The activity curve for component 1 is not shown, but it is similar to that for component 2 with $a_1 = 1$ at $x_2 = 0$, and $a_1 = 0$ at $x_2 = 1$, and having a shape as dictated by the Gibbs–Duhem relation.

nonexisting, except in the unlikely case when the solution is ideal in the Raoultian scale. Further, there is no need for the reference state for the Raoultain scale, but for the Henrian scale, the reference state is the infinitely dilute solution of i in a solvent. The activities based on these scales are illustrated in Fig. 1. The deviation from Raoult's law is positive for component i when γ_i is greater than 1, and negative when γ_i is smaller than 1. Similarly, the deviation from Henry's law is positive when f_i is greater than 1, and negative when f_i is smaller than 1. Generally, when $\gamma_i > 1$, then $f_i < 1$, and vice versa in the vicinity where the deviation from Henry's law starts.

Variation of Activity with Pressure and Temperature

Equation (1.13) may be written in the same form for either $\bar{G}_i$ or G_i°; hence, at constant T and composition,

$$\frac{\partial(\bar{G}_i - G_i^\circ)}{\partial P} = \bar{V}_i - V_i^\circ = RT\frac{\partial \ln a_i}{\partial P} = RT\frac{\partial \ln \gamma_i}{\partial P} \tag{1.40}$$

Both $\bar{V}_i$ and V_i° are small for condensed phases, and their difference is even smaller; therefore, a_i and γ_i for real solutions vary insignificantly at pressure changes of a few bars. For an ideal solution, $\gamma_i = 1$; hence, $\bar{V} = V_i^\circ$.

The activity a_i varies with temperature according to

$$\ln a_i = \frac{\bar{G}_i - G_i^\circ}{RT} = \frac{\bar{H}_i - H_i^\circ}{RT} - \frac{\bar{S}_i - S_i^\circ}{R} \tag{1.41}$$

where $\bar{G}_i = \bar{H} - T\bar{S}_i$ and $G_i^\circ = H_i^\circ - TS_i^\circ$. Differentiation of $\bar{G}_i/T$ with respect to temperature gives

$$\frac{\partial(\bar{G}_i/T)}{\partial T} = -\frac{\bar{H}_i}{T^2} + \frac{\partial \bar{H}_i}{T\partial T} - \frac{\partial \bar{S}_i}{\partial T} = -\frac{\bar{H}_i}{T^2} \tag{1.42}$$

where the second and third terms after the first equal sign cancel out because $\partial\bar{H}_i/\partial T = \bar{C}_{pi}$, and $\partial\bar{S}_i/\partial T = \bar{C}_{pi}/T$. We write a similar equation for G_i° and subtract it from equation (1.42) to obtain

$$\frac{\partial \ln a_i}{\partial T} = \frac{\partial \ln \gamma_i}{\partial T} = -\frac{\bar{H}_i - H_i^\circ}{RT^2}, \quad \text{or} \quad \frac{R\partial \ln a_i}{\partial(1/T)} = \bar{H}_i - H_i^\circ \tag{1.43}$$

For an ideal solution $a_i = x_i$ and the left side of this equation is zero because x_i is a variable independent of T in $\bar{G}_i = \bar{G}_i(P, T, x_1, x_2, \ldots)$; hence, $\bar{H}_i = H_i^\circ$, and the substitution of this equality in equation (1.41) gives

$$\bar{G}_i - G_i^\circ = RT \ln x_i = -T(\bar{S}_i - S_i^\circ) \quad \text{(ideal)} \tag{1.44}$$

Depression and Elevation of Freezing Point

The freezing point of a dilute solution is defined as the beginning or the onset of freezing. We designate a solvent as component 1, its freezing point in pure state as T_{fr}, a solute as component 2, and the onset of freezing of solution as T, so that $T_{fr} - T = \Delta T_{fr}$ is the depression or the elevation of the freezing point. We let x_2 be the mole fraction of component 2 in the binary liquid phase and y_2, the mole fraction in the binary crystal phase. For dilute solutions, the solvent obeys Raoult's law; hence,

$$\bar{G}_1(l) \equiv G_1^\circ(l) + RT \ln(1 - x_2) = \bar{G}_1(c) \equiv G_1^\circ(c) + RT \ln(1 - y_2) \quad (1.45)$$

Rearrangement of $G_1^{\circ}(c)$ and $G_1^{\circ}(l)$ of this equation gives $G_1^{\circ}(c) - G_1^{\circ}(l) \equiv H_1^{\circ}(c) - H_1^{\circ}(l) - T[S_1^{\circ}(c) - S_1^{\circ}(l)] \equiv \Delta H_{fr}^{\circ} - T\Delta S_{fr}^{\circ}$ where $\Delta H_{fr}^{\circ} = H_1^{\circ}(c) - H_1^{\circ}(l)$ is the enthalpy of freezing, and $\Delta S_{fr}^{\circ} = \Delta H_{fr}^{\circ}/T_{fr}$ is the entropy of freezing, and then the substitution for $G_1^{\circ}(c) - G_1^{\circ}(l)$ from equation (1.45) yields

$$\ln(1 - x_2) - \ln(1 - y_2) = \frac{\Delta H_{fr}^{\circ}}{RT_{fr}}\left[\frac{\Delta T_{fr}}{T}\right] \tag{1.46}$$

When x_2 and y_2 are very small, the left side becomes $y_2 - x_2$ and further, T becomes very close to T_{fr}; therefore,

$$\ln\frac{1 - x_2}{1 - y_2} \approx y_2 - x_2 \approx \frac{\Delta H_{fr}^{\circ}}{R}\left(\frac{\Delta T_{fr}}{T_{fr}^2}\right) \tag{1.47}$$

If the solubility y_2 is nearly zero, we obtain the familiar equation for the depression of freezing point for many ambient solutions. For alloys, however, when y_2 is greater than x_2, the melting point is elevated, and when y_2 is smaller than x_2, the melting point is depressed. Equation (1.45) can be rewritten for liquid–vapor equilibria and a similar procedure can be followed to obtain an equation identical in form with (1.47) for the elevation of the boiling point.

Molar Properties

A solution of c components may be considered as formed according to

$$x_1(\text{pure 1}) + x_2(\text{pure 2}) + \cdots \rightarrow 1 \text{ mole of solution} \tag{1.48}*$$

For the solution, the molar Gibbs energy is $G = \sum x_i \bar{G}_i$, and for the pure components, $G = \sum x_i G_i^{\circ}$, so that the Gibbs energy of formation of solution ΔG for reaction (1.48) is

$$\Delta G = \sum x_i \bar{G}_i - \sum x_i G_i^{\circ} = \sum x_i(\bar{G}_i - G_i^{\circ})$$
$$= RT \sum x_i \ln a_i; \qquad (\bar{G}_i - G_i^{\circ} = RT \ln a_i) \tag{1.49}$$

Likewise, for ΔH, by using equation (1.43), we derive

$$\Delta H = \sum x_i(\bar{H}_i - H_i^{\circ}) = -RT^2 \sum x_i \frac{\partial \ln \gamma_i}{\partial T}; \qquad \left(\bar{H}_i - H_i^{\circ} = -RT^2 \frac{\partial \ln \gamma_i}{\partial T}\right)$$
$$\tag{1.50}$$

---

*For alloys, 1 mole of solution is identical with 1 gram atom of solution.

The corresponding relationship for ΔS is obtained by substituting equations (1.49) and (1.50) into $\Delta S = (\Delta H - \Delta G)/T$:

$$\Delta S = \sum x_i(\bar{S}_i - S_i^\circ) = -R \sum \left(x_i \ln a_i + Tx_i \frac{\partial \ln \gamma_i}{\partial T} \right);$$

$$\bar{S}_i - S_i^\circ = -R \left(\ln a_i + T\frac{\partial \ln \gamma_i}{\partial T} \right) \tag{1.51}$$

Generally, the accuracy of data on γ_i for alloys does not justify a rigorous partial derivative in equations (1.50) and (1.51); instead, $\partial \ln \gamma_i/\partial T \approx \Delta \ln \gamma_i/\Delta$ is often used for solutions of metals to obtain ΔH and ΔS. A much more accurate procedure is to obtain ΔH by calorimetry at various concentrations and then to calculate ΔS from the data on a_i by using $\Delta S = R \sum x_i \ln a_i - (\Delta H/T)$. Figure 1.2 illustrates ΔG, $\bar{G}_1 - G_1^\circ$, $\bar{G}_2 - G_2^\circ$, ΔH, and ΔG for the binary system Al (component 1) and Cu (component 2) obtained by plotting the data compiled by Hultgren *et al.*[5]

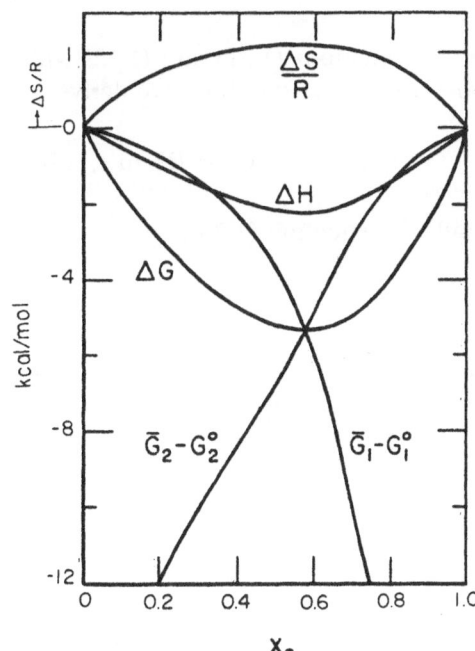

Figure 1.2. Selected thermodynamic properties of liquid Al-Cu system at 1373 K. Subscript 1 is for Al; 2, for Cu. Curves for ΔG, $\bar{G}_1 - G_1^\circ$, and $\bar{G}_2 - G_2^\circ$ coincide where ΔG is minimum. Lower left vertical scale is for ΔH, ΔG, and $\bar{G}_i - G_i^\circ$.

The foregoing equations can be rewritten for an ideal solution by observing that $\gamma_i = 1$, independent of temperature and composition; therefore, for reaction (1.48),

$$\Delta G(\text{ideal}) = RT \sum x_i \ln x_i; \quad [\bar{G}_i(\text{ideal}) - G_i^\circ = RT \ln x_i] \quad (1.49a)$$

$$\Delta H(\text{ideal}) = 0; \quad [\bar{H}_i(\text{ideal}) - H_i^\circ = 0] \quad (1.50a)$$

$$\Delta S(\text{ideal}) = -R \sum x_i \ln x_i; \quad [\bar{S}_i(\text{ideal}) - S_i^\circ = -R \ln x_i] \quad (1.51a)$$

Excess Molar Properties

The activity coefficient is a measure of deviation from ideality; therefore, certain thermodynamic properties related to the activity coefficients are called excess thermodynamic properties as introduced by Scatchard (1932). The excess Gibbs energy G^e and the excess partial molar Gibbs energy $\bar{G}_i^e$ are defined by

$$G^e = G(\text{real}) - G(\text{ideal}); \quad \bar{G}_i^e = \bar{G}_i(\text{real}) - \bar{G}_i(\text{ideal}) \quad (1.52)$$

From equations (1.31) and (1.32) with $\gamma_i \neq 1$ for real solutions, and with $\gamma_i = 1$ for ideal solutions, we derive

$$\bar{G}_i^e = RT \ln a_i - RT \ln x_i = RT \ln \gamma_i \quad (1.53)$$

and with equation (1.26),

$$G^e \equiv H^e - TS^e = RT \sum x_i \ln \gamma_i \quad (1.54)$$

We found earlier that $\bar{H}_i - H_i^\circ$ was zero for an ideal solution in conjunction with equations (1.43) and (1.44), or $\bar{H}_i = H_i^\circ$; therefore, $\bar{H}_i(\text{real}) - \bar{H}_i(\text{ideal})$ is the same as $\bar{H}_i(\text{real}) - H_i^\circ$; and from equations (1.43) and (1.50),

$$\bar{H}_i^e = \bar{H}_i - H_i^\circ = -RT^2 \frac{\partial \ln \gamma_i}{\partial T} \quad (1.55)$$

$$H^e = \sum x_i \bar{H}_i^e = \Delta H = -RT^2 \sum x_i \frac{\partial \ln \gamma_i}{\partial T} \quad (1.56)$$

where the molar enthalpy of formation ΔH is the same as the excess enthalpy of solution, H^e. Equations (1.53)-(1.56) and $\bar{G}_i^e = \bar{H}_i^e - T\bar{S}_i^e$ lead to

$$\bar{S}_i^e = -R \ln \gamma_i - RT \ln \frac{\partial \ln \gamma_i}{\partial T} \tag{1.57}$$

$$S^e = -R \sum \left(x_i \ln \gamma_i + Tx_i \frac{\partial \ln \gamma_i}{\partial T} \right) \tag{1.58}$$

Again a much more accurate procedure is to obtain $H^e = \Delta H$ and $\bar{H}_i^e$ by calorimetry at various concentrations and then use the activity coefficient in $S^e = (H^e/T) - R \sum x_i \ln \gamma_i$, and $\bar{S}_i^e = (\bar{H}^e/T) - R \ln \gamma_i$ to calculate S^e and $\bar{S}_i^e$.

Analytical Forms of Excess Partial Molar Properties

Binary Systems

The extensive excess Gibbs energy $\mathscr{G}^e = \mathscr{G}^e(P, T, n_1, n_2)$ for a binary system is related to $\bar{G}_1^e$ by equation (1.22), i.e.,

$$\bar{G}_1^e = \left(\frac{\partial \mathscr{G}^e}{\partial n_1} \right)_{P,T,n_2} \tag{1.59}$$

where P, T, and n_2 are explicitly shown as held constants during differentiation. It is convenient to use the molar property G^e instead of the extensive property $\mathscr{G}$ by using $\mathscr{G}^e = (n_1 + n_2)G^e$ in equation (1.59) to obtain

$$\bar{G}_1^e = G^e + (n_1 + n_2) \left(\frac{\partial G^e}{\partial n_1} \right) \tag{1.60}$$

Differentiation of $x_1 = n_1/(n_1 + n_2)$ at constant n_2 gives

$$dx_1 = \frac{n_2 \, dn_1}{(n_1 + n_2)^2} = \frac{x_2 \, dn_1}{n_1 + n_2}$$

Substitution of x_2/dx_1 for $(n_1 + n_2)/dn_1$ from this equation into equation (1.60) gives

$$\bar{G}_1^e = G^e + x_2 \left(\frac{\partial G^e}{\partial x_1} \right)_{P,T,n_2} \tag{1.61}$$

Since G^e is a function of one composition variable, e.g., x_1 for a binary solution, it is independent of n_2; therefore, the restriction required by the constancy of n_2 is unnecessary; hence,

$$\bar{G}_1^e = G^e + (1 - x_1)\left(\frac{\partial G^e}{\partial x_1}\right)_{P,T} \tag{1.62}$$

where $1 - x_1$ has been substituted for x_2 to obtain symmetry in the subscripts. The equation for $\bar{G}_2^e$ is obtained by replacing the subscripts 1 with 2 in equation (1.62), i.e.,

$$\bar{G}_2^e = G^e + (1 - x_2)\left(\frac{\partial G^e}{\partial x_2}\right)_{P,T} \tag{1.62a}$$

Compiled values[5] of G^e, $\bar{G}_1^e$, and $\bar{G}_2^e$ for Al(subscript 1)–Cu(subscript 2) at 1373 K are plotted in Fig. 1.3. All three curves intersect one another

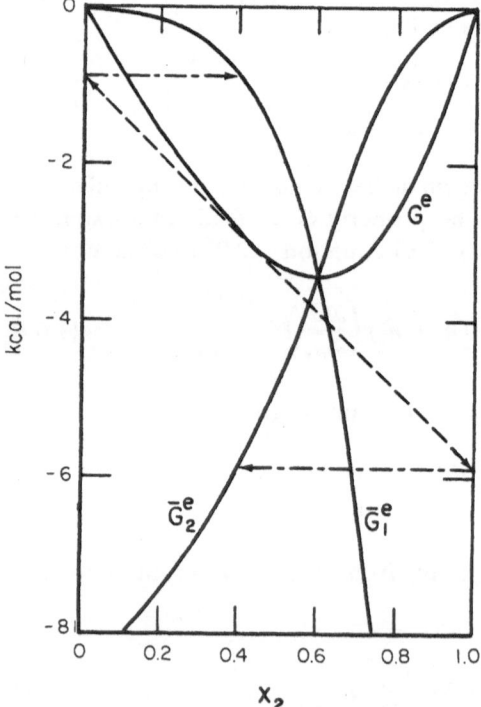

Figure 1.3. Excess molar Gibbs energy, G^e, and excess partial molar Gibbs energies, $\bar{G}_1^e$ and $\bar{G}_2^e$, for Al–Cu system at 1373 K. Subscript 1 is for Al; 2, for Cu. The tangent line on the curve for G^e at $x_2 = 0.4$ intersects the vertical coordinate at $x_2 = 0$ to yield $\bar{G}_1^e$, and at $x_2 = 1$ to yield $\bar{G}_2^e$, as shown by the dashed lines with arrows.

Note that $\bar{G}_2^e$ is equal to G^e plus $(1 - x_2)$ times the slope, $\partial G^e/\partial x_2$, as required by equation (1.62a), and similarly for $\bar{G}_1$, but the slope in this case is positive because $\partial G^e/\partial x_1 = -\partial G^e/\partial x_2$.

where G^e is a minimum as required by equations (1.62) and (1.62a). When G^e (or other excess molar properties) is plotted versus x_2, the method of tangent-intercepts can be used to obtain $\bar{G}_1^e$ and $\bar{G}_2^e$ as indicated in Fig. 1.3, but this method is now obsolete because a polynomial data fitting on a computer is superior (see Ref. 1). Equations (1.59) through (1.62) are general in the sense that G, H, S, or other related thermodynamic properties may be substituted for G^e to obtain the corresponding equations for them. The power series most often used for alloys are the Margules equation in which ε is the order of series; thus,

$$G^e = x_1 x_2(\alpha_1 + \alpha_2 x_2 + \alpha_3 x_2^2 + \cdots + \alpha_3 x_2^{\varepsilon-1}) = \sum_{q=2}^{\varepsilon} A_q x_2(1 - x_2^{q-1}) \quad (1.63)$$

where the relationships between α_i and A_q can be obtained by eliminating x_1 and setting the coefficients of x_2^i equal for equal values of i on the left and on the right sides; however, the result is of no great importance for our purposes, and we shall use only the equations involving A_q as follows:

$$\bar{G}_1^e = \sum_{q=2}^{\varepsilon} (q-1)A_q x_2^q \quad (1.64)$$

$$\bar{G}_2^e = \sum_{q=2}^{\varepsilon} A_q[1 - qx_2^{q-1} + (q-1)x_2^q] \quad (1.65)$$

The boundary conditions for the Raoultian activity scale require that for $x_1 = 1$, $\bar{G}_1^e = 0$, and for $x_2 = 1$, $\bar{G}_2^e = 0$, and therefore G^e is zero for $x_1 = 1$ or $x_2 = 1$. Traditionally, $\bar{G}_2^e$ is often written in symmetry with $\bar{G}_1^e$ so that

$$\bar{G}_2^e = \sum_{t=2}^{\varepsilon} (t-1)B_t x_1^t \quad (1.66)$$

The coefficients A_q and B_t must satisfy either the Gibbs–Duhem relation $x_1 \, d\bar{G}_2^e + x_2 \, d\bar{G}_2 = 0$ [cf. equation (1.27)] or equation (1.63); therefore,

$$B_t = (-1)^t \sum_{q=t}^{\varepsilon} \frac{q! A_q}{t!(q-t)!} \quad (1.67)$$

where the values of t start from 2 and end with ε. We stress here that when any set of power series is selected for $\bar{G}_1^e$ and $\bar{G}_2^e$, they must satisfy the Gibbs–Duhem relation, but when a power series satisfying the boundary conditions is selected for G^e, then $\bar{G}_1^e$, and $\bar{G}_2^e$ derived from such a G^e automatically satisfy the Gibbs–Duhem relation. For $\varepsilon = 6$, the results for B_t in terms of A_q from equation (1.67) are

$$B_2 = A_2 + 3A_3 + 6A_4 + 10A_5 + 15A_6$$

$$2B_3 = -2A_3 - 8A_4 - 20A_5 - 40A_6$$

$$3B_4 = 3A_4 + 15A_5 + 45A_6$$

$$4B_5 = -4A_5 - 24A_6$$

$$5B_6 = 5A_6 \tag{1.68}$$

It is evident that for $\varepsilon = 2$, $B_2 = A_2$ and we have

$$G^e = A_2 x_1 x_2; \qquad \bar{G}_1^e = A_2 x_2^2; \qquad \bar{G}_2^e = A_2 x_1^2 \tag{1.69}$$

Another set of forms[6,7] of Margules equations with $\varepsilon = 4$ is

$$G^e = x_1 x_2 (A_{21} x_1 + A_{12} x_2 - D_{12} x_1 x_2) \tag{1.70}$$

$$\bar{G}_1^e = x_2^2 [A_{12} + 2x_1 (A_{21} - A_{12} - D_{12}) + 3x_1^2 D_{12}] \tag{1.71}$$

$$\bar{G}_2^e = x_1^2 [A_{21} + 2x_2 (A_{12} - A_{21} - D_{12}) + 3x_2^2 D_{12}] \tag{1.72}$$

[In general, A_{12} is not equal to A_{21}; and the term $D_{12} x_1 x_2$ may be written as $x_1 x_2 (D_{21} x_1 + D_{12} x_2 - E_{12} x_1 x_2)$ if additional terms are required.] It is clear that $\bar{G}_2^e$ can be obtained from $\bar{G}_1^e$ by interchanging the subscripts 1 and 2 and agreeing that $D_{12} = D_{21}$ since D_{21} does not appear in equation (1.70). For $D_{12} = 0$, we have the cubic equations, and for $D_{12} = 0$ and $A_{12} = A_{21}$, we have the familiar quadratic equations of the type given by equation (1.69) since $A_{21} x_1 + A_{12} x_2 = A_{12}$. Various additional equations are summarized by Hala *et al.*[6]; however, no specific equation can be claimed to have much greater advantages over other equations in accurate representations of data.

The effect of temperature on G^e and $\bar{G}_1^e$ is adequately accounted for by assuming that each coefficient A_{ij}, D_{ij}, and so on is a linear function of temperature, e.g., $A_{12} = \alpha_{12} - \beta_{12} T$. The constant term α_{12} times the compositional variables contribute to H^e and $\bar{H}_i^e$, and β_2 times the same variables contribute to S^e and $\bar{S}_i^e$. This is self-evident since $\partial(G^e/T)/\partial(1/T) = H^e$, and $\partial G^e/\partial T = -S^e$, and similarly for $\bar{H}_i^e$ and $\bar{S}_i^e$.

EXAMPLE. Thermodynamic properties of the Mg–Cd system have been determined by Castanet *et al.*[8] at various temperatures and concentrations, by using a galvanic cell for measurements of $\bar{G}_1^e(Mg)$. The results have been fitted with the following equation:

$$\bar{G}_1^e(Mg) = (-25,413 + 43.949\,T)x_2^2 + (9494 - 176.000\,T)x_2^3$$
$$+ (-25,457 + 241.507\,T)x_2^4$$
$$+ (22,421 - 106.594\,T)x_2^5 \qquad (J/mole)$$

where subscript 1 = Mg and 2 = Cd. Since $\partial \bar{G}_1^e / \partial T = -\bar{S}_1^e$, it is evident that the coefficients of T in each term belong to $\bar{S}_1^e$, leaving the preceding temperature-independent term for $\bar{H}_1^e$. Therefore,

$$\bar{S}_1^e = -43.949x_2^2 + 176.000x_2^3 - 241.507x_2^4 + 106.594x_2^5$$

$$\bar{H}_1^e = -25,413x_2^2 + 9494x_2^3 - 25,457x_2^4 + 22,421x_2^5$$

This equation corresponds to equation (1.64) with $\varepsilon = 5$:

$$\bar{G}_1^e = A_2 x_2^2 + 2A_3 x_2^3 + 3A_4 x_2^4 + 4A_5 x_2^5$$

Hence, $A_2 = -25,413 + 43.549\,T$, $A_3 = 4747 - 88.000\,T$, and so on. It is therefore possible to express $\bar{G}_2 = B_3 x_1^3 + \cdots$ of equation (1.66) from equation (1.68); the result is

$$\bar{G}_2^e(Cd) = (-6043.5 - 3.522\,T)x_1^2 + (-53,713.7 + 64.9513\,T)x_1^3$$
$$+ (58,621.8 - 158.2205\,T)x_1^4 + (-22,421 + 106.594\,T)x_1^5$$

The corresponding equations for $\bar{H}_2^e$ and $\bar{S}_2^e$ can readily be obtained from this equation. In addition, the equation for the molar Gibbs energy G^e can be obtained by using equation (1.63) from which H^e and S^e can be obtained. Therefore, the measurements of $\bar{G}_1^e = RT \ln \gamma_1$ at various temperatures and compositions yield $\bar{G}_1^e$, $\bar{H}_1^e$, $\bar{S}_1^e$, $\bar{G}_2^e$, $\bar{H}_2^e$, $\bar{S}_2^e$, G^e, H^e, and S^e.

Ternary Systems

The equations selected for G^e of a ternary system must again satisfy the boundary conditions in that $G^e = 0$ when $x_i = 1$ ($i = 1, 2, 3$). When this condition is satisfied, the power series beginning with the second-order terms automatically satisfy the Gibbs–Duhem relation. The derivation of $\bar{G}_i^e$ from G^e requires $\mathcal{G}^e = (n_1 + n_2 + n_3)G^e$ and again writing

$$\bar{G}_1^e = \left(\frac{\partial \mathcal{G}^e}{\partial n_1}\right)_{n_i'} = G^e + (n_1 + n_2 + n_3)\left(\frac{\partial G^e}{\partial n_1}\right)_{n_i'} \qquad (1.73)$$

where n_i' signifies that all n_i other than n_1 are held constant during differentiation. The partial differential of x_1 is

$$(dx_1)_{n_i} = d\left(\frac{n_1}{n_1 + n_2 + n_3}\right) = \frac{(n_2 + n_3)\, dn_1}{(n_1 + n_2 + n_3)^2} = \frac{(1 - x_1)\, dn_1}{n_1 + n_2 + n_3} \quad (1.74)$$

The variable x_1 is an intensive property that may be written as $(n_1/n_3)/[(n_1/n_3) + n_2/n_3) + 1]$ where the restriction of n_i', or the constancy of n_2 and n_3, is simply equivalent to the constancy of ratio $r = n_2/n_3 = x_2/x_3$ in terms of mole fractions. Substitution of $(1 - x_1)/dx_1$ from equation (1.74) into (1.73) for $(n_1 + n_2 + n_3)/\partial n_1$ gives

$$\bar{G}_1^e = G^e + (1 - x_1)\left(\frac{\partial G^e}{\partial x_1}\right)_{P,T,x_2/x_3} \quad (1.75)$$

If this process is repeated for a multicomponent system, the additional restrictions in equation (1.74) are x_3/x_4, x_4/x_5, ... In order to obtain the analytical equation for $\bar{G}_1^e$ for a ternary system, a power series such as the following equation is needed for G^e:

$$G^e = x_1 x_2 (A_{21} x_1 + A_{12} x_2 - D_{12} x_1 x_2)$$
$$+ x_1 x_3 (A_{31} x_1 + A_{13} x_3 - D_{13} x_1 x_3)$$
$$+ x_2 x_3 (A_{32} x_2 + A_{23} x_3 - D_{23} x_2 x_3)$$
$$+ x_1 x_2 x_3 (A_{21} + A_{13} + A_{32} - C_1 x_1 - C_2 x_2 - C_3 x_3) \quad (1.76)$$

For $x_3 = 0$, this equation becomes identical with equation (1.70); likewise, for $x_2 = 0$, it becomes G^e for the binary system 1-3, and for $x_1 = 0$, G^e for the binary system 2-3. In addition, there are three ternary coefficients C_1, C_2, and C_3. The order of equation (1.76) is four, i.e., $\varepsilon = 4$. To derive the equation for $\bar{G}_1^e$, it is necessary to substitute equation (1.76) into equation (1.75) and carry out the differentiation with respect to x_1. Care must be exercised in this procedure because complications arise in obtaining the correct derivatives unless the sum of the mole fractions is written as $x_1 + x_2 + x_3 = x_1 + x_2(r + 1)/r = 1$, and $x_1 + (r + 1)x_3 = 1$, from $x_2/x_3 = r$; then from $dx_1 + [(r + 1)/r]\, dx_2 = 0$, and from $dx_1 + (r + 1)\, dx_3 = 0$, the derivatives are correctly written as

$$\left(\frac{\partial x_2}{\partial x_3}\right)_r = -\frac{x_2}{x_2 + x_3}; \quad \left(\frac{\partial x_3}{\partial x_1}\right)_r = -\frac{x_3}{x_2 + x_3} \quad (1.77)$$

These equations may be rewritten in the following convenient forms:

$$(1-x_1)\left(\frac{\partial x_2}{\partial x_1}\right)_r = -x_2; \qquad (1-x_1)\left(\frac{\partial x_3}{\partial x_1}\right)_r = -x_3 \qquad (1.78)$$

The second of these equations can be obtained from the first by replacing the subscript 2 with 3. The resulting equations for $\bar{G}_i$ ($i = 1, 2, 3$) by this procedure are as follows:

$$\begin{aligned}
\bar{G}_1^e = {} & x_2^2[A_{12} + 2x_1(A_{21} - A_{12} - D_{12}) + 3x_1^2 D_{12}] \\
& + x_3^2[A_{13} + 2x_1(A_{31} - A_{13} - D_{13}) + 3x_1^2 D_{13}] \\
& + x_2 x_3[A_{21} + A_{13} - A_{32} + 2x_1(A_{31} - A_{13}) \\
& + 2x_3(A_{32} - A_{23}) + 3x_2 x_3 D_{23} - C_1 x_1(2 - 3x_1) \\
& - C_2 x_2(1 - 3x_1) - C_3 x_3(1 - 3x_1)]
\end{aligned} \qquad (1.79)$$

$$\begin{aligned}
\bar{G}_2^e = {} & x_3^2[A_{23} + 2x_2(A_{32} - A_{23} - D_{23}) + 3x_2^2 D_{23}] \\
& + x_1^2[A_{21} + 2x_2(A_{12} - A_{21} - D_{12}) + 3x_2^2 D_{12}] \\
& + x_1 x_3[A_{32} + A_{21} - A_{13} + 2x_2(A_{12} - A_{21}) \\
& + 2x_1(A_{13} - A_{31}) + 3x_1 x_3 D_{13} - C_2 x_2(2 - 3x_2) \\
& - C_3 x_3(1 - 3x_2) - C_1 x_1(1 - 3x_2)]
\end{aligned} \qquad (1.80)$$

$$\begin{aligned}
\bar{G}_3^e = {} & x_1^2[A_{31} + 2x_3(A_{13} - A_{31} - D_{13} + 3x_3^2 D_{13})] \\
& + x_2^2[A_{32} + 2x_3(A_{23} - A_{32} - D_{23} + 3x_3^2 D_{23})] \\
& + x_1 x_2[A_{13} + A_{32} - A_{21} + 2x_3(A_{23} - A_{32}) \\
& + 2x_2(A_{21} - A_{12}) + 3x_1 x_2 D_{12} - C_3 x_3(2 - 3x_3) \\
& - C_1 x_1(1 - 3x_3) - C_2 x_2(1 - 3x_3)]
\end{aligned} \qquad (1.81)$$

An exchange of subscripts cannot yield $\bar{G}_i^e$ from $\bar{G}_j^e$ because a circular rotation of subscripts is necessary. For example, to obtain $\bar{G}_2^e$ from $\bar{G}_1^e$ it is necessary to use 1(old) → 2(new), 2(old) → 3(new), 3(old) → 1(new), and to observe that $D_{ij} = D_{ji}$ because there is only one D in equation (1.76). The patterns of coefficients in equations (1.76) through (1.81) are now established so that it is a simple matter to add the terms corresponding to increasing values of ε. If it is desirable to check the results, an additional thermodynamic relationship can be used. For this purpose, it is necessary to differentiate $G^e = x_1\bar{G}_1^e + x_2\bar{G}_2^e + x_3\bar{G}_3^e$ and use the Gibbs–Duhem relation $x_1\, d\bar{G}_1^e + x_2\, d\bar{G}_2^e + x_3\, d\bar{G}_3^e = 0$ to get $dG^e = \bar{G}_1^e\, dx_1 + \bar{G}_2^e\, dx_2 + \bar{G}_3^e\, dx_3$. At a fixed value of x_3, $dx_3 = 0$, then $dx_1 = -dx_2$, and the equation of dG^e becomes $dG^e = -\bar{G}_1^e\, dx_2 + \bar{G}_2^e\, dx_2$; from this equation, it is evident that

$$\bar{G}_2^e = \bar{G}_1^e + \left(\frac{\partial G^e}{\partial x_2}\right)_{x_3} \tag{1.82}$$

The exchange of subscripts 2 and 3 yields a similar equation for $\bar{G}_3^e$. The aditional terms in equation (1.76) contribute to $\bar{G}_i^e$ as required by equation (1.75).

The third-order Margules equation can be obtained from the foregoing equations by setting $D_{ij} = 0$ and $C_1 = C_2 = C_3 = C$. The results are given here for convenience and as a check on equations (1.79)–(1.81):

$$G^e = x_1 x_2 (A_{21}x_1 + A_{12}x_2) + x_1 x_3 (A_{31}x_1 + A_{13}x_3)$$
$$+ x_2 x_3 (A_{32}x_2 + A_{23}x_3) + x_1 x_2 x_3 (A_{21} + A_{13} + A_{32} - C) \tag{1.83}$$

$$\bar{G}_1^e = x_2^2 [A_{12} + 2x_1(A_{21} - A_{12})] + x_3^2 [A_{13} + 2x_1(A_{31} - A_{13})]$$
$$+ x_2 x_3 [A_{21} + A_{13} - A_{32} + 2x_1(A_{31} - A_{13})$$
$$+ 2x_3(A_{32} - A_{23}) - C(1 - 2x_1)] \tag{1.84}$$

$$\bar{G}_2^e = x_3^2 [A_{23} + 2x_2(A_{32} - A_{23})] + x_1^2 [A_{21} + 2x_2(A_{12} - A_{21})]$$
$$+ x_1 x_3 [A_{32} + A_{21} - A_{13} + 2x_2(A_{12} - A_{21})$$
$$+ 2x_1(A_{13} - A_{31}) - C(1 - 2x_2)] \tag{1.85}$$

$$\bar{G}_3^e = x_1^2 [A_{31} + 2x_3(A_{13} - A_{31})] + x_2^2 [A_{32} + 2x_3(A_{23} - A_{32})]$$
$$+ x_1 x_2 [A_{13} + A_{32} - A_{21} + 2x_3(A_{23} - A_{32})$$
$$+ 2x_2(A_{21} - A_{12}) - C(1 - 2x_3)] \tag{1.86}$$

Equation (1.83) contains only one ternary term, C, which must be determined from ternary data. It is possible to set $C = 0$ and use only the binary coefficients to obtain $\bar{G}_i^e$ for the ternary alloys with success for liquid alloys, but marginal results for solid alloys.

The second-order Margules equations can be obtained by setting $A_{12} = A_{21}$, $A_{13} = A_{31}$, $A_{23} = A_{32}$, and $C = 0$ in the preceding equations; the results are

$$G^e = A_{12}x_1 x_2 + A_{13}x_1 x_3 + A_{23}x_2 x_3 \tag{1.87}$$

$$\bar{G}_1^e = A_{12}x_2^2 + A_{13}x_3^2 + x_2x_3(A_{12} + A_{13} - A_{23}) \tag{1.88}$$

$$\bar{G}_2^e = A_{23}x_3^2 + A_{12}x_1^2 + x_1x_3(A_{23} + A_{12} - A_{13}) \tag{1.89}$$

$$\bar{G}_3^e = A_{13}x_1^2 + A_{23}x_2^2 + x_1x_2(A_{13} + A_{23} - A_{12}) \tag{1.90}$$

Equation (1.87) contains only the constants from the binary systems; therefore, it contains no terms with $x_1x_2x_3$. The interchange of any set of two subscripts does not change equation (1.87) since it assumed that $A_{ij} = A_{ji}$; therefore, only in such sets of equations, $\bar{G}_2^e$ can be obtained from $\bar{G}_1^e$ by interchanging the subscripts 1 and 2, and likewise, $\bar{G}_3^e$, by interchanging the subscripts 1 and 3. Various other types of equations and computational methods exist, as listed elsewhere in detail.[1,6,9,10] The foregoing equations are also valid when other molar properties such as ΔH, ΔS, and S^e replace G^e, after which the related partial molar properties can be derived by the same procedure.

Determination of G^e and $\bar{G}_i^e$ of Ternary Systems

Determination of the coefficients of the equation for $\bar{G}_i^e$ of one selected component is sufficient for obtaining the equations for the remaining $\bar{G}_i^e$, as well as for G^e as in binary systems. Consider the usual type of measurements for a ternary system in which one of the components, labeled as component 1, lends itself to convenient and accurate determinations of $\bar{G}_1^e$, e.g., the vapor pressure of P_1 over the system is accurately measurable from which $\bar{G}_1^e = RT \ln \gamma_1 = RT \ln[P_1/(x_1 P_1^*)]$ where P_1^* refers to the pure component. If, for example, the equation necessary for adequately representing $\bar{G}_1^e$ is that corresponding to $\varepsilon = 4$, i.e., equation (1.79), then there are 12 coefficients which require a minimum of six well-spaced ternary values of $\bar{G}_1^e$, the balance being three binary values for system 1–2, and three for system 1–3 because component "1" can form only two binary systems over which its partial pressure can be measured; if, however, at least two components out of three are volatile so that their vapor pressures can be measured accurately, a minimum of three data from the ternary system, and the remaining nine, with three from each binary system, would be sufficient for $\varepsilon = 4$. All the foregoing equations may also be evaluated entirely from well-spaced ternary data alone. Such ternary data are capable of yielding the values of G^e and $\bar{G}_i^e$ for the binary systems, but the data on the binary systems alone cannot usually yield good values of G_i^e and $\bar{G}_i^e$ for the ternary system although this approximation appears to yield fairly good values for a few ternary systems in which the deviations from the ideality are not large.

The coefficients of equation (1.79) can be determined by computer if the data are available for the activity coefficient of only one component. Various computer programs are available[10,11] for this purpose, but the essential point is simply solving a set of simultaneous equations in which the unknowns are the coefficients in equation (1.79). If, however, all three activity coefficients are known from experimental measurements, it is best to compute G^e at each set of compositions by $G^e = x_1\bar{G}_1^e + x_2\bar{G}_2^e + x_3\bar{G}_3^e$ and then solve for the unknown coefficients in equation (1.76). The foregoing analytical method is far superior to the graphical integration of the Gibbs–Duhem equation.[1,11]

Interaction Parameters

Equations (1.88)–(1.90) for $\bar{G}_1^e$, $\bar{G}_2^e$, and $\bar{G}_3^e$ become considerably more reliable as x_2 and x_3 approach zero, or x_1 approaches unity. For such dilute solutions of components 2 and 3 in solvent 1, differentiation of these equations yields

$$\left(\frac{\partial \bar{G}_2^e}{\partial x_3}\right)_{x_2} = \varepsilon_2^{(3)} = \left(\frac{\partial \bar{G}_3^e}{\partial x_2}\right)_{x_3} \equiv \varepsilon_3^{(2)} = A_{23} - A_{13} - A_{12}; \qquad (x_2 \text{ and } x_3 \to 0)$$

$$(1.91)$$

where the identity signs define $\varepsilon_i^{(j)}$ called the Wagner interaction parameters or coefficients,[1,12] $\varepsilon_2^{(3)}$ giving the effect of component 3 on $\bar{G}_2^e$, and $\varepsilon_3^{(2)}$, that of component 2 on $\bar{G}_3^e$. This differentiation is easily carried out after eliminating x_1 in equations (1.89) and (1.90), and then setting x_2 and x_3 to zero. To generalize $\varepsilon_i^{(j)}$ for all values of i and j, including $i = j$, it is sufficient to write $\varepsilon_i^{(i)}$ as $(\partial \bar{G}_i^e / \partial x_i)$; the results are

$$\left(\frac{\partial \bar{G}_2^e}{\partial x_2}\right)_{x_3} \equiv \varepsilon_2^{(2)} = -2A_{12}; \qquad (x_2 \text{ and } x_3 \to 0) \qquad (1.92)$$

$$\left(\frac{\partial \bar{G}_3^e}{\partial x_3}\right)_{x_2} \equiv \varepsilon_3^{(3)} = -2A_{13}; \qquad (x_2 \text{ and } x_3 \to 0) \qquad (1.93)$$

For a binary solution of components 1 and 2, $x_3 = 0$ in equation (1.89), and for $x_2 \to 0$ or $x_1 \to 1$,

$$\bar{G}_2^e(x_2 \to 0, x_3 = 0) = RT \ln \gamma_2^\circ = A_{12} = -\varepsilon_2^{(2)}/2 \qquad (1.94)$$

Equation (1.89), after substituting $1 - x_2 - x_3$ for x_1 and neglecting the second- and higher-order terms, gives

$$\bar{G}_2^e = RT \ln \gamma_2^\circ + \varepsilon_2^{(2)} x_2 + \varepsilon_2^{(3)} x_3 + \cdots + \varepsilon_2^{(n)} x_n$$

$$= RT \ln \gamma_2^\circ + \sum_{i=2}^{n} \varepsilon_2^{(i)} x_i \tag{1.95}$$

where $\varepsilon_2^{(n)} x_n$ is the nth term in an n-component system. Similarly,

$$\bar{G}_3^e = RT \ln \gamma_3 = RT \ln \gamma_3^\circ + \sum_{i=2}^{n} \varepsilon_3^{(i)} x_i \tag{1.96}$$

The Wagner interaction parameters and their generalized form have been the subject of numerous papers.[13-16] The usefulness of these coefficients lies in determining γ_2 for multicomponent dilute solutions from γ_2 for the binary and ternary systems. Thus, $\varepsilon_2^{(2)}/RT$ is the slope from a plot of $\ln \gamma_2$ versus x_2 for the binary system 1-2 and the linear portion of the plot near $x_2 \to 0$ gives $\partial \ln \gamma_2 / \partial x_2$; likewise, $\varepsilon_2^{(3)}/RT$ is the slope of $\ln \gamma_2$ versus x_3 at a selected fixed but low concentration of x_2. For example, in a given solution containing only $x_2 = 0.02$ as the solute, component 3 can be added in small increments of 0.01 from 0.00 to 0.05 and then $\ln \gamma_2$ is plotted versus x_3. The value of $\varepsilon_2^{(3)}$ is then obtained from

$$RT \left(\frac{\partial \ln \gamma_2}{\partial x_3} \right)_{x_2} = RT \frac{\ln \gamma_2 (\text{at } x_3 = 0.05) - \ln \gamma_2 (\text{at } x_3 = 0.00)}{0.05 - 0.00} = \varepsilon_2^{(3)} \tag{1.97}$$

where it is assumed that the plot in this range is linear. Similar equations for component 3 can be obtained by interchanging the subscripts 2 and 3. A more interesting method, when experimentally possible, is to determine $\bar{G}_i^e$ as a function of composition in sufficient ranges of concentration to determine A_{ij} in equations (1.88)-(1.90) and then use equations (1.91)-(1.93) to determine $\varepsilon_i^{(j)}$.

EXAMPLE. The activities a_i and activity coefficients and related thermodynamic properties for the binary systems of Hg(component 1), Sn(2), and Zn(3) have been evaluated and compiled by Hultgren et al.[5] The results are extrapolated to 673 K by using the listed values of H^e and S^e in $G^e = H^e - TS^e$ by assuming that H^e and S^e are independent of temperature. For example, the listed value for Hg-Sn at 450 K are $H^e = 211$, and $S^e = -0.142$ at $x_1 = 0.5$; therefore, $G^e = 211 + 673 \times 0.142 = 307$ cal/(g-atom of alloy) at 673 K, with a possible error of ± 100 cal/g-atom.

The values of G^e (in cal/g-atom) for the three binary alloys are fitted with equation (1.70) as follows:

$$G^e(\text{Hg-Sn}) = x_1x_2(2180x_1 + 1080x_2 - 1530x_1x_2) \text{ in cal/g-atom};$$

$$(A_{21} = 2180; \; A_{12} = 1080; \; D_{12} = 1530) \qquad (1.98)$$

$$G^e(\text{Hg-Zn}) = x_1x_3(850x_1 + 1100x_3); \qquad (A_{31} = 850; \; A_{13} = 1100; \; D_{13} = 0)$$

$$(1.99)$$

$$G^e(\text{Sn-Zn}) = x_2x_3(1160x_2 + 2500x_3 - 970x_2x_3);$$

$$(A_{32} = 1160; \; A_{23} = 2500; \; D_{23} = 970) \qquad (1.100)$$

The results for Hg–Zn by Kozin et al.[17] are in agreement with equation (1.99). The foregoing equations could have been represented as functions of temperature but the ternary data to be used later here are available for one temperature only,[18] i.e., 673 K; therefore, equations (1.98)–(1.100), valid for 673 K, are adequate for our purposes. The calculations show that the values of γ_i for each binary system vary by a factor of approximately 2 from $x_i = 0$ to $x_i = 1$; hence, the deviation from Raoult's law is not severe.

The values of a_1 for Hg in the ternary system Hg–Sn–Zn were determined by Nigmetova et al.[18] by measuring the vapor pressure of Hg over various liquid alloys at 673 K. We wish to use equations (1.76) and (1.79) with $C_i = 0$, i.e., to obtain ternary equations by using the binary coefficients, in order to test the resulting values with the experimental ternary data.[18] The results for G^e and $\bar{G}_1^e$ are as follows:

$$G^e = x_1x_2(2180x_1 + 1080x_2 - 1530x_1x_2) + x_1x_3(850x_1 + 1100x_3)$$

$$+ x_2x_3(1160x_2 + 2500x_3 - 970x_2x_3) + 4440x_1x_2x_3 \qquad (1.101)$$

$$\bar{G}_1^e = x_2^2(1080 - 860x_1 + 4590x_1^2) + x_3^2(1100 - 500x_1)$$

$$+ x_2x_3(2120 - 500x_1 - 2680x_3 + 2910x_2x_3) \qquad (1.102)$$

The result from this equation for $\bar{G}_1^e = RT \ln \gamma_1$ (at $x_1 = 0.4$, $x_2 = 0.3$, $x_3 = 0.3$) is 337.4 cal/g-atom, from which $\gamma_1 = 1.287$. The experimental result is $\gamma_1 = 1.285$ as read carefully from an appropriate figure by Nigmetova et al.[18] The agreement is good because the deviation from ideality is not severe.

Other Definitions of Partial Molar Gibbs Energy

The partial molar Gibbs energy $\bar{G}_i$ of component i is usually defined by $\bar{G}_i = (\partial \mathcal{G}/\partial n_i)_{P,T,n_1,n_2,\ldots}$, but there are four additional definitions. The energy of a *closed system*, $\underline{E}$, underlined to show its correspondence with $\mathcal{G}$, is a function of entropy and volume since $d\underline{E} = T\,d\underline{S} - P\,d\underline{V}$. For an *open system*, the numbers of moles of components are also variables of state; therefore,

$$\underline{E} = \underline{E}(\underline{S}, \underline{V}, n_1, n_2, n_3, \ldots) \tag{1.103}$$

The total differential of energy is then

$$d\underline{E} = T\,d\underline{S} - P\,d\underline{V} + \left(\frac{\partial \underline{E}}{\partial n_1}\right)_{s,v,n'} \cdot dn_1 + \left(\frac{\partial \underline{E}}{\partial n_2}\right)_{s,v,n'} \cdot dn_2 + \cdots \tag{1.104}$$

where n' means that the numbers of moles other than that inside the parentheses are regarded as constants. The total differential of $\mathcal{G} = \underline{E} + P\underline{V} - T\underline{S}$ is

$$d\mathcal{G} = d\underline{E} + P\,d\underline{V} + \underline{V}\,dP - T\,d\underline{S} - \underline{S}\,dT \tag{1.105}$$

Substitution of the right side of equation (1.104) for $d\underline{E}$ in equation (1.105) gives

$$d\mathcal{G} = \underline{V}\,dP - \underline{S}\,dT + \left(\frac{\partial \underline{E}}{\partial n_1}\right)_{s,v,n'} \cdot dn_1 + \left(\frac{\partial \underline{E}}{\partial n_2}\right)_{s,v,n} \cdot dn_2 + \cdots \tag{1.106}$$

From this equation, it is evident that an alternative definition of $\bar{G}_1$ is given by

$$\bar{G}_1 = \left(\frac{\partial \mathcal{G}}{\partial n_1}\right)_{p,T,n'} = \left(\frac{\partial \underline{E}}{\partial n_1}\right)_{s,v,n'} \tag{1.107}$$

Likewise, starting with $\underline{H} = f(\underline{S}, P, n_1, \ldots)$, $\underline{S} = f(\underline{E}, \underline{V}, n_1, \ldots)$, and $\underline{A} = f(\underline{V}, T, n_1, \ldots)$, and following a similar procedure, the following additional definitions of $\bar{G}_1$ can be derived:

$$\bar{G}_1 = \left(\frac{\partial \underline{H}}{\partial n_1}\right)_{S,P,n'} = -T\left(\frac{\partial \underline{S}}{\partial n_1}\right)_{E,V,n'} = \left(\frac{\partial \underline{A}}{\partial n_1}\right)_{V,T,n'} \tag{1.108}$$

The definition given by $\bar{G}_i = (\partial \mathscr{G}/\partial n_i)_{p,T,n'}$, is more convenient than the remaining definitions because it is a property obtained under conveniently attainable conditions of constant temperature and pressure. Therefore, the additional definitions given in this section are seldom used in thermodynamics.

References

1. N. A. Gokcen, *Thermodynamics*, Techscience, Hawthorne, California (1975).
2. H. Margenau and G. M. Murphy, *Mathematics of Physics and Chemistry*, Second Edition, Krieger, Huntington, New York (1976).
3. D. R. Gaskell, *Metallurgical Thermodynamics*, Second Edition, McGraw-Hill, New York (1981).
4. O. F. Devereux, *Topics in Metallurgical Thermodynamics*, Wiley-Interscience, New York (1983); see also R. A. Swalin, *Thermodynamics of Solids*, Wiley-Interscience, New York (1972).
5. R. Hultgren, P. D. Desai, D. T. Hawkins, M. Gleiser, and K. K. Kelley, *Selected Values of the Thermodynamic Properties of Binary Alloys*, ASM, Metals Park, Ohio (1973).
6. E. Hala, J. Pick, V. Fried, and O. Vilim, *Vapour-Liquid Equilibrium*, Second Edition, translated by G. Standart, Pergamon Press, Elmsford, New York (1967).
7. N. A. Gokcen, *High Temp. Sci.* **15**, 293 (1982).
8. R. Castanet, Z. Moser, and W. Gasior, *Calphad* **4**(4), 231 (1980).
9. E. Hala, E. Wichterle, J. Polak, and T. Boublik, *Vapor-Liquid Equilibrium Data at Normal Pressures*, Pergamon Press, Elmsford, New York (1968).
10. H. Renon and J. M. Prausnitz, *J. Am. Inst. Chem. Eng.* **14**, 135 (1968); see also J. M. Prausnitz, *Molecular Thermodynamics of Fluid Phase Equilibria*, Prentice-Hall, Englewood Cliffs, New Jersey (1969).
11. J. M. Prausnitz, C. A. Eckert, R. V. Orye, and J. P. O'Connell, *Computer Calcuations for Multicomponent Vapor-Liquid Equilibria*, Prentice-Hall, Englewood Cliffs, New Jersey (1967).
12. C. Wagner, *Thermodynamics of Alloys*, translated by S. Mellgren and J. H. Westbrook, Addison-Wesley, Reading, Massachusetts (1952).
13. M. Ohtani and N. A. Gokcen, *Trans. Metall. Soc. AIME* **218**, 533 (1960).
14. C. H. P. Lupis and J. F. Elliott, *Acta Metall.* **14**, 529 and 1019 (1966).
15. F. Neumann and H. Schenk, *Arch. Eisenhuettenwes.* **30**, 477 (1959).
16. Z.-I. Morita and T. Tanaka, *Iron Steel Inst. J.* **23**, 824 (1983).
17. L. F. Kozin, R. S. Nigmetova, and M. B. Dergache, *Izd. Nauka, Alma-Ata* (1977).
18. R. S. Nigmetova, N. A. Golovanova, and L. F. Kozin, *Z. Fiz. Khim.* **53**, 1450 (1979).

2

Phase Equilibria and Phase Diagrams

Single-Component Equilibria

The phase equilibria constitute a vast area of interest in metallurgy, geology, chemistry, physics, and related fields. In a single-component system, such equilibria occur between two phases, or three phases. The two-phase equilibria and the corresponding Gibbs energy relations for a component A are

$$A(\text{Phase I}) = A(\text{Phase II}); \quad \Delta G = G^{\text{II}} - G^{\text{I}} = 0, \text{ or } G^{\text{I}} = G^{\text{II}} \tag{2.1}$$

A rigorous derivation of equation (2.1) will be given later in this chapter. Two phases may coexist over a wide range of temperature and the corresponding range of pressure. Three phases may also coexist at a single temperature and pressure for a pure component; therefore,

$$A(\text{I}) = A(\text{II}); \quad A(\text{II}) = A(\text{III}); \quad G^{\text{I}} = G^{\text{II}} = G^{\text{III}} \tag{2.2}$$

The equality of Gibbs energies in equation (2.1) requires that $dG^{\text{I}} = dG^{\text{II}}$, and since $dG = V\,dP - S\,dT$, it is evident that

$$V^{\text{I}}\,dP - S^{\text{I}}\,dT = V^{\text{II}}\,dP - S^{\text{II}}\,dT \tag{2.3}$$

Rearrangement of this equation with the substitution of $S^{\text{II}} - S^{\text{I}} = \Delta H / T$ gives

$$\frac{dT}{dP} = \frac{T(V^{\text{II}} - V^{\text{I}})}{\Delta H}; \quad \text{or} \quad \frac{dP}{dT} = \frac{\Delta H}{T\Delta V} \tag{2.4}$$

The numerical values of ΔH and T are positive; therefore, if V^{II} is larger than V^{I}, as in melting of most metals and vaporization of all elements and compounds, then dT/dP is positive, i.e., an increase in P causes an increase in T. On the other hand, when V^{II} is less than V^{I}, as in melting of As, Sb, Bi, and H_2O, an increase in P causes a decrease in T.

A phase diagram for a single-component system is shown in Fig. 2.1. Solid and vapor phases coexist along the curve AB; liquid and vapor phases, along BC; and solid and liquid phases, along BD. The areas represent the regions of stability for the single phases, and the lines, the coexisting two phases. At B, all three phases are in equilibrium, and B is thus called a triple point. The relative slopes of the lines at B are significant in thermodynamics. The slope of BD is usually very steep because ΔV is very small; in Fig. 2.1 ΔV, and hence the slope, is taken to be positive. At the triple point B, the temperature is designated as T_B and the slope of the sublimation curve AB at T_B is

$$\frac{dP}{dT} = \left(\frac{\Delta H_{subl}}{T_B R T_B}\right) P_B \tag{2.5}$$

where ΔH_{subl} is the molar enthalpy of sublimation, and $V^{II} - V^{I} = V(\text{vapor}) - V(\text{solid}) \approx V(\text{vapor}) = RT/P$, because the volume of vapor is

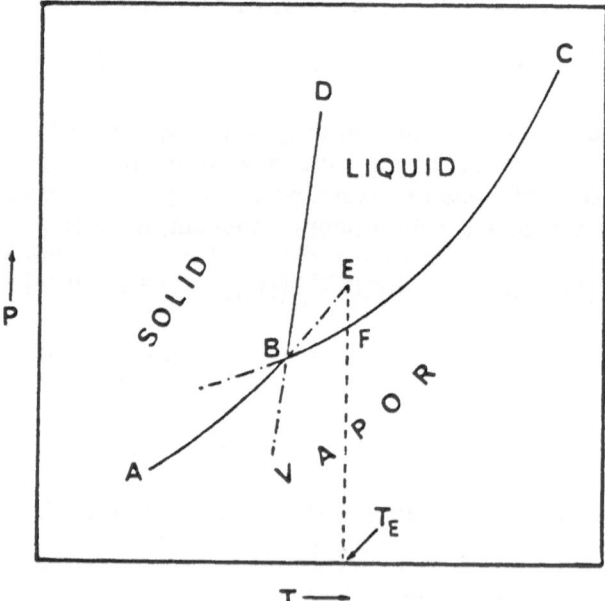

Figure 2.1. Phase equilibria for one-component system.

usually about 1000-fold larger than that of a condensed phase under equilibrium conditions. Similarly, for the vaporization curve BC,

$$\frac{dP}{dT} = \left(\frac{\Delta H_{vap}}{RT_B^2}\right) P_B \tag{2.6}$$

where ΔH_{vap} is the enthalpy of vaporization. Equations (2.5) and (2.6) are also valid at points other than B, i.e., at all possible values of P and T. According to the first law of thermodynamics, the enthalpy of sublimation at B is equal to the sum of the enthalpies of melting ΔH_m and vaporization ΔH_{vap}, or $\Delta H_{subl} = \Delta H_m + \Delta H_{vap}$; hence, $\Delta H_{subl} > \Delta H_{vap}$. Therefore, equations (2.5) and (2.6) show that the slope of AB is greater than the slope of BC at their point of intersection. This requirement necessitates that the phase boundaries intersect each other at angles less than 180° as shown in Fig. 2.1. Consequently, the supercooled liquid and superheated solid must have higher vapor pressures than the corresponding equilibrium phases. For example, at T_E, P_E is greater than P_F, and since $G^g = G^{o,g} + RT \ln P$, then G^g(over solid at E) > G^g(over liquid at F) and for the coexisting phases G^g(at E) = G^s and G^g(at F) = G^l; therefore, the preceding inequality is identical with $G^s > G^l$ at T_E. The solid at E must therefore transform into liquid at F either directly or by vaporization and condensation, because the phase with the lower Gibbs energy is the liquid phase. Thus, the extended portions of all the curves terminate in regions where they represent nonequilibrium conditions. A pure substance may have several additional triple points among its solid allotropes at various temperatures and considerably higher pressures than the solid–liquid–vapor triple point. It will be seen later that the intersecting curves at a triple point in composition–temperature phase diagrams also obey the requirement that they intersect one another at angles less than 180°.

Multicomponent Equilibria

Equilibria in multicomponent systems require the definition of degrees of freedom and conditions of equilibrium. It was seen in Chapter 1 that the Gibbs energy $\mathcal{G}$ for an open multicomponent system consisting of one phase is a function of pressure, temperature, and c numbers of moles, i.e.,

$$\mathcal{G} = \mathcal{G}(P, T, n_1, n_2, \ldots, n_c) \tag{2.7}$$

The partial molar Gibbs energy of component i was defined to be $\bar{G}_i = \partial G/\partial n_i$, and $\bar{G}_i$ is a function of P, T, and $c - 1$ composition variables which may be taken as mole fractions, x_i; thus,

$$\bar{G}_i = \bar{G}_i(P, T, x_1, x_2, \ldots, x_{c-1}) \qquad (2.8)$$

The number of independent intensive variables in this equation is called *the independent variables of states for a single phase*, which consists of P, T, and $c - 1$ composition variables, or simply $c + 1$ variables. Under specified conditions, such as fixing P, T, and any number of composition variables, and the coexistence of several phases, the number of independent variables of state is reduced. The number of independent variables of state of any system under equilibrium is called *the degree of freedom, or variance.*

If magnetic, electric, and gravitational fields were also variables of state in equation (2.7), the number of independent variables would increase by 3 and the degrees of freedom would be $4 + c$ instead of $1 + c$. Additional variables of state do not present particular difficulties, and for simplicity it is sufficient to consider the variables of state as given in equation (2.7). The total differential of $\mathcal{G}$ for a phase is

$$d\mathcal{G} = \frac{\partial \mathcal{G}}{\partial P} dP + \frac{\partial \mathcal{G}}{\partial T} dT + \bar{G}_1\, dn_1 + \bar{G}_2\, dn_2 + \cdots$$

The first two terms represent the contributions to the Gibbs energy when the number of moles of all components is constant, or when the system is a closed system; hence, from $d\mathcal{G} = \underline{V}\, dP - \underline{S}\, dT$, $\partial\mathcal{G}/\partial P = \underline{S}$, it follows that

$$d\mathcal{G} = \underline{V}\, dP - \underline{S}\, dT + \bar{G}_1\, dn_1 + \bar{G}_2\, dn_2 + \cdots \qquad (2.9)$$

where $\underline{V}$ and $\underline{S}$ are the volume and the entropy, respectively, both of which are extensive properties.

The Gibbs energy $\mathcal{G}$ for a system consisting of ϕ phases is the sum of Gibbs energies of all phases, i.e.,

$$\mathcal{G} = \mathcal{G}^{\mathrm{I}} + \mathcal{G}^{\mathrm{II}} + \cdots + \mathcal{G}^{\phi} \qquad (2.10)$$

where the superscripts refer to the coexisting phases. The total differential of $\mathcal{G}$ is then the sum of $d\mathcal{G}^{\mathrm{I}}$, $d\mathcal{G}^{\mathrm{II}}$, $\ldots$, each of which is written out according to equation (2.9); therefore,

$$\begin{aligned}
d\mathcal{G} = {}& (\underline{V}^{\mathrm{I}} + \underline{V}^{\mathrm{II}} + \cdots + \underline{V}^{\phi})\, dP - (\underline{S}^{\mathrm{I}} + \underline{S}^{\mathrm{II}} + \cdots + \underline{S}^{\phi})\, dT \\
& + \bar{G}_1^{\mathrm{I}}\, dn_1^{\mathrm{I}} + \bar{G}_1^{\mathrm{II}}\, dn_1^{\mathrm{II}} + \cdots + \bar{G}_1^{\phi}\, dn_1^{\phi} \\
& + \bar{G}_2^{\mathrm{I}}\, dn_2^{\mathrm{I}} + \bar{G}_2^{\mathrm{II}}\, dn_2^{\mathrm{II}} + \cdots + \bar{G}_2^{\phi}\, dn_2^{\phi} \\
& \;\;\vdots \\
& + \bar{G}_c^{\mathrm{I}}\, dn_c^{\mathrm{I}} + \bar{G}_c^{\mathrm{II}}\, dn_c^{\mathrm{II}} + \cdots + \bar{G}_c^{\phi}\, dn_c^{\phi}
\end{aligned} \qquad (2.11)$$

The system is in equilibrium under four specific conditions dictated by

$$d\varnothing_{P,T,n} = 0 \tag{2.12}$$

where (1) the change in Gibbs energy is zero, (2) pressure and (3) temperature are fixed and uniform from one phase to another, and (4) the system must be closed, and therefore the amount of each component $n_1, n_2, \ldots, n_c$ is fixed as indicated by the subscript n. The last condition requires that

$$dn_1 = 0 = dn_1^{\mathrm{I}} + dn_1^{\mathrm{II}} + \cdots + dn_1^{\phi}$$

$$dn_2 = 0 = dn_2^{\mathrm{I}} + dn_2^{\mathrm{II}} + \cdots + dn_2^{\phi}$$

$$\vdots$$

$$dn_c = 0 = dn_c^{\mathrm{I}} + dn_c^{\mathrm{II}} + \cdots + dn_c^{\phi}$$

Multiplication of the first equality in this set of equations by $\bar{G}_1^{\mathrm{I}}$, the second by $\bar{G}_2^{\mathrm{I}}, \ldots$, the last by $\bar{G}_c^{\mathrm{I}}$, and subtraction of the results from equation (2.11) after setting $d\varnothing$, dP, and $dT = 0$, yields

$$0 = (\bar{G}_1^{\mathrm{II}} - \bar{G}_1^{\mathrm{I}})\, dn_1^{\mathrm{II}} + (\bar{G}_1^{\mathrm{III}} - \bar{G}_1^{\mathrm{I}})\, dn_1^{\mathrm{III}} + \cdots + (\bar{G}_1^{\mathrm{P}} - \bar{G}_1^{\mathrm{I}})\, dn_1^{\phi}$$

$$+ (\bar{G}_2^{\mathrm{II}} - \bar{G}_2^{\mathrm{I}})\, dn_2^{\mathrm{II}} + (\bar{G}_2^{\mathrm{III}} - \bar{G}_2^{\mathrm{I}})\, dn_2^{\mathrm{III}} + \cdots + (\bar{G}_2^{\mathrm{P}} - \bar{G}_2^{\mathrm{I}})\, dn_2^{\phi}$$

$$\vdots$$

$$+ (\bar{G}_c^{\mathrm{II}} - \bar{G}_c^{\mathrm{I}})\, dn_c^{\mathrm{II}} + (\bar{G}_c^{\mathrm{III}} - \bar{G}_c^{\mathrm{I}})\, dn_c^{\mathrm{III}} + \cdots + (\bar{G}_c^{\mathrm{P}} - \bar{G}_c^{\mathrm{I}})\, dn_c^{\phi} \tag{2.13}$$

For every possible value of each dn_i^j, which is generally nonzero, equation (2.13) is zero if, and only if, the coefficient of each dn_i^j is zero; therefore,

$$\bar{G}_1^{\mathrm{I}} = \bar{G}_1^{\mathrm{II}} = \cdots = \bar{G}_1^{\phi}; \qquad (\phi \text{ terms})$$

$$\bar{G}_2^{\mathrm{I}} = \bar{G}_2^{\mathrm{II}} = \cdots = \bar{G}_2^{\phi}$$

$$\vdots$$

$$\bar{G}_c^{\mathrm{I}} = \bar{G}_c^{\mathrm{II}} = \cdots = \bar{G}_c^{\phi}$$

$$\left. \begin{array}{c} \\ \\ \\ \\ \\ \end{array} \right\} [c(\phi - 1) \text{ equations}] \tag{2.14}$$

The method used to derive this equation is called Lagrange's method of undetermined multipliers.[1] Equation (2.14) contains a very important thermodynamic statement first derived by J. W. Gibbs (1875), that the chemical potential of a component is the same in all phases under equilibrium conditions. The term "chemical potential" is still used, but the appropriate term is the "partial molar Gibbs energy". It is interesting to note that for a single-component system, or for $c = 1$, $\bar{G}_1^I = G_1^I$, and $\bar{G}_1^{II} = G_1^{II}$, and this immediately gives a rigorous derivation of equation (2.1).

Phase Rule

The total number of intensive variables that can define each phase is $c - 1$ composition variables, and pressure and temperature. For ϕ coexisting phases, therefore, the total number of intensive variables defining the system is

$$\text{Total number of variables} = 2 + \phi(c - 1) \qquad (2.15)$$

the second term is the total number of composition variables, $c - 1$, for each of the ϕ phases. According to equations (2.14), the value of $\bar{G}_i$ for any selected component i is the same in all the ϕ phases, and each line in equations (2.14) represents $\phi - 1$ independent equations, and further, c lines represent $c(\phi - 1)$ independent equations, i.e.,

$$\text{Total number of independent equations} = c(\phi - 1) \qquad (2.16)$$

The number of independent variables, called the degrees of freedom, and denoted by Y, is equal to the total number of variables, minus the number of independent equations, i.e., $2 + \phi(c - 1) - c(\phi - 1)$, or simply

$$Y = c - \phi + 2 \qquad (2.17)$$

This equation is the formal statement of the well-known phase rule originally derived by J. W. Gibbs (1875). The degrees of freedom, or the variance, Y, represents the number of unrestricted variables of state which may be a set of variables out of P, T, and $c - 1$ compositions. It must be emphasized that Y can be zero or a positive number, but never a negative number. It should be remembered that the composition variables refer to each individual phase and not to the bulk composition of a heterogeneous system of two or more phases. The permissible number of composition restrictions could be all in one phase or in several phases; for example, fixing the mole fractions x_i and x_j of components i and j in one phase, or fixing x_i in one phase and x_j in another phase, or fixing only x_i in two coexisting phases decreases the degrees of freedom by two.

All the components were assumed to be soluble in all phases in the derivation of the phase rule. However, in some cases the solubility of certain components in some phases is very small or practically zero. If, for example, the solubilities of k_1 components in Phase I, and k_2 components in Phase II are zero, the number of composition variables for Phase I decreases by k_1, and for Phase II, by k_2. Therefore, when k_i components are insoluble in ϕ_j phases, the total number of variables in equation (2.14) decreases by $k_i\phi_j$ so that the right side becomes $2 + \phi(c - 1) - k_i\phi_j$. The Gibbs energy $\mathscr{G}^q$ is independent of n_i in a phase q in which i is insoluble; hence, for $\bar{G}_i = \partial\mathscr{G}^q/\partial n_i = 0$. Consequently, the number of independent equations in equation (2.14) decreases by ϕ_j on each line of equation (2.14) and by $k_i\phi_j$ on all lines; the phase rule then becomes

$$2 + \phi(c - 1) - k_i\phi_j - [c(\phi - 1) - k_i\phi_j] = c - \phi + 2 \qquad (2.18)$$

Therefore, the absence of some components in some of the phases does not alter the phase rule. It is often convenient to fix the pressure and decrease the degrees of freedom by one in dealing with condensed phases such as for substances with low vapor pressures. The phase rule then becomes

$$Y = c - \phi + 1 \qquad \text{(constant pressure)} \qquad (2.19)$$

A few examples illustrate some applications of the phase rule represented by equation (2.17). In a single-component system, for one phase, $Y = 1 - 1 + 2 = 2$, and hence pressure and temperature may vary independently; for two phases, $Y = 1 - 2 + 2 = 1$, either pressure or temperature may vary independently, and if pressure or temperature is fixed $Y = 0$, or the system is completely defined. If pressure and temperature are both fixed in the last case, $Y = -1$; therefore, either the phase rule is violated, or an unnecessarily large number of restrictions are imposed on the system. For three phases, Y is zero and again the system is completely defined. If either temperature or pressure is varied, one of the three phases must disappear to permit this degree of freedom.

In a system consisting of three components, the maximum value of Y is 4, since the minimum number of phases for any system is 1. When $\phi = 3$, then Y is 2 and Y may be chosen out of the following total number of variables: 2(pressure and temperature) + 3(3 − 1)(composition variables) = 8.

Phase Diagrams

A phase diagram represents a map of coexisting phase boundaries as affected by the variables of state. The relationships between the phase diagrams and the molar Gibbs energy diagrams are presented in detail in the remaining sections of this chapter. We limit our discussion to condensed phase diagrams in which pressure is held constant. The removal of the restriction on pressure does not require a special treatment, but complicates the phase diagrams. We shall consider the temperature–composition phase diagrams, involving the solids and liquids, and limit our presentation largely to binary phase diagrams.[2-5]

A condensed binary phase diagram represented on the composition-temperature coordinates consists of the single-phase regions separated by the two-phase regions as shown in Figs. 2.2–2.4. The curves separating the single phases from two phases represent the compositions of single phases. When a homogeneous liquid phase freezes to form a homogeneous solid, the phase diagram may be as in Fig. 2.2(A) with the phase boundary curves spaced as indicated, or as in Fig. 2.2(B) with a common maximum point for both boundary curves, or as in Fig. 2.2(C) with a common minimum

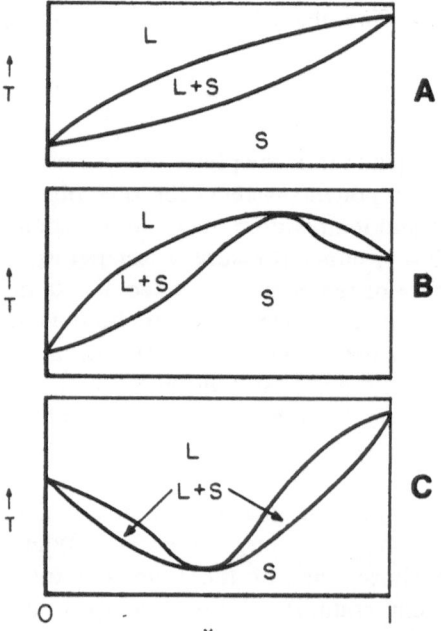

Figure 2.2. Hypothetical binary phase diagrams for liquid (L), solid (S), and S + L phase fields. Upper curve in each panel is "liquidus," and lower curve, "solidus."

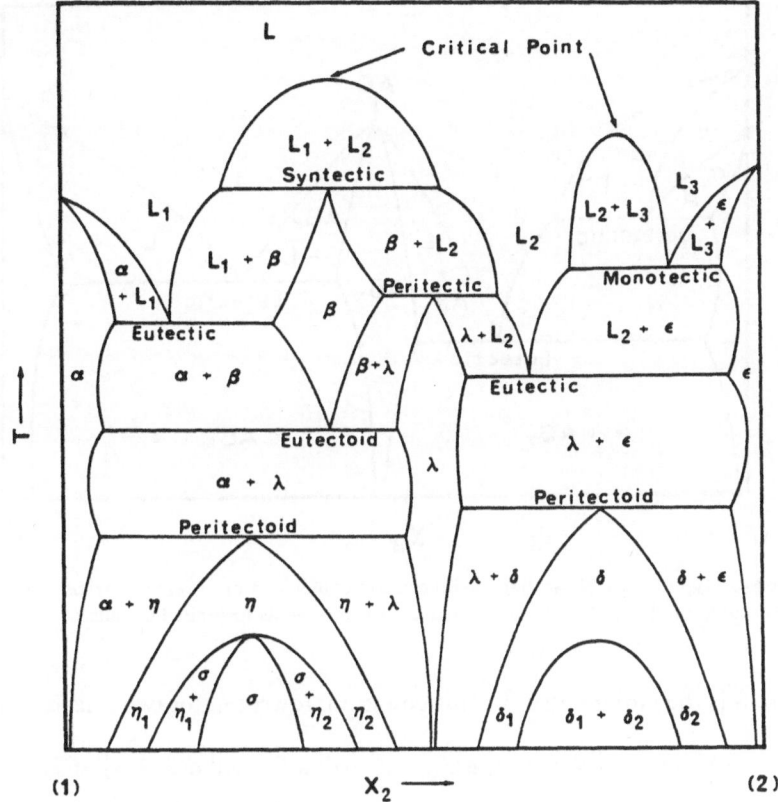

Figure 2.3. Hypothetical phase diagram for various transformations.

point. The curve showing the start of solidification, called the liquidus, is tangent to the curve showing the completion of freezing, called the solidus, at the same extremum point in Fig. 2.2(B) and (C). This requirement is necessitated by the Gibbs–Konovalow theorem,[1-6] which can be proved by using the Gibbs energy $\mathcal{G}$ written as $\mathcal{G} = n_1 \bar{G}_1 + n_2 \bar{G}_2$, and functionally as $\mathcal{G} = \mathcal{G}(P, T, n_1, n_2)$. The total differentials of these equations are

$$d\mathcal{G} = n_1 \, d\bar{G}_1 + n_2 \, d\bar{G}_2 + \bar{G}_1 \, dn_1 + \bar{G}_2 \, dn_2$$

$$d\mathcal{G} = \underline{V} \, dP - \underline{S} \, dT + \bar{G}_1 \, dn_1 + \bar{G}_2 \, dn_2$$

These equations yield $\underline{V} \, dP - \underline{S} \, dT = n_1 \, d\bar{G}_1 + n_2 \, d\bar{G}_2$, and after division with $(n_1 + n_2)$, and then imposing the constant-pressure restriction, we obtain

$$-S \, dT = x_1 \, d\bar{G}_1 + x_2 \, d\bar{G}_2 \qquad (2.20)$$

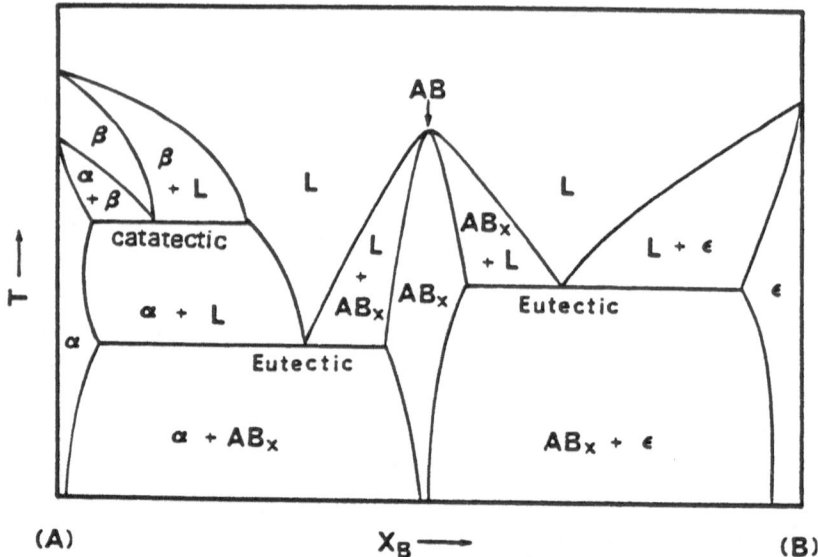

Figure 2.4. Hypothetical phase diagram for catatectic (metatectic) and eutectic transformations, and for compound AB capable of dissolving its components to limited extents.

where S is the molar entropy. This equation, rewritten for two phases, yields

$$-\underline{S}^I\, dT = x_1^I\, d\bar{G}_1 + x_2^I\, d\bar{G}_2; \qquad -S^{II}\, dT = x_1^{II}\, d\bar{G}_1 + x_2^{II}\, d\bar{G}_2$$

where $\bar{G}_i$ is the same for both phases for a given temperature. Rearrangement of the preceding equations and division by dx_2^I yields

$$(S^I - S^{II})\frac{dT}{dx_2^I} = (x_1^{II} - x_1^I)\frac{d\bar{G}_1}{dx_2^I} + (x_2^{II} - x_2^I)\frac{d\bar{G}_2}{dx_2^I} \qquad (2.21)$$

It is important to note that phases I and II may be liquid and solid, respectively. This equation leads to the proof of the Gibbs–Konovalow theorem in its concise and most useful form.

THEOREM. *If a phase boundary curve for Phase I meets another phase boundary curve for Phase II at a point $(x_2^I = x_2^{II}, T_2)$, both curves must become horizontally tangent to each other.*

PROOF. The compositions of both phases are equal, hence $x_1^{II} - x_1^I$ and $x_2^{II} - x_2^I$ in equation (2.21) are both zero; further, $d\bar{G}_1/dx_2^I$ and $d\bar{G}_2/dx_2^I$ are both finite because $\bar{G}_1$ and $\bar{G}_2$ are continuous functions of T and x_2, and $S^I - S^{II}$ is finite and nonzero; therefore, dT/dx_2^I must be zero. Equation (2.21) can be divided by dx_2^{II} instead of dx_2^I, and the same argument can be followed to show that dT/dx_2^{II} is also zero. Therefore, both curves must be horizontally tangent to each other and this completes the proof.

Eutectic-Type Reactions

The liquidus and solidus lines are often depressed as shown in the upper left of Fig. 2.3 in such a way that at a particular composition and temperature, the liquid and two solids, i.e., three phases, coexist. The reaction upon cooling is liquid(L_1) → solid(α) + solid(β), which is called the eutectic reaction. Above and below the eutectic temperature, one phase must disappear according to the phase rule. There are other similar reactions involving three phases with various possible combinations. Each reaction is named according to the states of aggregation of phases and the number of reactant phases. The suffix "tectic" is used for reactions involving one

Table 2.1. Three-Phase Equilibria in Binary Systems

Name	Diagram	Reaction upon cooling
	Transitions involving liquid phases	
Eutectic		Liquid(L_1) → solid(α) + solid(β)
Monotectic		Liquid(L_3) → liquid(L_2) + solid(ε)
Peritectic		Solid(β) + liquid(L_2) → solid(λ)
Syntectic		Liquid(L_1) + liquid(L_2) → solid(β)
Catatectic[7] (metatectic)		Solid(β) → solid(α) + liquid(L)
	Transitions involving only solid phases	
Eutectoid		Solid(β) → solid(α) + solid(λ)
Peritectoid		Solid(α) + solid(λ) → solid(η)

or two liquid phases, and the suffix "tectoid" for three solid phases. Various types of three-phase reactions are presented in Figs. 2.3 and 2.4 and summarized in Table 2.1 with the coexisting phases marked as in these figures.

These transformations consist of (1) eutectic-type, in which the reactant is a single phase, and (2) peritectic-type, in which the reactants consist of two phases. Thus, there are four eutectic-type and three peritectic-type transitions in the preceding list.

A critical point exists when one phase dissociates into two phases at a point where the phase boundary has a horizontal inflection point, i.e., when $\partial T/\partial x_2$ and $\partial^2 T/\partial x_2^2$ are both zero as shown in Fig. 2.3 for $L \to L_1 + L_2$, $L \to L_2 + L_3$, and $\delta \to \delta_1 + \delta_2$. It should be noted that the maximum point in the η-region does not have an inflection point and the phase boundaries for η and σ are tangent to each other at their common maximum points. The maximum point in the center of Fig. 2.4 is for a congruently melting compound, AB, capable of dissolving its component elements to limited extents as indicated by the phase region AB_x.

The limits of solubility of one solid phase in another solid phase is sometimes called the solvus curve, or briefly, the solvus. Thus, the ε-phase boundary below the eutectic in Fig. 2.4 is the solvus, which is the limits of solubility of AB_x in ε. However, the ε-phase boundary above the eutectic is the solidus.

The preceding diagrams contain all the possible phase equilibrium types encountered in the condensed phase binary systems. Various compilations of phase diagrams exist,[8-11] and recently evaluated diagrams are published in the *Bulletin of Alloy Phase Diagrams* (a bimonthly journal begun in 1980, published by ASM-NBS).

Erroneous Diagrams

A number of important aspects of the phase equilibria summarized in Figs. 2.2-2.4 must be observed to avoid errors in drawing the phase boundaries. All such errors violate (a) the phase rule, (b) the Gibbs-Konovalow theorem, and (c) the requirement that the extended portions of the phase boundaries terminate in the two-phase regions. Examples of these violations are illustrated in Fig. 2.5 and discussed as follows.

a. Along BC, α, L_1, L_2, and β coexist and thus violate the phase rule since the degree of freedom Y with four coexisting phases is -1 and Y cannot assume a negative value. This error can be corrected by joining L_1 and L_2 at one point on the straight line BC. On D, α, ε, λ, and β coexist because at E there are more than two curves and one straight line intersecting one another; the phase boundary curve ME must therefore not terminate at E. Accordingly, there must be no more than one straight line and two

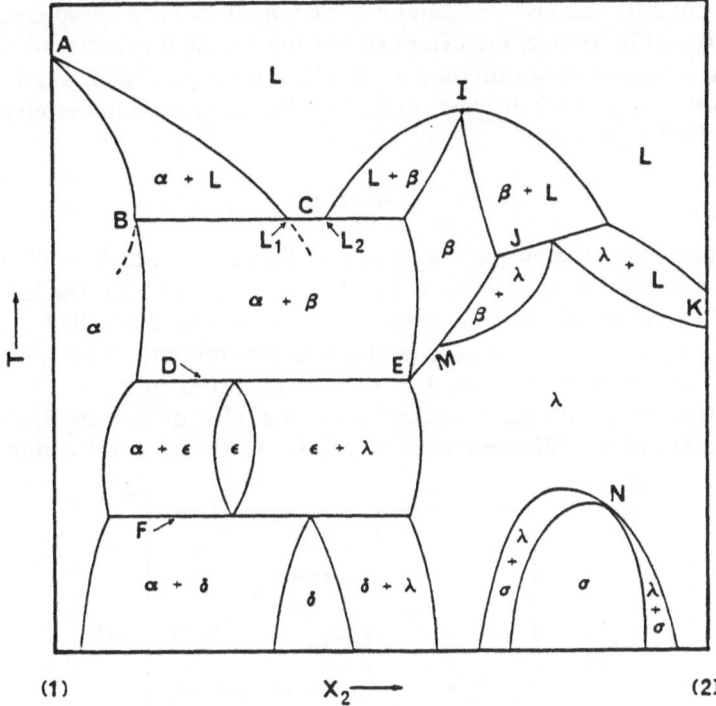

Figure 2.5. Errors in a hypothetical phase diagram violating phase rule or other thermodynamic principles.

curves at a point of intersection of phase boundaries. On F, a eutectoid and a peritectoid on the same horizontal line signifies that α, ε, δ, and λ coexist and violate the phase rule. There are three phases on the straight line J, and since J is not horizontal these phases coexist over a temperature range and violate the phase rule since $Y = -1$. At K, there are two phases λ and L over a temperature range for pure component 2, and again $Y = -1$; the phase boundaries must therefore intersect the vertical line for the pure component at the same point.

b. At I, M, and N, the Gibbs–Konovalow theorem is violated, i.e., each pair of phase boundaries does not meet at an extremum point. At M, the phase boundaries should not meet at all, and an entirely different construction must be made to eliminate the error, particularly because of the erroneous separation of β and λ along a curve EM instead of a two-phase field.

c. The extended portion of the phase boundary at B, shown by the dashed extended curve, and the unstable liquid on the extended portion of AC are in equilibrium. Since the liquid below C is unstable, the dashed curve below B must also be in an unstable region, not in a one-phase region which is stable. Hence, the extended portion below B must terminate in a two-phase region, or within the area BCDE. This is possible when the phase boundary curves and the horizontal line BC intersect one another at an angle smaller than 180°.

Lever Rule

Figure 2.6(a) shows a simple phase diagram in which solid phase λ dissociates at the critical point Q into two solids λ_1 and λ_2. The horizontal line ACB ending on the phase boundaries at A and B is called a tie-line. Point C represents the bulk composition x_c of a mixture of two phases. As point C moves from A toward B, the compositions of λ_1 and λ_2 remain unchanged as x_A and x_B, respectively, but the mass of λ_2 increases relative to the mass of λ_1. All compositions x_i refer to the atomic fraction of the

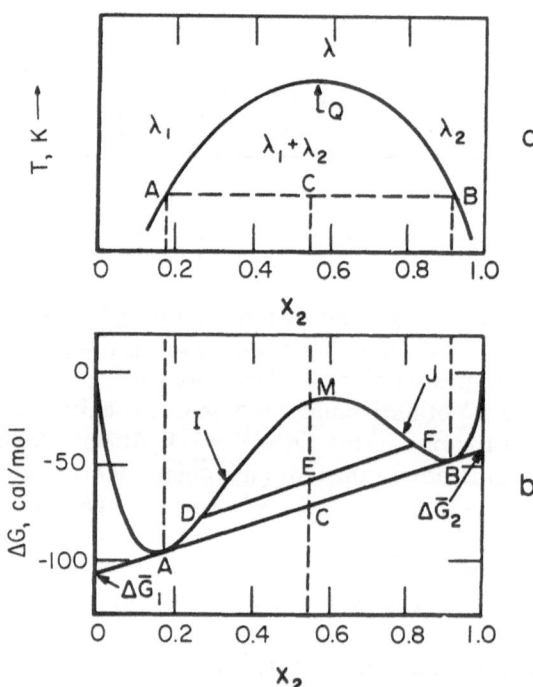

Figure 2.6. (a) Hypothetical binary phase diagram for dissociation of solid λ-phase into two solid phases, $\lambda_1 + \lambda_2$; critical point is Q; bulk composition is taken as C at temperature corresponding to AB; (b) ΔG versus x_2 for temperature corresponding to ACB. Equation (2.24) represents ΔG curve for 1 g-atom of alloy; A, C, and B represent corresponding compositions in both panels.

second component. Let $m(A)$ and $m(B)$ represent gram atoms of λ_1 and λ_2 at A and B, respectively, with the restriction that $m(A) + m(B) = 1$. Atomic balance requires that

$$x_A m(A) + x_B m(B) = x_C[m(A) + m(B)]$$

This equality can be used to derive

$$\frac{m(A)}{m(B)} = \frac{x_B - x_C}{x_C - x_A} = \frac{\overline{CB}}{\overline{AC}} \tag{2.22}$$

which is known as the lever rule. Note that x_c representing the bulk composition of two phases is not a composition variable for thermodynamic properties of each phase such as ΔG and $\Delta \bar{G}_i$. Equation (2.22) may be transformed into a useful form by adding 1 to both sides of the first equality and simplifying by using $m(A) + m(B) = 1$ so that

$$m(A) = \frac{x_B - x_C}{x_B - x_A}; \qquad m(B) = \frac{x_C - x_A}{x_B - x_A} \tag{2.23}$$

where the second relationship is obtained from the first by using $m(B) = 1 - m(A)$.

Molar Gibbs Energy of Mixing-Phase Diagrams

The molar Gibbs energy of mixing refers to the process prescribed by x_1 (pure component 1) + x_2 (pure component 2) → 1 gram atom of alloy. A simple phase diagram and the molar Gibbs energy of mixing ΔG are presented in Fig. 2.6. Variation of ΔG with x_2 refers to the temperature represented by the tie-line ACB in Fig. 2.6(a). The curve in Fig. 2.6(b) is drawn by assuming that ΔG is represented by

$$\Delta G = 1000x_2 + 700x_2^2 - 2800x_2^3 + 1100x_2^4$$
$$+ 600[(1 - x_2) \ln(1 - x_2) + x_2 \ln x_2] \qquad \text{(cal/mole)} \tag{2.24}$$

where the last term is the molar Gibbs energy of mixing of an ideal solution, $\Delta G(\text{ideal}) = RT(x_1 \ln x_1 + x_2 \ln x_2)$ with RT taken to be 600 cal/mole (301.93 K) as a convenient simple quantity. The remaining coefficients in equation (2.24) were obtained by using one maximum point M and two minimum points near A and near B. The tangent line ACB to the curve for

ΔG intersects the vertical axis at $x_2 = 0$ at the point corresponding to $\Delta \bar{G}_1 = \bar{G}_1 - G_1^\circ$, and similarly, $\Delta \bar{G}_2 = \bar{G}_2 - G_2^\circ$, according to the tangent intercept method. The tangent line ACB shows that $\Delta \bar{G}_1 = \bar{G}_1 - G_1^\circ$ and $\Delta \bar{G}_2 = \bar{G}_2 - G_2^\circ$ for λ_1 phase are the same as those for λ_2 phase.

The section of the straight line between A and B in Fig. 2.6(b) represents ΔG for the mixture of two phases; thus, $\Delta G(\text{at C})$ is

$$\Delta G(\text{at C}) = \frac{x_B - x_C}{x_B - x_A} \Delta G(\text{at A}) + \frac{x_C - x_A}{x_B - x_A} \Delta G(\text{at B}) \qquad (2.25)$$

which is in accord with the lever rule because ΔG of a mixture of two phases is the sum of ΔG of its constituent phases. Equation (2.25) is linear and it is represented by the straight line between A and B. Likewise, the straight line DEF represents ΔG of two phases having the compositions corresponding to D and F. The value of $\Delta G(\text{at E})$ is larger than that of $\Delta G(\text{at C})$; therefore, any straight line joining two points on the curve AMB has higher values of ΔG than those corresponding to the straight line ACB at the same values of x_c; hence, a phase mixture along DEF is unstable with respect to that along ACB for the same bulk composition.

The curve for ΔG in Fig. 2.6(b) has two inflection points, one at I $(x_2 = 0.331)$ and the other at J $(x_2 = 0.791)$ as can be shown by substituting these values of x_2 in the second derivatives of equation (2.24) with respect to x_2. These points are called the spinodes, which are important in kinetics of nucleation and growth of new phases from the supersaturated single phases. The equations for the activities of components can be derived and the results can be plotted versus x_2. It can be shown that the maximum and the minimum points in such activity versus composition diagrams for both components coincide with the spinodes.

The shape of the curve in Fig. 2.6(b) changes with increasing temperature as required by the phase diagram. Thus, A and B approach each other and finally minimum, maximum, and the inflection points coincide at the horizontal inflection point Q in Fig. 2.6(a) as required by the phase diagram.

ΔG Diagrams for Other Phases

The diagram shown in Fig. 2.6(a) is for a single solid phase decomposing into two solid phases. For all other types of transformation, it is customary to represent ΔG versus x_2 for each phase at each selected temperature on the same diagram. If the phases in equilibrium are solid and liquid, the convention for writing ΔG for each phase through the entire range of composition at a selected temperature T is as follows:

$$\Delta G(l) \equiv G(l) - x_1(l) G_1^\circ(\text{stable phase at } T)$$
$$- x_2(l) G_2^\circ(\text{stable phase at } T) \qquad (2.26)$$

$$\Delta G(s) \equiv G(s) - x_1(s) G_1^\circ(\text{stable phase at } T)$$
$$- x_2(s) G_2^\circ(\text{stable phase at } T) \qquad (2.27)$$

where $G(l)$ and $G(s)$ are the molar Gibbs energies of the liquid and solid solutions. Equation (2.26) at $x_2 = 1$ becomes $\Delta G(l) = G(l) - G_2^\circ(s)$ when the stable pure phase for component 2 is solid at T, and since for $x_2 = 1$, $G(l)$ is the same as $G_2^\circ(l)$ for the pure liquid 2, then $\Delta G^\circ(l)$ is identical with $\Delta G_{2,m}^\circ$ of melting for pure component 2. If, however, the stable phase for component 2 is liquid, $\Delta G(l)$ in equation (2.26) is then zero at $x_2(l) = 1$.

Reconsider equations (2.26) and (2.27) at a temperature T greater than the melting points of both components so that

$$\Delta G(l) = G(l) - x_1(l) G_1^\circ(l) - x_2(l) G_2^\circ(l) \qquad (2.28)$$

$$\Delta G(s) = G(s) - x_1(s) G_1^\circ(l) - x_2(s) G_2^\circ(l) \qquad (2.29)$$

Addition of $x_1(s) G_1^\circ(s) + x_2(s) G_2^\circ(s)$ to the right side of equation (2.29) and then subtraction of these terms from the same side, followed by a simple rearrangement, yields

$$\Delta G(s) = G(s) - x_1(s) G_1^\circ(s) - x_2(s) G_2^\circ(s)$$
$$- x_1(s) \Delta G_{1,m}^\circ - x_2(s) \Delta G_{2,m}^\circ \qquad (2.30)$$

where $\Delta G_{i,m}^\circ$ is the standard molar Gibbs energy of fusion of pure component i. The first three terms after the equal sign can be transformed into $RTx_1(s) \ln a_1(s) + RTx_2(s) \ln a_2(s)$ by using the definition of activity, i.e., $\bar{G}_i(s) = G_i^\circ(s) + RT \ln a_i(s)$ and observing that $G(s) = x_1(s)\bar{G}_1 + x_2(s)\bar{G}_2$. Likewise, the three terms on the right side of equation (2.28) can be rewritten; the results for equations (2.28) and (2.29) are therefore

$$\Delta G(l) = RTx_1(l) \ln a_1(l) + RTx_2(l) \ln a_2(l) \qquad (2.31)$$

$$\Delta G(s) = RTx_1(s) \ln a_1(s) + RTx_2(s) \ln a_2(s)$$
$$- x_1(s) \Delta G_{1,m}^\circ - x_2(s) \Delta G_{2,m}^\circ \qquad (2.32)$$

A simple example for these equations is now presented, with the following arbitrarily assigned values:

$$T_{1,m} = 900 \text{ K} \qquad \Delta H_{1,m}^\circ = 900R \qquad \Delta S_{1,m}^\circ = R$$

$$T_{2,m} = 300 \text{ K} \qquad \Delta H_{2,m}^\circ = 300R \qquad \Delta S_{2,m}^\circ = R$$

where $T_{i,m}$ is the melting point of i. From the definition of ΔG° as $\Delta G^\circ = \Delta H^\circ - T\Delta S^\circ$, with $\Delta C_p^\circ = 0$, it is evident that

$$\Delta G_{1,m}^\circ = 900R - RT; \qquad \Delta G_{2,m}^\circ = 300R - RT \qquad (2.33)$$

If, in addition, the solution is ideal so that the activities can be set equal to the mole fractions, equations (2.31) and (2.32) become

$$\Delta G(\mathrm{l}) = RTx_1(\mathrm{l}) \ln x_1(\mathrm{l}) + RTx_2(\mathrm{l}) \ln x_2(\mathrm{l}) \qquad (2.34)$$

$$\Delta G(\mathrm{s}) = RTx_1(\mathrm{s}) \ln x_1(\mathrm{s}) + RTx_2(\mathrm{s}) \ln x_2(\mathrm{s})$$
$$- x_1(\mathrm{s})\Delta G_{1,m}^\circ - x_2(\mathrm{s})\Delta G_{2,m}^\circ \qquad (2.35)$$

Equations (2.34) and (2.35) are plotted in Fig. 2.7 for $T = 1000$ K. We note that $\Delta G(\mathrm{s})$ for the solid in Fig. 2.7 is higher than $\Delta G(\mathrm{l})$ for the liquid;

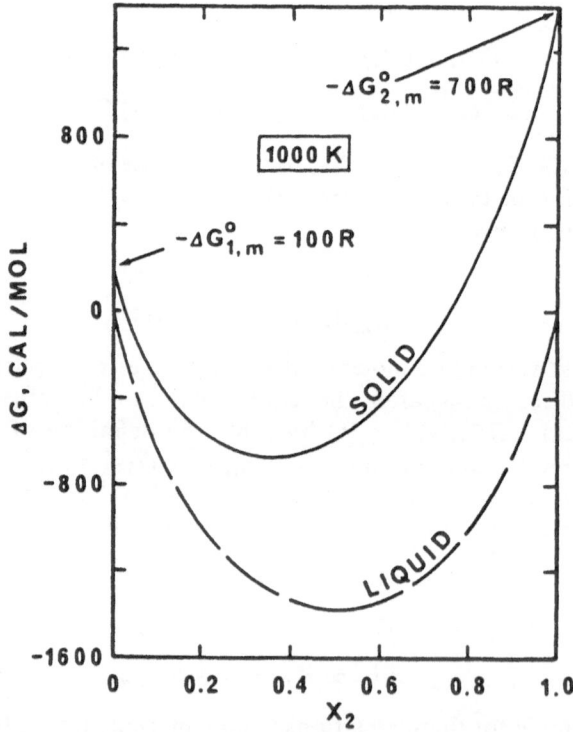

Figure 2.7. Diagram for $\Delta G(\mathrm{l})$ and $\Delta G(\mathrm{s})$ at 1000 K from equations (2.34) and (2.35), respectively. $\Delta G(\mathrm{s}) > \Delta G(\mathrm{l})$ throughout, and solid phase is thus unstable relative to liquid phase. Phase diagram for this system is in lower portion of Fig. 2.8.

hence, the liquid phase is stable relative to the solid phase at 1000 K. The difference between the vertical intercepts (at $x_1 = 1$ and at $x_2 = 1$) for those curves represent $\Delta G^\circ_{1,\text{frz}} = +100R$ and $\Delta G^\circ_{2,\text{frz}} = +700R$ where $\Delta G^\circ_{i,\text{frz}} = -\Delta G^\circ_{i,m}$ with the subscripts frz and m referring to freezing and melting, respectively. The positions of the curves below 300 K are reversed, i.e., the curve for $G(l)$ is above that for $\Delta G(s)$ because the solid phase is stable relative to the liquid phase below 300 K as can be shown by plotting a different but similar figure.

Next to be considered are a set of curves at 550 K for which the liquid and solid phases coexist at appropriate concentrations. These curves are given by

$$\Delta G(l) = G(l) - x_1(l)G^\circ_1(l) - x_2(l)G^\circ_2(l) + x_1(l)\Delta G^\circ_{1,m} \quad (2.36)$$

$$\Delta G(s) = G(s) - x_1(s)G^\circ_1(s) - x_2(s)G^\circ_2(s) - x_2(s)\Delta G^\circ_{2,m} \quad (2.37)$$

Substitution of $a_i = x_i$ and $\bar{G}_i = G^\circ_i + RT \ln x_i$ in these equations gives

$$\Delta G(l) = RTx_1(l) \ln x_1(l) + RTx_2(l) \ln x_2(l) + x_1(l)\Delta G^\circ_{1,m} \quad (2.38)$$

$$\Delta G(s) = RTx_1(s) \ln x_1(s) + RTx_2(s) \ln x_2(s) - x_2(s)\Delta G^\circ_{2,m} \quad (2.39)$$

Substitution of equation (2.33) in these equations gives

$$\Delta G(l) = 550Rx_1(l) \ln x_1(l) + 550Rx_2(l) \ln x_2(l) + x_1(l)350R \quad (2.40)$$

$$\Delta G(s) = 550Rx_1(s) \ln x_1(s) + 550Rx_2(s) \ln x_2(s) + x_2(s)250R \quad (2.41)$$

Equations (2.40) and (2.41) are represented in the upper portion of Fig. 2.8 by the curves marked LIQUID and SOLID respectively. The straight line tangent to both curves gives the compositions of solid and liquid at $x_2 = x_2(s) = 0.450$ and $x_2 = x_2(l) = 0.709$, respectively. Above the tangent, a pair of phases have a higher value of ΔG than the phases at the points of tangency as discussed in conjunction with Fig. 2.6. The liquid phase is stable from $x_2 = 0.709$ to $x_2 = 1$, and the solid phase is stable from $x_2 = 0.0$ to $x_2 = 0.450$, as required by the relatively lower values of ΔG represented by the lower sections of the curves. At the point of intersection of the curves, $\Delta G(l)$ and $\Delta G(s)$ are equal but $\bar{G}_1(l)$ and $\bar{G}_1(s)$, as well as $\bar{G}_2(l)$ and $\bar{G}_2(s)$, are not equal; therefore, there is no equilibrium at this point. The vertical intercepts of the curves in Fig. 2.8 correspond to $\Delta G^\circ_{1,m} = 350R$ and $-\Delta G^\circ_{2,m} = 250R$, as indicated in the figure. The solidus and liquidus points given by the upper portion of Fig. 2.8 are shown in the lower portion. The repetition of the foregoing procedure for $\bar{G}_i$ at various temperatures generates the entire phase diagram as shown in the lower portion of Fig. 2.8.

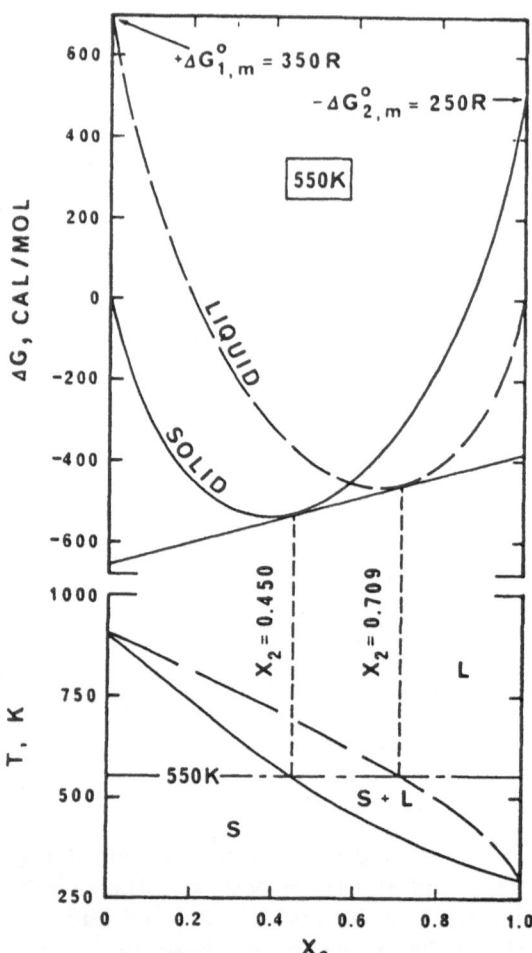

Figure 2.8. Upper diagram is for $\Delta G(l)$ and $\Delta G(s)$ at 550 K from equations (2.36) and (2.37), respectively. Phase diagram in lower portion is obtained from equations (2.43) and (2.44).

The phase boundaries in Fig. 2.8 have been drawn by equating $\bar{G}_i(l) = \bar{G}_i(s)$ and then substituting various values of T to solve for $x_i(s)$ and $x_i(l)$. In this region G_1° is solid but G_2° is liquid. The liquid phase is ideal, and for $\bar{G}_1(l)$ it is evident that $\bar{G}_1(l) = G_1^\circ(l) + RT \ln x_1(l)$, and likewise, $\bar{G}_1(s) = G_1^\circ(s) + RT \ln x_1(s)$, and the equality of $\bar{G}_1(l)$ and $\bar{G}_1(s)$ yields

$$G_1^\circ(l) - G_1^\circ(s) = \Delta G_{1,m}^\circ = -RT \ln \frac{x_1(l)}{x_1(s)} \qquad (2.42)$$

which is simply the equilibrium distribution ratio (or equilibrium constant) for (component 1 in solid phase $\rightleftarrows$ component 1 in liquid phase). Substitution of $\Delta G^\circ_{1,m}$ from equation (2.33) into equation (2.42) and simplification gives

$$1 - \frac{900}{T} = \ln \frac{x_1(l)}{x_1(s)} \tag{2.43}$$

A similar equation is obtained for component 2 by an identical procedure; the resulting final equation is

$$1 - \frac{300}{T} = \ln \frac{x_2(l)}{x_2(s)} \tag{2.44}$$

These equations represent the solidus and liquidus curves. For example, at 550 K, $x_1(l)/x_1(s) = 0.5292$, and $x_2(l)/x_2(s) = 1.5755$, and then substitution of $1 - x_1(l) = x_2(l)$ and $1 - x_1(s) = x_2(s)$ reduces the number of unknowns in the first set of equations by two, leading to $x_2(s) = 0.450$ and $x_2(l) = 0.709$; these points are indicated in Fig. 2.8.

Deviations from ideality modify the curves for ΔG represented in the preceding figures; however, the principles involved in the representations are basically the same. The curves similar to those in Figs. 2.6–2.8 for nonideal solutions require substitution of $a_1 = \gamma_1 x_1$ and $a_2 = \gamma_2 x_2$ in equations (2.31) and (2.32), and expansion to generate a set of terms expressed by $G^e = RT(x_1 \ln \gamma_1 + x_2 \ln \gamma_2)$. Appropriate analytical equations for $G^e(l)$ and $G^e(s)$ must then be added to equations (2.34) and (2.35), respectively, to represent them as functions of composition and temperature for nonideal solutions. A detailed example of such a procedure will be given in Chapter 6 in thermodynamic calculations of the iron–carbon phase diagram.

ΔG Diagrams for Complex Systems

The diagrams of ΔG versus x_2 for complex phase diagrams may be illustrated schematically by the eutectic system shown in Fig. 2.9. The ΔG diagram for the liquid and α phases between the melting points of components 1 and 2, e.g., at T_I, is, in principle, the same as the diagram in Fig. 2.8, and therefore not presented. At T_{II}, below the melting points of components and above the eutectic temperature T_e, the curves are shown in the upper portion of Fig. 2.9. The straight lines tangent to each pair of curves give the compositions of each pair of phases. Single phases are stable outside

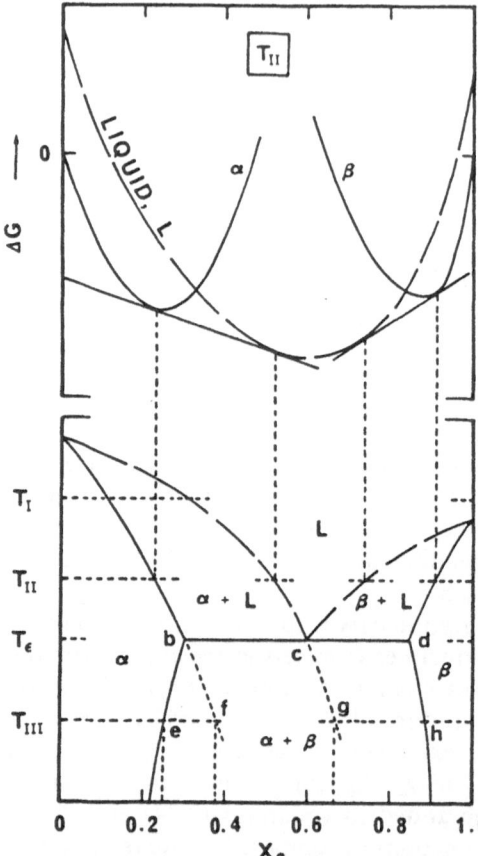

Figure 2.9. Upper figure schematically shows ΔG for liquid α and β phases at T_{II}. Lower figure is phase diagram.

the two-phase regions where ΔG is lower for the stable phases than the unstable phases. Figure 2.10 shows the diagrams at T_e and T_{III} indicated on the phase diagram in Fig. 2.9. At T_e, three phases coexist as shown by the single tangent line to the three curves. As the temperature decreases, the curve for the liquid, $\Delta G(l)$, moves up and the remaining curves move down so that the α-phase at e is in equilibrium with the β-phase at h for T_{III}. Assume that the liquid and α phrases are supercooled as shown by the extended dashed portions of the phase boundaries in Fig. 2.9. The liquid at g, and the α-phase at f must then coexist as shown in Fig. 2.9 and in Fig. 2.10, by the tangent line at f and g. The location of the curve for $\Delta G(l)$ at T_{III} necessitates that the tangent at f and g be in the two-phase region, and since the values of ΔG for the two phases at f and g are higher than

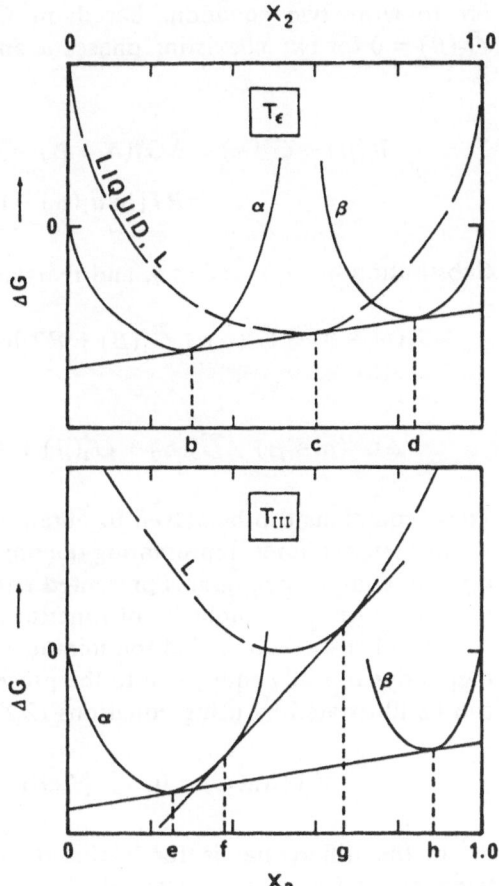

Figure 2.10. Upper figure schematically shows ΔG for liquid α and β phases at T_e, and lower figure, at T_{III}. For T_e, T_{III}, and compositions, see phase diagram of Fig. 2.9.

those at e and h, the phases at f and g are unstable. The extended portions of the phase boundaries in Fig. 2.9, terminating in the two-phase regions, are therefore properly constructed. This is possible when the angles between the adjoining curves and the eutectic line at b and at d are less than 180° as stressed earlier.

Calculation of Phase Diagrams from Thermodynamic Data

The criterion for equilibrium is expressed by the equality of $\bar{G}_1$ as well as $\bar{G}_2$ in the coexisting phases. The required calculations for a binary system are to write two equations based on $\bar{G}_1(\alpha) - \bar{G}_1(\beta) = 0$ and $\bar{G}_2(\alpha) - \bar{G}_2(\beta) = 0$ for two coexisting phases α and β. The result for component 1 is

$$G_1^\circ(\beta) - G_1^\circ(\alpha) = \Delta G_1^\circ(\alpha \to \beta) = \Delta H^\circ(\alpha \to \beta) - T\Delta S^\circ(\alpha \to \beta)$$
$$= RT[\ln a_1(\alpha) - \ln a_1(\beta)] \tag{2.45}$$

Substitution of $\bar{G}_i^e = RT \ln \gamma_i$ and rearrangement of the result gives

$$\Delta G_1^\circ(\alpha \to \beta) - \bar{G}_1^e(\alpha) + \bar{G}_1^e(\beta) + RT \ln\{[1 - x_2(\beta)]/[1 - x_2(\alpha)]\} = 0 \tag{2.46}$$

$$\Delta G_2^\circ(\alpha \to \beta) - \bar{G}_2^e(\alpha) + \bar{G}_2^e(\beta) + RT \ln[x_2(\beta)/x_2(\alpha)] = 0 \tag{2.47}$$

These equations can be solved to obtain the values of $x_2(\alpha)$ and $x_2(\beta)$ at various temperatures, representing the phase boundaries. There are various types of computer programs presented and discussed for this purpose,[12-17] depending on the complexity of functions chosen for $\bar{G}_i^e$.

Another method, called the minimization of Gibbs energy method, is thermodynamically equivalent to the preceding method. Its basic principle can be illustrated by using equations (2.24) and (2.25), and recalling that

$$d\Delta G(\text{at C}) = 0; \qquad [\Delta G(\text{at C}) \text{ is a minimum}] \tag{2.48}$$

where the differential is for a closed system at constant pressure and temperature. The total differential of a function $u = u(x_A, x_B)$ is zero when $\partial u/dx_A$ and $\partial u/\partial x_B$ are zero in $du = (\partial u/\partial x_A)\, dx_A + (\partial u/\partial x_B)\, dx_B$; therefore,

$$\frac{\partial \Delta G(\text{at C})}{\partial x_A} = 0; \qquad \frac{\partial \Delta G(\text{at C})}{\partial x_B} = 0 \tag{2.49}$$

For this purpose we label x_2 in equation (2.24) once as $x_A = x_2(A)$ and then as $x_B = x_2(B)$ to obtain the equations for $\Delta G(\text{at A})$ and $\Delta G(\text{at B})$, respectively. These equations must then be substituted in equation (2.25) and then in equation (2.49) to obtain two simultaneous equations with two unknown mole fractions x_A and x_B. The solution of these equations yields $x_A = 0.170$, $x_B = 0.919$ at $RT = 600$ cal/mole (301.93 K). The calculational detail[1] shows that x_c cancels out in setting equation (2.49) to zero.

The foregoing methods can be reversed to obtain thermodynamic properties of mixtures, in particular G^e, from the phase diagrams. The computational procedure in this process is simpler because it does not involve solving for the unknowns with the logarithmic terms. The accuracy of the experimental methods for the determination of the phase boundaries, particularly at high temperatures, is not sufficient for highly reliable values of G^e. The general principle based on experience in this respect can be summarized as follows: Moderately accurate data for G^e or G_i^e can generate an excellent phase diagram, but a very highly accurate phase diagram is necessary to obtain moderately reliable values of G^e or $\bar{G}_i^e$.

Empirical Rules on Phase Stability

Examination and generalization of binary phase diagrams have led to interesting, and sometimes controversial empirical rules set forth by a number of investigators. The most widely known such rules bear the names of Hume-Rothery (1934) for substitutional solid solutions, and Hägg (1929) for interstitial solid solutions. In substitutional solutions the atoms of metals A and B occupy equivalent lattice positions, whereas in interstitial solutions, the smaller atoms of B occupy the voids or the interstices in the lattice of the larger atoms A, often by small expansion of the A-lattice. As more binary phase diagrams became established in time, and more knowledge of electronic properties developed, these rules were modified, not only by the original proponents, but also by various investigators. The Hume-Rothery rules[18,19] have been stated and restated by himself and others, often with differing interpretations.[20] The main reason for this is the existence of various scales for electronegativities, atomic radii, and valencies. Therefore, it is appropriate to begin the Hume-Rothery rules by eliminating one of the original four rules, called the "relative valency effect," according to which a metal of lower valency was regarded as more likely to dissolve a metal of higher valency as a solute than vice versa.[18,20] This elimination, recommended by Massalski,[20] is also in accord with the more recent monograph by Hume-Rothery et al.[18] Further, a statistical analysis by Gschneidner[21] shows that this rule is not useful, though it works out well for the alloys of Cu, Au, and Ag with the B-subgroup elements. This leaves the following Hume-Rothery rules accepted by recent critical reviewers.[18,20,21]

Rule 1, size-factor. Solid solubility of a metal A in B (or B in A) is restricted to a few atomic percent (less than 5 at.% according to Gschneidner[22]) if the difference between the atomic radii of A and B is more than 15%. The solubilities generally decrease further with increasing size differences in excess of 15%. For example, the atomic radius of Cu is 1.28 Å, and that of Cd, 1.52 Å, and Cd is therefore about 19% larger than Cu, and only 1.7 at.% Cd is soluble in Cu. In contrast, the atomic radius of Zn is 1.37 Å, which is 7% larger than that of Cu, and Cu–Zn form extensive solid solutions. The radii recommended by Hume-Rothery are half of the closest distances of approach of atoms in crystals of the pure elements, but in the examples cited here, the conclusion is not significantly affected when other atomic radii are used.

Within the favorable range of 15%, known as the *favorable size-factor*, other properties of atoms play important roles; consequently, the favorable size-factor is *not sufficient* for substantial terminal solubilities. For example Mg (1.60 Å) and Sb (1.61 Å) have nearly identical atomic radii, but their mutual solubilities are less than 0.04 at.%. This rule is therefore a negative rule; i.e., if the size difference between two metals exceeds 15%, then the mutual solubilities are limited, but if the size difference is less than 15%, the mutual solubilities may or may not be extensive. Theoretical justification for this rule has been made by Friedel, Blandin, Eshelby, and others.[2]

Rule 2, electronegativity factor. Formation of stable intermetallic compounds, more appropriately stable intermediate phases, will restrict terminal solid solubilities. The possibility of formation of such compounds increases with increasing differences in the electronegativities of component metals. In general, when the difference in Pauling electronegativities exceeds ±0.4 volt, the solubilities are restricted even when the size-factor is favorable. This rule is also known as the *rule of electronegativity effect*.

Rule 3, electron concentration factor. In many alloy systems an important factor that determines the extent of solubilities in terminal and intermediate phases is the *electron concentration*, which is usually expressed in terms of electrons per atom of alloy, e/at. The compositional range of existence of each phase having a particular crystal structure is frequently observed to correspond to a narrow range of electron concentration.

Computation of the e/at. ratios is made by adding the electrons contributed by each pure component element and then dividing the sum by the number of atoms in the alloy. The number of electrons contributed to e/at. by each element is not a universally accepted quantity. Usually, but not always, the outermost s and p orbitals are considered to contribute to the electron concentration.[24] Indeed, in the B-subgroup elements with paired d-electrons, it is generally agreed that only the s- and p-orbitals contribute electrons. (d-electrons are completely paired when there are ten such electrons in the same orbital.) However, the convention is by no means universal for the transition and noble elements in which the inner orbitals are assumed to contribute to the electron concentration. Any method of selection is likely to be controversial, but the main objective is to develop a consistent empirical picture correlating the phase regions with the e/at. ratio. A set of assigned values for selected elements, commonly used for computing $e/\text{at.}$ ratios, are listed in Table 2.2.

For a better fit of experimental results, numerous investigators have presented arguments for assigning variable valencies to Cr, Mn, Fe, Co, and Ni. When phase diagrams are plotted as temperature versus $e/\text{at.}$, several interesting features and similarities among various systems are exhibited. Unfortunately, there is no general agreement on the electron concentration ranges' for various phases encountered in various binary diagrams. In addition, frequently a phase of known crystal structure in one binary diagram does not appear in another binary diagram in the same range of $e/\text{at.}$ ratio. Evidently, this as well as other rules encounter greater degrees of success for the elements in particular groups in the periodic chart. Greater degrees of divergences and greater numbers of exceptions occur when greater numbers of metals are considered. It is therefore advisable to apply these rules with appropriate modifications to limited selected groups of elements for greater degrees of success.

The Cu–Zn diagram is shown in Fig. 2.11, in which Cu and Zn contribute one and two electrons per atom, respectively. Two simple structures among six encountered in the Cu–Zn system are (1) the bcc β-phase that exists at 1.36 to 1.55 in $e/\text{at.}$ (this phase is sometimes written as CuZn to show its roughly equiatomic composition), and (2) the cph ε-phase that occurs at 1.78 to 1.87 in $e/\text{at.}$ (sometimes designated as $CuZn_3$). Such phases are frequently called *electron phases*, and they often show no definite stoichiometry to be called *electron compounds*. A few examples of electron phases are listed in Table 2.3 where (1) the limit of maximum solid solubility of Cu corresponds to $e/\text{at.} \approx 1.4$, (2) bcc structure appears as Zn is added in Cu when $e/\text{at.} \approx 1.5$, (3) γ-phase boundary corresponds to $e/\text{at.} \approx 1.6$ to 1.65, and (4) cph boundary for the ε-phase is variable but $e/\text{at.}$ lies about 1.8 for most binary alloys of the noble metals. For each group of elements a similar scheme can be devised with different values of $e/\text{at.}$ for

Table 2.2. Electron Contributions of Selected Elements to Their Alloys

Group	$e/\text{at.}$
IB: Cu, Ag, Au	1
IIA: Be, Mg, Ca,...	2
IIB: Zn, Cd, Hg	2
IIIB: B, Al, Ga,...	3
IVB: Si, Ge, Sn, Pb	4
VB: P, As, Sb, Bi	5
VIIIA: Fe, Co, Ni,...	0

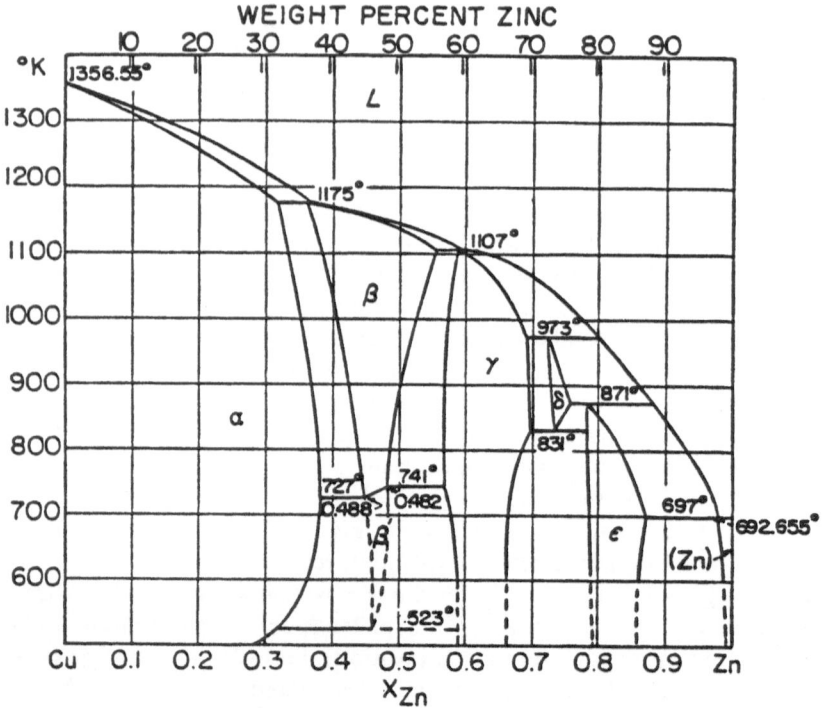

Figure 2.11. Cu–Zn phase diagram. (Adapted from Hultgren *et al.*[11] with permission.)

Table 2.3. Ranges of Electron/Atom Ratios for Selected Electron Phases

Alloy	fcc upper boundary	Minimum bcc boundary	γ-phase[a] boundary	cph boundary (ε)
Cu–Zn	1.38[b]	1.48	1.58–1.66	1.78–1.87[c]
Cu–Al	1.41	1.48	1.63–1.77	—
Cu–Si	1.42	1.49	—	—
Cu–Sn	1.27	1.49	1.60–1.63	1.63–1.75
Ag–Zn	1.38	—	1.58–1.63	1.67–1.90
Ag–Al	1.41	—	—	1.55–1.80
Au–Zn	1.31	—	—	—

[a] Complex cubic.
[b] Lower limit is 1.0.
[c] Terminal Zn-rich η-phase is also cph, and contains only 3 at.% maximum Cu, with e/at. = 1.97 minimum; see Cu–Zn phase diagram (Fig. 2.11).

various phases. The electron concentration factor is therefore not a definite factor, but instead is a scheme that attempts to predict the existence of various phases at certain ranges of values for electron concentration. Based on the foregoing analysis, an acceptable form of electron concentration factor may be stated as follows: the boundaries of each phase in the alloys formed by each group of metals with the remaining metals occur at approximately the same ranges of electron concentration.

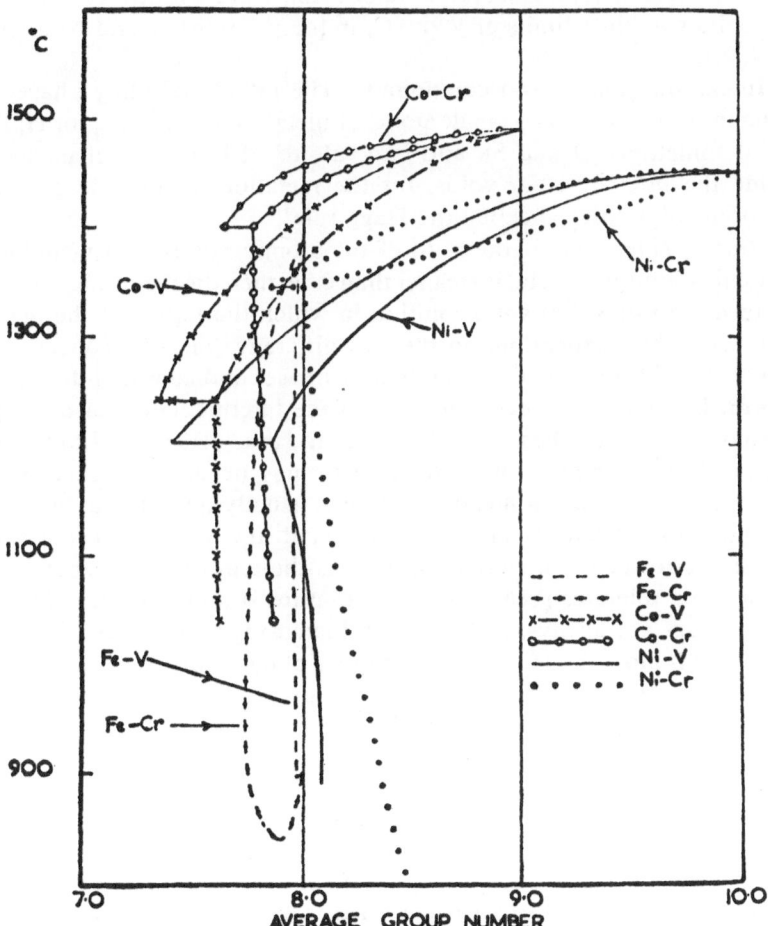

Figure 2.12. fcc solid solubility limits as solidus and solvus, plotted in terms of average group number (AGN) values for Ni–Cr, Ni–V, Co–Cr, Co–V, Fe–Cr, Fe–V. In the systems Ni–Co, Ni–Fe, and Co–Fe, continuous fcc solid solutions exist. (From Hume-Rothery et al.[18] with permission.)

A different interpretation of the electron concentration effects is presented by Engel and Brewer as will be discussed in Appendix A. Other methods, such as the average group number (AGN) versus temperature diagrams,[18] have been proposed to avoid the controversy regarding the values of e/at. for the elements. For this purpose, AGN is taken to be the number of electrons outside the inert gas shell of each metal; e.g., for an equiatomic alloy of Ti-V, AGN is 4.5 since Ti contributes four electrons, and V, five electrons per atom. The range of AGN is 1 to 10 for groups headed by K to Ni. An interesting plot[18] of AGN versus temperature for the face-centered cubic solid solubility limits of V and Cr in fcc γ-Fe, β-Co, and Ni is shown in Fig. 2.12.

Interstitial phases and compounds: The interstitial alloy phases and compounds are formed by small atoms of metalloids consisting of H, B, C, N, and sometimes O and Si, and large atoms of transition elements that provide the interstices, or voids, for the metalloids. The rule governing interstitial solubilities is called the Hägg rule.

Rule 4, Hägg rule. If the ratio of the atomic radius of a metalloid to the atomic radius of a metal is smaller than 0.59, then the metal and metalloid may form an interstitial solid solution in which the lattice of the metal is usually stretched depending on the size difference and the concentration of the metalloid. Frequently, intermediate phases and compounds may also be formed. The stoichiometry of some of the intermediate phases roughly corresponds to $A_n B$ where A is the metal and n is often equal to 0.5, 1, 2, or 4, but large deviations in stoichiometry may occur in all such phases. The lattice strain about a metalloid atom is usually quite large; hence, the solubilities of metalloids are generally small and usually decrease with increasing sizes of the metalloid atoms. As the temperature increases, the solvent metal tolerates greater degrees of lattice strains and the solubilities generally, but not always, increase with increasing temperature. When the radius ratio is more than 0.59, the solubilities of metalloids in metals become extremely small, but intermetallic compounds may be formed, wherein the crystal structures of solvent metals become highly distorted or modified.

References

1. N. A. Gokcen, *Thermodynamics*, Techscience, Hawthorne, California (1975).
2. A. Alper, editor, *Phase Diagrams*, three volumes, Academic Press, New York (1970); H. C. Yeh in Volume I, p. 167; T. B. Massalski and H. Pops in Volume II, p. 221.
3. A. G. Guy and J. J. Hren, *Elements of Physical Metallurgy*, Third edition, Addison–Wesley, Reading, Massachusetts (1974).
4. A. Reisman, *Phase Equilibria*, Academic Press, New York (1970).
5. *Bulletin of Alloy Phase Diagrams*, Vols. 1–5, ASM, Metals Park, Ohio (1978–1984).

6. I. Prigogine and R. Defay, *Chemical Thermodynamics*, translated by D. H. Everett, Longmans, Green, New York (1954).
7. S. Wagner and D. A. Rigney, *Metall. Trans.* **5**, 2115 (1974).
8. M. Hansen and K. Anderko, *Constitution of Binary Alloys*, McGraw-Hill, New York (1958).
9. R. P. Elliott, *Constitution of Binary Alloys, First Supplement*, McGraw-Hill, New York (1965).
10. F. A. Schunk, *Constitution of Binary Alloys, Second Supplement*, McGraw-Hill, New York (1969); J. F. Smith and Z. Moser, *J. Nuclear Mater.*, **59**, 158 (1976).
11. R. Hultgren, P. D. Desai, D. T. Hawkins, M. Gleiser, and K. K. Kelley, *Selected Values of the Thermodynamic Properties of Binary Alloys*, ASM, Metals Park, Ohio (1973).
12. M. Hillert, *Physics* **103B**, 31 (1981).
13. See, e.g., the articles by (a) M. Hillert, (b) P. J. Spencer, and (c) P. L. Lin, C. W. Bale, and A. D. Pelton, in *Calculation of Phase Diagrams and Thermochemistry of Alloy Phases*, edited by Y. A. Chang and J. F. Smith, Metall. Soc. AIME (1979).
14. L. Kaufman and H. Bernstein, *Computer Calculation of Phase Diagrams*, Academic Press, New York (1970).
15. H. Gaye and C. H. P. Lupis, *Metall. Trans.* **6A**, 1049 (1975).
16. C. W. Bale, A. D. Pelton, and W. T. Thompson. An on-line computer program (F*A*C*T) is available for phase-diagram calculations through McGill University/Ecole Polytechnique de Montreal, Canada.
17. I. Ansara, *Int. Metals Rev.* No. 1, p. 20, Metals Society and ASM (1979).
18. W. Hume-Rothery, R. E. Smallman, and C. W. Haworth, *The Structure of Metals and Alloys*, The Institute of Metals, London (1969).
19. W. Hume-Rothery, in *Phase Stability in Metals and Alloys*, edited by P. S. Rudman, J. Stringer, and R. I. Jaffee, McGraw-Hill, New York, p. 3 (1967).
20. T. B. Massalski, in *Theory of Alloy Phase Formation*, edited by L. H. Bennett, AIME, Warrendale, Pennsylvania (1980).
21. K. A. Gschneidner, Ref. 20, p. 36.
22. K. A. Gschneidner, Ref. 20, p. 1.
23. J. Friedel, *Trans. Metall. Soc. AIME* **230**, 632 (1964); see also pp. 8-9, and 38-39, in Ref. 20.
24. T. B. Massalski, in *Physical Metallurgy*, edited by R. W. Cahn, North-Holland, Amsterdam (1965); see also C. S. Barrett and T. B. Massalski, *Structure of Metals*, McGraw-Hill, New York (1980).

3

Statistical Thermodynamics

Statistical thermodynamics is the discipline that develops thermodynamic relationships from the mechanics of submicroscopic particulates (for brevity, particles) coupled with the laws of statistics.[1,2] The particles of interest in this book are atoms and molecules, but we shall also have brief remarks on photons, electrons, and nuclear components. Since classical mechanics does not always lead to the most elegant results, we shall use a limited number of rules and axioms from quantum mechanics. The reader interested in basic quantum mechanics is referred to standard texts[3,4]; proficiency in this field is helpful but not essential for our purposes. This chapter is intended to be a thorough background for acquiring the important procedures of statistical mechanics as related to thermodynamics.

Distribution of Independent Particles

The distribution of independent particles in various ways depends on the following general axiom, based on actual enumeration of all the possible arrangements for a given set of particles and conditions.

GENERAL AXIOM. *If the first arrangement (e.g., selection, ordering) can be made N_1 different ways, after which the second arrangement can be made N_2 different ways, independently of the first arrangement, then both arrangements can be made in $N_1 N_2$ different ways.*

The result is based on actual construction of independent arrangements with a few particles and then generalization to any number of particles. For example, if we have two sets of letters, (A, B, C) and (E, F), then the number of ways we can select one letter from the first set is 3, and from the second set, 2, so that there are 3×2 ways of making both selections in the stated sequence, i.e., AE, AF, BE, BF, CE, and CF. An obvious extension of this axiom is for more than two sets when there are N_1, N_2, N_3, ... independent particles; the total number of arrangements is then $N_1 \times N_2 \times N_3 \times \cdots$. The number of distinct ways of arranging N different objects all taken together is $N!$, and this result is called the permutation of N different objects. The deduction is easy for three letters A, B, C, because the first letter can be selected three different ways, after which the second letter can be selected two ways, leaving one letter to be selected last so that we have ABC, ACB, BAC, BCA, CAB, and CBA, or $3! = 6$. Among the N objects, if N_1 were alike, then $N_1!$ of all the arrangements would be indistinguishable; therefore, the total number of distinct arrangements (or distribution) is

$$D = \frac{N!}{N_1! N_2!} \tag{3.1}$$

We now consider a system containing N independent particles which are identical. Let there be g_1, g_2, ..., g_i boxes, each set of boxes corresponding to each set of energies ε_1, ε_2, ..., ε_i, respectively, called the energy levels. The occupation number N_1, N_2, ..., N_i of each box is the number of identical particles contained in each box. Thus, we have

Boxes: g_1, g_2, ..., g_i

Energy levels: ε_1, ε_2, ..., ε_i

Occupation numbers: N_1, N_2, ..., N_i

Each set of g_i of the ith set of identical boxes differing from other boxes represents the degeneracy of that level as given by quantum mechanics, i.e., the number of single-particle wavefunctions that yield the corresponding single energy level. Note that g_1, g_2, ... are the degeneracies, but ε_1, ε_2, ... are themselves not degenerate. The set N_1, N_2, ..., N_i taken altogether is called a distribution. A change in any or all of selected N_i represents a new distribution.

Fermi–Dirac Statistics

Let there be N_1 identical particles in g_1 identical boxes with the restrictions that $N_1 \le g_1$ and that each box has no more than one particle in it. The number of boxes, each containing one particle, is N_1, and the number of empty boxes is $g_1 - N_1$. The number of ways, D_1, for arranging g_1 boxes, of which N_1 are alike and each contains one particle, and $g_1 - N_1$, which are empty and also alike, is given by equation (3.1), i.e.,

$$D_1 = \frac{g_1!}{N_1!(g_1 - N_1)!} \tag{3.2}$$

For a given set of occupation numbers, the total number of distributions is the product of all D_i; hence,

$$D = \prod_{i=1}^{i} D_i = \prod_{i=1}^{i} \frac{g_i!}{N_i!(g_i - N_i)!} \tag{3.3}$$

We use the Stirling approximation, $\ln g_1! \approx g_1 \ln g_1 - g_1$ and similarly for $N_1!$, to write equation (3.3) as follows:

$$\ln D = \sum_{i=1}^{i} \left[g_i \ln \left(\frac{g_i}{g_i - N_i} \right) + N_i \ln \left(\frac{g_i - N_i}{N_i} \right) \right] \tag{3.4}$$

The particles obeying equation (3.3) or (3.4) are called fermions, typical examples of which are electrons, neutrons, and protons which, according to the Pauli exclusion principle, cannot occupy the same state with another particle, or no two such particles can have the same quantum numbers.

Bose–Einstein Statistics

If we remove the restriction in the preceding section that each box may contain no more than one particle, then the distribution becomes completely different. Each box may now contain any number of particles; thus, for five particles and four boxes, one of the possible arrangements is

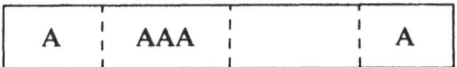

Four boxes are generated by inserting the dashed, movable partitions, and since there are three of these partitions, the number of movable partitions is one fewer than the number of boxes so that we have $g_1 - 1$ partitions. The total number of objects to be arranged is 8, or $g_1 - 1 + N_1$. The number of ways of arranging $g_1 - 1 + N_1$ objects into $g_1 - 1$ and N_1 objects is also given by equation (3.1), i.e.,

$$D_1 = \frac{(g_1 - 1 + N_1)!}{N_1!(g_1 - 1)!} \tag{3.5}$$

Again D is the product of all D_i so that

$$D = \prod_{i=1}^{i} \frac{(g_i - 1 + N_i)!}{N_i!(g_i - 1)!} \tag{3.6}$$

and when the Stirling approximation is used after neglecting 1 in $(g_i - 1)$, we obtain

$$\ln D = \sum_{i=1}^{i} \left\{ -g_i \ln \left(\frac{g_i}{g_i + N_i} \right) + N_i \ln \left(\frac{g_i + N_i}{N_i} \right) \right\} \tag{3.7}$$

The particles (e.g., photons) that obey equation (3.6) are called bosons.

Boltzmann Statistics

Consider equation (3.2) for fermions when g_i is very much greater than N_i, i.e., $g_i \gg N_i$; hence, most of the boxes are not occupied and equation (3.2) can be rewritten as

$$D_1 = \frac{g_1(g_1 - 1)(g_1 - 2) \cdots (g_1 - N_1 + 1)[(g_1 - N_1)!]}{N_1!(g_1 - N_1)!}$$

$$= \frac{g_1(g_1 - 1) \cdots (g_1 - N_1 + 1)}{N_1!} \tag{3.8}$$

There are N_1 factors in the numerator, and since $g_1 - N_1 + 1$ and the preceding factors are very close to g_1 because g_1 is very much greater than N_1, equation (3.8) becomes

$$D_1 = \frac{g_1^{N_1}}{N_1!} \tag{3.9}$$

The total number of distributions, D, is therefore

$$D = \prod_{i=1}^{i} \frac{g_i^{N_i}}{N_i!} \tag{3.10}$$

Again for ln D, we obtain

$$\ln D = \sum_{i=1}^{i} [N_i \ln(g_i/N_i) + N_i] \tag{3.11}$$

The particles obeying equation (3.10) are called the corrected boltzons, and for the most part we shall be concerned with these particles. In the historical development of equation (3.10) the right-hand side contained $N!$ as a factor immediately after the equal sign and such particles were simply called the uncorrected boltzons. We shall have no specific use for uncorrected boltzons. The preceding equations for D are functions of N_i only since degeneracies are generally not variables.

Distribution Laws

Consider a system consisting of N fixed number of particles with a fixed total energy, E, and a fixed volume, V. Such a system was called a microcanonical ensemble by J. W. Gibbs. Statistical thermodynamic properties of this system are functions of N, E, and V. The requirements that

$$\sum_{i=1}^{i} N_i = N \qquad \text{(first restrictive condition)} \tag{3.12}$$

and

$$\sum_{i=1}^{i} N_i E_i = E \qquad \text{(second restrictive condition)} \tag{3.13}$$

are called the restrictive conditions wherein E_i is the energy of particle i. We proceed to solve for the maximum value of ln D within the restrictive conditions specified by equations (3.12) and (3.13). We rewrite the differentials of these equations as

$$\sum_{i=1}^{i} \frac{\partial N}{\partial N_i} dN_i = 0, \qquad \sum_{i=1}^{i} \frac{\partial E}{\partial N_i} dN_i = \sum_{i=1}^{i} \varepsilon_i \, dN_i = 0 \tag{3.14}$$

where $\partial E/\partial N_i = \varepsilon_i$.

We next assume that the expression for D, within the constraints imposed by the restrictive conditions and within a very close approximation, is the same as the expression we shall obtain for the maximum term for D.

This assumption states that only the maximum term of D in very close proximity of E contributes significantly to D, or stated differently, if D were plotted versus energy for a closed system, D would sharply peak out at the prescribed value of E for the system. The justification for using the maximum term for D is that the resulting statistical thermodynamic equations accurately describe the observable thermodynamic properties. For the maximum value of D, $d \ln D$ is zero; hence,

$$d \ln D = 0 = \sum_{i=1}^{i} \frac{\partial \ln D}{\partial N_i} dN_i \qquad (3.15)$$

Multiplication of the first and second relationships in equation (3.14) with the Lagrangian multipliers α and $-\beta$, respectively, and combination of the result with equation (3.15) yields

$$\sum_{i=1}^{i} \left(\frac{\partial \ln D}{\partial N_i} + \alpha \frac{\partial N}{\partial N_i} - \beta \varepsilon_i \right) dN_i = 0 \qquad (3.16)$$

Since each dN_i is finite and nonzero, equation (3.16) is satisfied if, and only if, the coefficient of each dN_i is zero; hence,

$$\frac{\partial \ln D}{\partial N_i} + \alpha \frac{\partial N}{\partial N_i} - \beta \varepsilon_i = 0 \qquad (3.17)$$

From equation (3.12), $\partial N / \partial N_i = 1$, and

$$\frac{\partial \ln D}{\partial N_i} + \alpha - \beta \varepsilon_i = 0; \qquad (i = 1, 2, \ldots) \qquad (3.18)$$

Equations (3.12), (3.13), and (3.18) provide the simultaneous equations for solving the unknowns N_i, α, and β.

For fermions, the first term on the left side of equation (3.18) is equal to $\ln(g_i - N_i) - \ln N_i$; because g_i is independent of N_i, therefore,

$$\ln\left(\frac{g_i - N_i}{N_i} \right) = -\alpha + \beta \varepsilon_i; \qquad \text{or} \qquad \frac{N_i}{g_i} = \frac{1}{e^{-\alpha} e^{\beta \varepsilon_i} + 1}; \qquad \text{(fermions)} \qquad (3.19)$$

Likewise,

$$\frac{N_i}{g_i} = \frac{1}{e^{-\alpha} e^{\beta \varepsilon_i} - 1}; \qquad \text{(bosons)} \qquad (3.20)$$

and

$$\frac{N_i}{g_i} = e^{\alpha} e^{-\beta \varepsilon_i}; \qquad \text{(boltzons)} \qquad (3.21)$$

Equations (3.19)–(3.21) are called the distribution laws for systems consisting of independent particles.

The values of α and β can now be obtained for boltzons. For this purpose, we rewrite equation (3.12) by using equation (3.21) so that

$$\sum N_i = \sum g_i e^{\alpha} e^{-\beta \varepsilon_i} = e^{\alpha} \sum g_i e^{-\beta \varepsilon_i} = N \qquad (3.22)$$

For compact writing we use

$$q \equiv \sum g_i e^{-\beta \varepsilon_i} \qquad (3.23)$$

and call q the molecular partition function. If the energy, ε_i, is separable to ε_i and ε_j, then $q = q_i q_j = (\sum g_i e^{-\beta \varepsilon_i})(\sum g_j e^{-\beta \varepsilon_j})$. For example, ε_i and ε_j may refer to kinetic and potential energy, respectively. Equation (3.23) can be substituted in equation (3.22) to write

$$e^{\alpha} q = N \qquad (3.24)$$

We solve for e^{α} from this equation and substitute it into equation (3.21) to obtain N_i and then use equation (3.13) to obtain

$$E = \sum N_i \varepsilon_i = \frac{N}{q} \sum \varepsilon_i g_i e^{-\beta \varepsilon_i} = -N \left(\frac{\partial \ln q}{\partial \beta} \right)_V \qquad (3.25)$$

where we assume that ε_i is a function of volume and therefore q is a function of β and V. Since E is known from measurements, g_i and ε_i are known from molecular properties of a system such as an ideal gas, and β can then be determined from equation (3.25). For this purpose we substitute q and the known value of N to obtain α from equation (3.24). The value of β can be obtained by using a simple ideal gas such as neon or argon. According to the kinetic theory of gases, the molecules of an ideal gas are point particles randomly moving in space without exerting attractive or repulsive forces on one another. Consider that there is 1 mole of an ideal monatomic gas consisting of N molecules and occupying V cm^3 of volume. An isolated cube of 1 cm contains N/V molecules of this gas. Let m be the mass of each molecule and u its average velocity and assume that one-third of all molecules move in each of the x, y, and z directions as shown in Fig. 3.1.

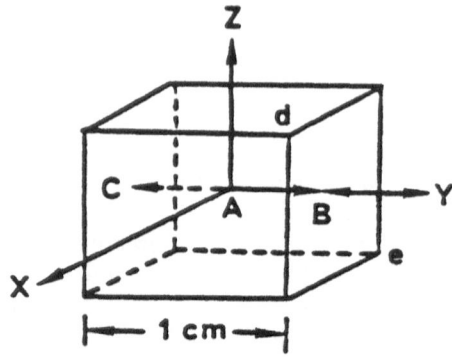

Figure 3.1. One-centimeter cube containing an ideal gas.

A molecule at "A" may move toward "B" to collide with the surface at "B" and then rebound and travel to "C" and return to "A" again to repeat its journey. In this process, each molecule travels a distance of 2 cm for each collision on the selected surface dBe. When a molecule rebounds from a surface, its average velocity changes from $+u$ to $-u$ and the change of its momentum at the surface becomes $mu - (-mu) = 2mu$. According to Newton's laws of motion, the force exerted on a surface by a moving body is equal to the rate of change of momentum and since this force acts on dBe which has an area of 1 cm^2, it also represents the pressure. The time, t, necessary for each collision is $t = (2 \text{ cm})/(u \text{ cm/sec}) = 2/u$, and thus the rate of change of momentum is $2mu/(2/u) = mu^2$ which is the pressure due to each molecule. For all the molecules moving in the direction of y, i.e., $N/3V$, the pressure is

$$P = (N/3V)mu^2 \tag{3.26}$$

Multiplying both sides by V and observing that, from experiments, $PV = RT$, it is readily seen that

$$PV = (2N/3)(mu^2/2) = RT = NkT \tag{3.27}$$

where $k = R/N$ is the gas constant per molecule, i.e., the Boltzmann constant.

The kinetic theory states that the internal energy of a monatomic gas consists entirely of the translations, or kinetic energy of its molecules; therefore,

$$E = N(mu^2/2) \tag{3.28}$$

Combination with equation (3.27) gives

$$E = 1.5RT; \qquad \varepsilon_i(\text{average}) = E/N = mu^2/2 \qquad (3.29)$$

Consequently, ε_i in $q = \sum g_i e^{-\beta \varepsilon_i}$ is equal to $mu_i^2/2$ where the velocity u_i is the quantized individual velocity component. The exponent of e must be dimensionless; hence, β must be proportional to $(mu^2)^{-1}$, and since $(kT)^{-1} = (mu_i^2/3)^{-1}$ from equation (3.27), then β is proportional to $(kT)^{-1}$, and in fact a more rigorous treatment beyond our scope also shows that $\beta = (kT)^{-1}$.

Entropy and Related Properties

The energies, ε_i, are functions of volume only. The reversible work (dw) done on a system is given by $dw = \sum N_i(d\varepsilon_i/dV)\, dV = \sum N_i\, d\varepsilon_i$. From $E = N_i\varepsilon_i$, $dE = \sum (\varepsilon_i\, dN_i + N_i\, d\varepsilon_i)$; however, from the first law of thermodynamics, $dE = dq_{th} + dw$ where q_{th} is the thermal energy exchange. Consequently,

$$dq_{th} = dE - dw = \sum \varepsilon_i dN_i \qquad (3.30)$$

The second law of thermodynamics gives the existence of entropy S, by the infinitesimal reversible thermal energy dq_{th} divided by T, $dS = dq_{th}/T$; therefore,

$$dS = \sum \frac{\varepsilon_i}{T} dN_i \qquad (3.31)$$

It was shown by equation (3.17) that

$$\sum \frac{\partial \ln D}{\partial N_i} dN_i = \sum (\beta \varepsilon_i\, dN_i - \alpha\, dN_i) \qquad (3.32)$$

The last term is zero because for a closed system $\alpha \sum dN_i = \alpha\, dN = 0$. The left side of this equation is equal to $\sum d \ln D$ where D is a function of N_i only. If the most probable value of D is the only significant contribution to $\sum d \ln D$, then the summation can be replaced by the most probable value of D_{mp} which we designated again as D so that

$$d \ln D = \beta \sum \varepsilon_i\, dN_i = \frac{1}{k} \sum \frac{\varepsilon_i}{T} dN_i \qquad (3.33)$$

Thus, from equations (3.31) and (3.33),

$$dS = kd \ln D \tag{3.34}$$

The integratin of this equation yields the definition of the entropy, i.e.,

$$S = k \ln D \tag{3.35}$$

where it is assumed that for $D = 1$, $S = 0$. For boltzons, we have

$$S = k \ln D = k \sum \left(N_i \ln \frac{g_i}{N_i} + N_i \right) \tag{3.36}$$

From equation (3.21), $N_i = g_i e^\alpha e^{-\beta \varepsilon_i}$, and from equation (3.24), $e^\alpha = N/q$, so that $g_i/N_i = (q/N) e^{\beta \varepsilon_i}$; hence, with $\beta = (kT)^{-1}$,

$$S = k \left[\left(\ln \frac{q}{n} \right) \sum N_i + \sum (\beta N_i \varepsilon_i + N_i) \right] = kN \ln \frac{q}{N} + \frac{E}{T} + kN \tag{3.37}$$

where the last equality was obtained by using equations (3.12) and (3.13). The Helmholtz energy is given by

$$A = E - TS = -kNT \ln \frac{q}{N} - kNT \tag{3.38}$$

and the Gibbs energy by $G = A + PV = A + kNT$; hence,

$$G = -kNT \ln \frac{q}{N} \tag{3.39}$$

where G (as well as A) at 0 K is taken to be zero for simplicity. It should be noted from equations (3.24) and (3.39) that

$$\alpha = G/(kNT) = \bar{G}/kT \tag{3.40}$$

where $\bar{G} = G/N$ is the Gibbs energy per particle. Similar equations can be obtained for fermions, and bosons, starting with the appropriate D but the resulting equations for G and α are identical with equations (3.39) and (3.40).

Fermions, Bosons, and Boltzons

The electrons in metals and semiconductors obey the Fermi-Dirac statistics, i.e.,

$$\frac{N_i}{g_i} = \frac{1}{e^{-\alpha} e^{\beta \varepsilon_i} + 1} = \frac{e^{\alpha} e^{-\beta \varepsilon_i}}{1 + e^{\alpha} e^{-\beta \varepsilon_i}} \qquad (3.41)$$

where $\alpha = \bar{G}/kT$ and $\bar{G}$ here is often called the Fermi energy by physicists, and denoted as E_f by them. The equations corresponding to fermions are

$$E = \sum N_i \varepsilon_i = \sum \frac{g_i \varepsilon_i e^{\alpha} e^{-\beta \varepsilon_i}}{1 + e^{\alpha} e^{-\beta \varepsilon_i}} \qquad (3.42)$$

$$S/k = \beta E - N\alpha + \sum g_i \ln(1 + e^{\alpha} e^{-\beta \varepsilon_i}) \qquad (3.43)$$

$$A = E - (S/k\beta) = NkT\alpha - kT \sum \ln(1 + e^{\alpha} e^{-\beta \varepsilon_i}) \qquad (3.44)$$

For fermions, $\alpha = \bar{G}/kT$ can be substituted into equation (3.41) to write

$$\frac{N_i}{g_i} = \frac{1}{[e^{\beta(\varepsilon_i - \bar{G})}] + 1} \qquad (3.45)$$

As T approaches zero, or as β approaches infinity, we have three possibilities, i.e., $\varepsilon_i > \bar{G}$ for which $N_i/g_i \to 0$; $\varepsilon_i = \bar{G}$, for which $N_i/g_i \to 0.5$; and for $\varepsilon_i < \bar{G}$, $N_i/g_i \to 1$. Therefore, all the levels for the first case are empty, and those for the last case are full, but those for the second case are half full. Equation (3.45) will be discussed in detail in conjunction with semiconductors. Neglecting the small contribution from the second case, the value of N is given by

$$N = \sum N_i = \sum_{\varepsilon_i=0}^{\varepsilon_i = \bar{G}} g_i; \qquad (T \to 0) \qquad (3.46)$$

If for the fermions in the following equation

$$\frac{N_i}{g_i} = \frac{e^{\beta(\bar{G}-\varepsilon_i)}}{1 + e^{\beta(\bar{G}-\varepsilon_i)}} \qquad (3.47)$$

ε_i is much larger than $\bar{G}$, and β is finite and small, or the temperature is sufficiently high, then the second term in the denominator is negligible and we obtain equation (3.21) for boltzons, i.e.,

$$\frac{N_i}{g_i} = e^{\beta(\bar{G}-\varepsilon_i)} = \frac{N}{q} e^{-\beta\varepsilon_i} \qquad (3.48)$$

As a result of $\varepsilon_i \gg \bar{G}$, we have $N_i \ll g_i$, which is the requirement for applicability of the Boltzmann statistics.

The plus sign preceding the exponential term in the denominator of equations (3.42) and (3.45), and after ln in equations (3.43) and (3.44), must be changed to minus to obtain the corresponding equations for bosons. These results can be obtained by starting with equation (3.20) and following an identical procedure; the result is

$$\frac{N_i}{g_i} = \frac{1}{e^{\beta(\varepsilon_i-\bar{G})} - 1}; \qquad \text{(bosons)} \qquad (3.49)$$

Again for $\varepsilon_i \gg \bar{G}$ and high temperatures, this equation becomes the same as equation (3.48) for boltzons.

Gibbsian Ensembles

Equations (3.12), (3.13), and (3.23) through (3.25) show that q refers to a closed system described by constant E, N, and V. A microcanonical ensemble is therefore a closed system defined by E, N, and V. (Canonical = standard.)

Consider equation (3.38), which can be rewritten as

$$A = -kT \ln \frac{q^N}{N!} \qquad (3.50)$$

From thermodynamics, A is a function of V and T and if we let the $Q = q^N/N!$, then we have

$$A = -kT \ln Q; \qquad (Q \equiv q^N/N!) \qquad (3.51)$$

Therefore, Q now refers to a system defined by N, V, and T and such a system is called a canonical ensemble.

Subtraction of equation (3.51) from (3.39) yields

$$G - A = PV = kT \ln(N^N Q/q^N) \equiv kT \ln \Xi \qquad (3.52)$$

where Ξ is called the grand partition function, considered to be a function of V, T, and $\bar{G}$. The system to which Ξ refers is an open system, called the grand canonical ensemble, although a more useful concept is the isothermal isobaric ensemble $\mathbb{I}$ defined by

$$G = -kT \ln \mathbb{I} = -kT \ln \left(\frac{q}{N}\right)^N ; \qquad \left[\mathbb{I} = \left(\frac{q}{N}\right)^N\right] \qquad (3.53)$$

where equation (3.39) is used on the right side, and $\mathbb{I}$, like G, is a function of N, P, and T. We shall generally have very little use of Ξ and $\mathbb{I}$ in the topics considered in this book, and frequently use q and Q.

Quantum Mechanics of Free Particles

Ideal gases, electrons or other charged particles in rarified spaces, and electrons in semiconductors can be treated as free particles. We consider first an ideal gas, and then briefly the electron gas.

The partition function of an ideal gas is given by a simple form of equation (3.23), i.e.,

$$q = \sum_i e^{-\beta \varepsilon_i} \qquad (3.54)$$

where i is the index that counts the distinguishable and degenerate states individually, one by one. According to quantum mechanics, the energy of a free particle is given by

$$\varepsilon_{abc} = \frac{h^2}{8m} \left(\frac{a^2}{A^2} + \frac{b^2}{B^2} + \frac{c^2}{C^2}\right) \qquad (3.55)$$

where h is Planck's constant, a, b, and c are the quantum numbers, and A, B, and C are the dimensions of the container or box; hence, this system is often called "particles in a box." Resemblance to the kinetic energy in terms of the momentum, $p = mu$, i.e., $\varepsilon = p^2/2m$, is self-evident. The details of intermediate steps leading to equation (3.55) involve postulates that are accepted without question because the end results always justify the means used in quantum mechanics. Therefore, we accept equation (3.55) without presenting the details of postulates and procedures leading to this equation. The values of the quantum numbers in equation (3.55) are so large and close enough that they may be considered as continuous from zero to infinity. The partition function given by equation (3.54) can therefore be written as

$$q = \sum_{a=0}^{\infty} e^{-\beta h^2 a^2/8mA^2} \sum_{b=0}^{\infty} e^{-\beta h^2 b^2/8mB^2} \sum_{c=0}^{\infty} e^{-\beta h^2 c^2/8mC^2} = q_a q_b q_c \qquad (3.56)$$

where $q_a = \sum \exp(-\beta h^2 a^2 / 8mA^2)$, and so on. This summation can be obtained as follows:

$$q_a = \sum_{a=0}^{\infty} e^{-\beta h^2 a^2/8mA^2} = \int_0^{\infty} e^{-(\beta h^2/8mA^2)a^2} \, da = \left(\frac{2\pi m}{\beta h^2}\right)^{0.5} A \quad (3.57)$$

where the integral is listed in standard mathematical tables. The replacement of the sum by the integral is permissible when the coefficient of a^2 in the exponent is much smaller than unity, and this requirement is always met for $A \approx 1$ cm, and practically all the experimentally attainable temperatures. The result for q is

$$q = q_a q_b q_c = \left(\frac{2\pi m}{\beta h^2}\right)^{1.5} ABC = \left(\frac{2\pi m}{\beta h^2}\right)^{1.5} V; \quad \text{(ideal gas)} \quad (3.58)$$

where $ABC = V$ is the volume of the box. The energy E is obtained from

$$E = -N \left(\frac{\partial \ln q}{\partial \beta}\right)_V = \frac{1.5N}{\beta} \quad (3.59)$$

The energy of an ideal gas is given by $E = 1.5NkT$, hence, again $\beta = 1/kT$.

The free electrons obey the same treatment except a spin degeneracy factor of 2 (electron spin up and spin down) must appear in equation (3.58); thus,

$$q = 2 \left(\frac{2\pi m_- kT}{h^2}\right)^{1.5} V; \quad \text{(electron gas)} \quad (3.60)$$

where m_- is the mass of a freely moving electron. This equation will be used in conjunction with electrons and holes in semiconductors.

Rotation and Vibration of Diatomic Molecules

Two unsymmetrical masses m_i and m_j, separated by a fixed distance of r_{ij}, have a moment of inertia of $I = r_{ij}^2 m_i m_j / (m_i + m_j)$. According to quantum mechanics, the energy of two such masses rotating in three dimensions is

$$\varepsilon_j = \frac{J(J+1)h^2}{8\Pi^2 I}; \quad (J = 0, 1, \ldots) \quad (3.61)$$

where J is the rotational quantum number. The equation for q_{rot} is similar to equation (3.57) and can be integrated with J as the variable from zero to infinity for sufficiently large values of T to obtain

$$q_{rot} = \frac{8\pi^2 IkT}{h^2} \qquad (3.62)$$

For a symmetric molecule this equation must be divided by 2. The Gibbs energy and the enthalpy per mole from equation (3.62) are

$$G_{rot} = -RT \ln q_{rot}, \qquad H_{rot} = RT \qquad (3.63)$$

where G_{rot}(at 0 K) and H_{rot}(at 0 K) are taken to be zero for simplicity. We present these equations for completeness in the treatment of particles since there will be only a brief reference to them later in this book.

Two particles, i and j, may vibrate with respect to each other by forming a harmonic oscillator. The vibrational energy of such an oscillator with a reduced mass of $m_r = m_i m_j/(m_i + m_j)$ is also quantized. The energy is given in quantums of energy $L(h\nu)$, where L is the quantum number and ν is the frequency of vibration. The partition function is again

$$q_{vib} = \sum_{L=0}^{\infty} e^{-Lh\nu/kT} = \frac{1}{1 - e^{-h\nu/kT}} \qquad (3.64)$$

The last equality is obtained easily because the summation represents the sum of a geometrical series. For sufficiently high temperatures and small frequencies, $h\nu/kT$ is small and q_{vib} is very closely approximated by

$$q_{vib} = kT/h\nu; \qquad (h\nu/kT < 0.1) \qquad (3.65)$$

The Gibbs energy and enthalpy per mole from equation (3.65) are

$$'G = -RT \ln q_{vib}; \qquad H = RT \qquad (3.66)$$

The vibrational frequency, ν, of a harmonic oscillator is proportional to the inverse square root of the reduced mass, $a/\sqrt{m_r}$, where a is a proportionality constant; hence, for small values of m_r, e.g., for an electron and a nucleus, m_r is very small and ν is large and $q_{vib} \approx 1$, and the contribution to the related thermodynamic properties is negligible.

Vibrations of Atoms in Lattices

Consider a simple crystal consisting of N atoms of a single element. Each lattice point is occupied by one atom vibrating in three possible geometrical directions with a constant frequency of ν. The restoring force to the mean position of the atom is proportional to the displacement from the mean position. Therefore, each atom is regarded as three independent but localized harmonic oscillators having the same frequency in all three directions. One might also picture the atoms attached by a lattice of short rubber bands to their sites, and vibrating with increasing amplitude, with increasing temperature. The energy of the crystal of N atoms is assumed to be due entirely to vibration. At 0 K, the amplitude of vibration is so small that the vibrational energy and the heat capacity are zero.

Each atom in a crystal acts as a three-dimensional oscillator, i.e., there are $3N$ oscillators for a crystal consisting of N atoms. Therefore, equations (3.66) need to be multiplied by 3; thus,

$$G = -3RT \ln q_{\text{vib}}, \qquad H = 3RT \tag{3.67}$$

The heat capacity of a solid at high temperatures is $3R$ from dH/dT, in accord with the law of Dulong and Petit. Application of these equations to alloys requires a realistic representation of the frequency ν as a function of composition and temperature. Attempts to accomplish this have been largely empirical.

Potential Energy Functions

The potential energy functions generally attempt to correlate the energy as a function of the distance of approach of molecules from the gases to the condensed states. Such functions are usually empirical or at best semi-empirical. Two symmetric and neutral molecules, or atoms, attract each other to decrease their energy per particle, E'/N, by $-A/r^6$, where r is the intermolecular distance. The force causing this attraction is called the London dispersion force [F. London (1930)] arising from an attraction between mutually induced dipoles in the interacting molecules. However, when r becomes smaller, repulsive forces come into play because of intermolecular repulsion. An empirical approximation for the repulsive energy is given by $E''/N_0 = B/r^{12}$; therefore, the total energy per particle is

$$\frac{E}{N_0} = -\frac{A}{r^6} + \frac{B}{r^{12}} \tag{3.68}$$

This equation is known as the Lennard-Jones potential (1924). The parameters A and B are related to r_0, the minimum point in Fig. 3.2, by setting dE/dr to zero; the result is $A = 2B/r_0^6$. The minimum energy is designated as $-\varepsilon = E(\text{min})/N_0$, and substituted in equation (3.68) with $A = 2B/r_0^6$, to obtain

$$-\varepsilon = \frac{E(\text{min})}{N_0} = -\frac{2B}{r_0^{12}} + \frac{B}{r_0^{12}} = -\frac{B}{r_0^{12}}$$

from which $B = \varepsilon r_0^{12}$, and then $A = 2\varepsilon r_0^6$. Therefore, equation (3.68) becomes

$$\frac{E}{N_0} = -2\varepsilon \left(\frac{r_0}{r}\right)^6 + \varepsilon \left(\frac{r_0}{r}\right)^{12} \tag{3.69}$$

This equation, known as the Lennard-Jones 6–12 potential, has been used extensively for gases and liquids (see Refs. 5 and 6). It represents an intermolecular pair potential in gases, but the concept has been extended to the condensed phases.

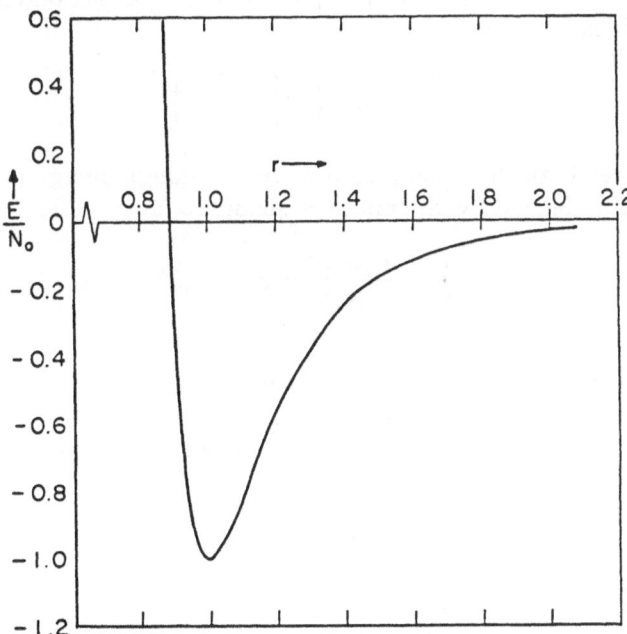

Figure 3.2. Lennard-Jones 6–12 potential for a pair of molecules. Vertical scale is based on $\varepsilon = 1$, and horizontal scale, $r_0 = 1$ in equation (3.69). Minimum point is at $r = r_0 = 1$ and $E(\text{min})/N_0 = -\varepsilon = -1$.

Machlin[7] has used a function similar to equation (3.68) for the structural stability of alloy phases, lattice parameter defects, and other related properties:

$$E = -\frac{\alpha}{r^4} + \frac{\beta}{r^8} \tag{3.70}$$

The parameters α and β are related to physically and mechanically measurable quantities. The pseudopotentials[8] will not be discussed here since brief summaries are presented elsewhere.[8] They are not unique and not greatly successful for the transition elements[9] but often used by physicists.

Ionically bound crystals, such as NaCl, are also considered to have a potential energy consisting of two terms. The attractive ionic forces contribute $E'/N_0 = -Me^2/r$ where E'/N_0 is the energy per ion, M is the Maedelung constant, and e is the electronic charge. The values of M are in the neighborhood of 1.7. The repulsive forces cause an increase in energy, expressed by $E''/N_0 = R/r^m$ where m is an exponent that varies from compound to compound, e.g., $m \approx 6$ for LiF and $m \approx 8$ for NaCl. The parameter R is related to M through the minimum point r_0 on the curve for E/N_0 versus r by $R = Me^2 r^{m-1}/m$; therefore, the net potential energy per particle is

$$\frac{E}{N_0} = Me^2\left(-\frac{1}{r} + \frac{r^{m-1}}{mr^m}\right) \tag{3.71}$$

In general, all the potentials near the minimum energy point can be closely approximated by an empirical quadratic function, i.e.,

$$\frac{E}{N_0} - \frac{E(\min)}{N_0} = \sigma(r - r_0)^2 \tag{3.72}$$

where σ is an empirical constant.

The overwhelming majority of potential energy functions are empirical in nature. The validity of assumptions used in these potentials may often be questionable but success has been obtained in a number of cases for certain types of metals and their alloys.[9]

References

1. D. A. McQuarrie, *Statistical Mechanics*, Harper & Row, New York (1983).
2. F. C. Andrews, *Equilibrium Statistical Mechanics*, Wiley–Interscience, New York (1975).
3. A. S. Davydov, *Quantum Mechanics*, translated by D. Ter Haar, Pergamon Press, Elmsford, New York (1976).

4. D. A. McQuarrie, *Quantum Chemistry*, University Science Books, Mill Valley, California (1983).
5. J. O. Hirschfelder, editor, *Intermolecular Forces*, Interscience, New York (1967).
6. J. O. Hirschfelder, C. F. Curtiss, and R. B. Bird, *Molecular Theory of Gases and Liquids*, Wiley-Interscience, New York (1964).
7. E. S. Machlin, in *Interatomic Potentials and Crystalline Defects*, edited by J. K. Lee, Metall. Soc. AIME, p. 33 (1981).
8. D. Stroud, in *Theory of Alloy Phase Formation*, edited by L. H. Bennett, Metall. Soc. AIME, p. 84 (1980); see also W. A. Harrison, *Pseudopotentials in the Theory of Metals*, Benjamin, New York (1966).
9. R. Taylor, in *Interatomic Potentials and Crystalline Defects*, edited by J. K. Lee, Metall. Soc. AIME, p. 71 (1981).

4

Theories of Solutions

The most general and simple theory intended to fit liquid and solid solutions is the regular solution theory developed by vanLaar and Lorenz (1925), Heitler (1926), and Hildebrand (1928). The term regular solution was proposed by Hildebrand (1929) for solutions described by random and yet nonideal behavior. This proposed behavior was properly called the zeroth approximation to the regular solutions by Guggenheim,[1] as distinct from the first approximation, to be discussed later in this chapter. Early developments in this field are presented in a number of publications.[1-3]

Regular Solutions—Zeroth Approximation

The treatment involved in the zeroth approximation is based on the assumptions that (1) the molecules* in solution are distributed randomly despite nonzero enthalpy of mixing, (2) the component molecules are not greatly different in size from one another so that the number of nearest neighbors Z is the same for the pure components and for their binary solutions, (3) the molecular interaction is limited to the nearest neighbors, and hence the interaction beyond the nearest molecules is neglected, and (4) the bond energy e_{ij} for a bond between unlike molecules i and j is independent of composition and of temperature. The bond energy e_{ij} is negative and refers to

$$N_i(\text{gas}) + N_j(\text{gas}) = N(\text{solution with } ZN/2 \text{ bonds}) \qquad (4.1)$$

where N_i and N_j are the number of molecules of i and j, respectively, and

*Molecules and atoms are identical for solutions of metals on pp. 81–111.

81

N is the sum of N_i and N_j, and for convenience, N is taken to be Avogadro's number. There are $Z/2$ bonds per molecule and Z is the number of nearest neighbors, which is also called the coordination number. Likewise, when $N_i(\text{gas})$ and $N_j(\text{gas})$ molecules condense into pure N_i and N_j in the solid or liquid state, they form $ZN_i/2$ bonds of i-i type and $ZN_j/2$ bonds of j-j type, respectively. We assume that we can assign Z values to the liquid in the same way that we assign Z values to the crystals. The probability of finding a molecule of i at a selected site is proportional to its mole fraction x_i in solution, and that of finding another i next to the selected i is $x_i x_i$. Likewise, the probability of finding a molecule of j and the probability of finding a molecule i next to j is $x_j x_i$, and the sum of all probabilities for all possible pairs is unity; thus,

$$x_i x_i + x_i x_j + x_j x_i + x_j x_j = 1 \tag{4.2}$$

For a multicomponent solution of b and c components, equation (4.2) is written as

$$\sum_{p=1}^{b} \sum_{q=1}^{c} x_p x_q = 1 \tag{4.3}$$

The normalization of all probabilities to unity is therefore automatically satisfied when the composition is expressed in mole fractions. The total number of bonds, $ZN/2$, from equation (4.1) is distributed among the probability terms in equation (4.2) and each term has its bond energy e_{ii}, e_{ij}, and e_{jj}, so that the energy terms for the respective bonds are $(ZN/2)x_i x_i e_{ii}$, $ZN x_i x_j e_{ij}$, and $(ZN/2)x_j x_j e_{jj}$, and the total energy of 1 mole of solution is

$$E = (ZN/2)(x_i^2 e_{ii} + 2x_i x_j e_{ij} + x_j^2 e_{jj}) \tag{4.4}$$

We assume that the energy E is nearly identical with the enthalpy H, because the pressure–volume product PV in the definitional relationship $H = E + PV$ is negligibly small for condensed phases at ordinary pressures. Likewise, the enthalpy for $x_i N$ molecules of pure condensed i is $x_i H_i^\circ = (ZN/2)x_i e_{ii}$ where H_i° refers to 1 mole of pure i. The corresponding enthalpy for pure j is $x_j H_j^\circ = (ZN/2)x_j e_{jj}$, and for the mixing process, i.e.,

$$x_i \text{ mole of pure } i + x_j \text{ mole of pure } j = 1 \text{ mole of } i\text{-}j$$

the molar enthalpy of mixing (or solution) is

$$\Delta H = H - x_i H_i^\circ - x_j H_j^\circ = N W_{ij} x_i x_j \tag{4.5}$$

where equation (4.4) is substituted for H to obtain the last equality, and W_{ij}, called the exchange energy, is defined by

$$W_{ij} = (Z/2)(2e_{ij} - e_{ii} - e_{jj}); \qquad (W_{ij} = \text{exchange energy}) \qquad (4.6)$$

The enthalpy change, ΔH, is also called the molar enthalpy of formation of solution. The arguments presented for the foregoing equations can be readily extended to multicomponent solutions to show that

$$\Delta H = NW_{ij}x_ix_j + NW_{ik}x_ix_k + NW_{jk}x_jx_k + \cdots \qquad (4.7)$$

where each term originates from each pairwise interaction. In equations (4.5) and (4.7) ΔH is also the excess molar energy of solution, H^e. This equation is identical in form with the second-order Margules equation (1.87). Therefore, each term in equation (4.7) originates from each binary system. The equation for $\Delta \bar{H}_i$, which is equal to $\bar{H}_i^e$, is identical to equation (1.69), and for a binary solution,

$$\bar{H}_1^e = NW_{12}x_2^2; \qquad \bar{H}_2^e = NW_{12}x_1^2 \qquad (4.8)$$

The exchange energy W_{ij} can be determined from a single experimental value of ΔH or $\bar{H}_i^e$ at a convenient concentration x_j in equation (4.5) or (4.8). The equations for a multicomponent solution can thus be determined from the values of W_{ij}, $W_{ik}, \ldots$ for the constituent binary systems.

The first assumption for the simple regular behavior, i.e., the random distribution, signifies that the excess molar entropy S^e be zero; therefore,

$$G^e = \Delta H = H^e = NW_{12}x_1x_2 = x_1RT \ln \gamma_1 + x_2RT \ln \gamma_2 \qquad (4.9)$$

$$\bar{G}_1^e = \bar{H}_1^e = NW_{12}x_2^2 = RT \ln \gamma_1;$$

$$\bar{G}_2^e = \bar{H}_2^e = NW_{12}x_1^2 = RT \ln \gamma_2 \qquad (4.10)$$

The equality of $\bar{G}_i^e$ and $\bar{H}_i^e$ signifies that, while $\bar{H}_i^e$ is not zero, $\bar{S}_i^e$ is zero, and we observe immediately that $\bar{S}_i^e = 0$ in $\bar{G}_i^e = \bar{H}_i^e - T\bar{S}_i^e$ is thermodynamically suspect since $\bar{H}_i^e$ cannot be equal to $\bar{G}_i^e$ unless $\bar{H}_i^e$ is also zero, i.e., unless the solution is ideal. Further, a solution for which $\bar{H}_i^e$ is not zero cannot have its molecules distributed randomly. The first approximation to the regular solutions, to be discussed in the next section, confirms that indeed $\bar{S}_i^e$ cannot be equal to zero unless $\bar{H}_i^e$ is also zero.

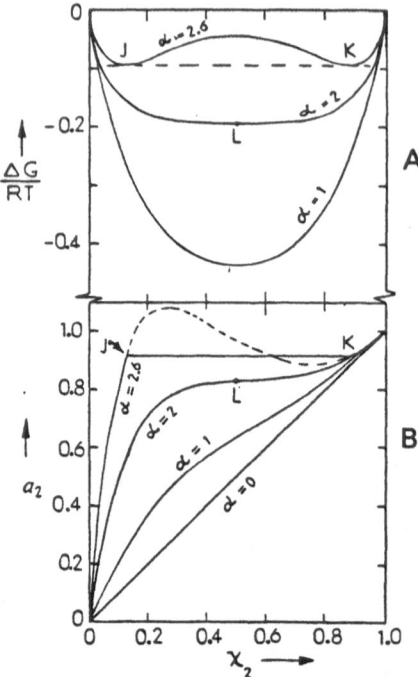

Figure 4.1. Regular solutions for various values of $\alpha = NW/RT$. Upper curves are for $\Delta G/RT$. Dashed horizontal line is tangent at J and K to curve for $\Delta G/RT$, with $\alpha = 2.6$. L is critical point. Lower curves are for activity a_2 of component 2; activity a_1 is symmetric to a_2 about vertical line through $x_2 = 0.5$.

Equation (4.7) may also be treated according to equations (1.87)–(1.90) for multicomponent solutions, after writing $G^e = \Delta H = H^e$, to obtain

$$\bar{G}_i^e = \bar{H}_i^e = RT \ln \gamma_i = -H^e + NW_{ij}x_j + NW_{ik}x_k + \cdots \quad (4.11)$$

We return to equation (4.9) for a binary system and use $\Delta G = G^e + G(\text{ideal})$, with $\Delta G(\text{ideal})^* = RTx_1 \ln x_1 + RTx_2 \ln x_2$, to write

$$\Delta G/RT = \alpha x_1 x_2 + x_1 \ln x_1 + x_2 \ln x_2 \;;$$

$$(\alpha = NW/RT; \quad W = W_{12}) \quad (4.12)$$

where we use W instead of W_{12} for simplicity in notation when we consider only a binary system. It was assumed that W is independent of temperature; consequently, α is inversely proportional to T, and the solution should tend to ideality with increasing T. The dimensionless property $\Delta G/RT$ is plotted as a function of x_2 at three different values of α in Fig. 4.1(A). The curve

---

*$\Delta G(\text{ideal})$ is sometimes unnecessarily called the configurational Gibbs energy and $\Delta S(\text{ideal})$, configurational entropy.

for $\Delta G/RT$ when $\alpha = 1$ is concave up without maximum and inflection points. The curve for $\alpha = 0$ is very similar to that for $\alpha = 1$, and therefore is not shown. When $\alpha = 2$, $\Delta G/RT$ has an inflection point at L where the first and the second derivatives of $\Delta G/RT$ become zero at $x_2 = 0.5$; thus,

$$\frac{\partial(\Delta G/RT)}{\partial x_2} = \alpha - 2\alpha x_2 + \ln\frac{x_2}{1 - x_2}; \qquad (4.13)$$

$$\frac{\partial^2(\Delta G/RT)}{\partial x_2^2} = -2\alpha + \frac{1}{x_2(1 - x_2)} \qquad (4.14)$$

are both zero for $\alpha = 2$ at $x_2 = 0.5$; this is the critical point. At $\alpha > 2$, the curves for $\Delta G/RT$ show two minima; e.g., for $\alpha = 2.6$, the minimum points are at J with $x_2 = 0.124$ and at K with $x_2 = 0.876$ and J and K are located symmetrically with respect to each other. In the range of composition between J and K, the solution separates into two phases, where the composition of each phase is constant but the relative proportions of these phases vary from J to K. The tangent line to the curve at J and K is horizontal and the intercepts with the vertical axes are given by $\Delta\bar{G}_1/RT = \Delta\bar{G}_2/RT = \Delta G/RT$, with the values of $\Delta\bar{G}_i$ calculated either at J or at K. The negative values of α yield curves similar to that for $\alpha = 1$ but with sharper dips.

Equation (4.14) for $\alpha > 2$ is zero at the inflection points; e.g., for $\alpha = 2.6$, $x_2 = 0.252$ and $x_2 = 0.748$ are the inflection points, which are called the spinodes.

The equation for $\Delta\bar{G}_2$ can be derived from equation (4.12); the result is

$$\Delta\bar{G}_2/RT \equiv \ln a_2 = \alpha x_1^2 + \ln x_2 \qquad (4.15)$$

where a_2 is the activity of component 2. The values of a_2 from this equation are plotted in Fig. 4.1(B) corresponding to the selected values of α. For the positive values of α the deviation from Raoult's law is positive, and for $\alpha = 0$ the solution is ideal (Raoultian). The horizontal inflection point for $\alpha = 2$ is also indicated by L in Fig. 4.1(B). The curve for the activity of component 1, a_1, is symmetric to that for a_2 for each corresponding value of α. The deviation from ideality is negative for the negative values of α as expected, but not shown in Fig. 4.1(B). It can be shown that for $\alpha > 2$, the maximum and minimum values of a_2 for x_2 versus a_2 correspond to the inflection points of the curve for x_2 versus $\Delta G/RT$.

Preliminary Concepts

The first approximation to the regular solutions has been obtained in its correct form by permutation of molecules.[4] The earlier approximation by Bethe[5] and by Guggenheim[1] were obtained by permutation of bonds, leading to physically meaningless spliced molecules as shown in Fig. 4.2. Thus, for example, for tetragonal crystals, such as certain silicon alloys for which $Z = 4$, permutation of bonds of $i-j$ type in Fig. 4.2(a) may form $j-i$ bonds in Fig. 4.2(b) and create physically impossible spliced molecules. A critique of the method based on the permutation of bonds is given later in this chapter after the presentation of the correct approximation based on the permutation of molecules. A binary system $i-j$ will be considered first and then the results will be extended to multicomponent systems.

The total numbers of like and unlike bonds are useful in the treatment of all types of solutions. In an ideal solution, the component molecules are distributed randomly, and we recall in conjunction with equation (4.4) that

$$(ZN/2)x_i^2 = \text{number of } i-i \text{ bonds}$$

$$(ZN/2)2x_ix_j = Z(Nx_ix_j) = \text{number of } i-j \text{ bonds}$$

$$(ZN/2)x_j^2 = \text{number of } j-j \text{ bonds}$$

For this distribution, we denote for brevity that

$$Y^* = Nx_ix_j \tag{4.16}$$

hence, the number of $i-j$ bonds is equal to ZY^* in an ideal solution. Likewise, the total number of unlike bonds in a nonideal solution may be denoted by

$$\text{Total } i-j \text{ bonds} = ZY \tag{4.17}$$

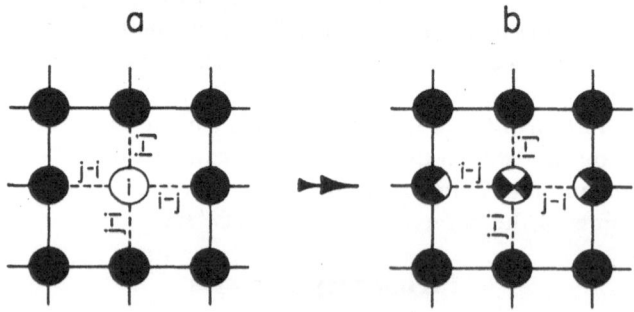

Figure 4.2. Permutation of bonds in two-dimensional crystals. Bonds of $i-j$ type in (a) form spliced molecules of the types in (b) after permutation.

where Y is the parameter that gives the correct number of i–j bonds when multiplied by Z. If we accept the concept of half-bonds for the moment, then there are $2ZY$ half-bonds, half of which, i.e., ZY, have to be i half-bonds emanating from the molecules of i, and the other half, ZY, have to be j half-bonds emanating from the molecules of j. However, prior to mixing, the pure i molecules have ZN_i half-bonds of i type; therefore, after mixing, the i half-bonds, not connected to the j half-bonds, are $ZN_i - ZY$, and these half-bonds must meet with each other to form i–i bonds. Likewise, the total number of j half-bonds connected to the j half-bonds is $ZN_j - ZY$. The reader can verify these relationships by using unidimensional crystals, i.e., beads in necklaces as in Fig. 4.3. The total number of whole bonds of each type in the mixture is therefore

$$i\text{–}j \text{ bonds} = ZY$$

$$i\text{–}i \text{ bonds} = (Z/2)(N_i - Y) \qquad (4.18)$$

$$j\text{–}j \text{ bonds} = (Z/2)(N_j - Y)$$

It is very important to remember that the total number of unlike bonds is equal to ZY. The total number of bonds before and after mixing is evidently the same, i.e., $NZ/2$. Since the number of bonds per molecule is $Z/2$, then the net number of molecules having unlike neighbors is $2Y$, the net numbers

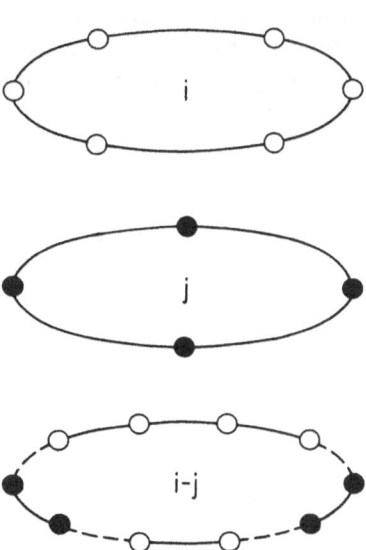

Figure 4.3. Molecules in unidimensional pure crystals i and j, and their solution i–j; $Z = 2$. i–j bonds = $ZY = 4$ in the bottom configuration; hence, $Y = 2$, $N_i - Y = 4$, and $N_j - Y = 2$.

of i and j molecules surrounded by like neighbors are $(N_i - Y)$ and $(N_j - Y)$, respectively. Figure 4.3 shows the various types of bonds for unidimensional crystals for which $Z = 2$. To eliminate the error from a limited number of molecules, the dangling terminal bonds in a linear crystal are tied to form crystals similar to the beads in a necklace. For a large number of molecules in a linear crystal, the end effect is negligible. The upper drawings show pure i and pure j, and the lower drawing shows one of the possible resulting alloys wherein the solid lines are the like bonds and the dashed lines are the unlike bonds. We observe that $ZY = 4$ since there are four i-j bonds, but $Z = 2$; hence, $Y = 2$. Therefore, the i-i bonds are $(6 - 2)Z/2 = 4$, and the j-j bonds are $(4 - 2)Z/2 = 2$. In this case, $N_i - Y = 4$ and $N_j - Y = 2$ since $Z = 2$. It is also very easy to construct such crystals for $Z = 4$ and obtain the same relationships as will be seen later.

The equilibrium among the net numbers of molecules having the like and unlike bonds is

$$-\overset{|}{\underset{|}{i}}- + -\overset{|}{\underset{|}{j}}- = \cdots\overset{:}{\underset{:}{i}}\cdots + \cdots\overset{:}{\underset{:}{j}}\cdots \qquad (4.19)$$

where one molecule of i with $Z/2$ of i-i bonds (or Z half-bonds) and one molecule of j with $Z/2$ of j-j bonds (as shown by the solid lines for $Z = 4$ on the left side) react to form one molecule of i with $Z/2$ of i-j bonds and one molecule of j with $Z/2$ bonds (as shown by the dotted lines on the right side), so that altogether two molecules with opposite neighbors are formed. Further justification for this reaction will be presented later. For a solution formed by N_i and N_j molecules of pure components i and j, the foregoing reaction has the following numbers of species at equilibrium:

$$[(N_i - Y)] + [(N_j - Y)] \leftrightharpoons 2Y \qquad (4.20)$$

where $N_i - Y$ and $N_j - Y$ are the net numbers of reactant molecules in reaction (4.19), and $2Y$ are the net numbers of product molecules.

Distribution of Molecules

One-Dimensional Crystals

We shall first obtain the distribution of molecules for $Z = 2$ and then extend it to $Z = 4$ to 6, where $Z = 6$ occurs in simple cubic and two-dimensional hexagonal crystals. We take another set of arrangements shown in Fig. 4.4 to derive the total number of arrangements for given values of

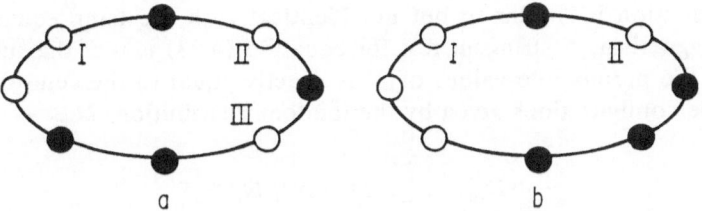

Figure 4.4. Arrangements of molecules in unidimensional crystals ($Z = 2$). $Y = 3$ for (a), $Y = 2$ for (b). Open circles are i molecules; black circles, j molecules.

Y, and $N_i = N_j = 4$. In Fig. 4.4(a), $Y = 3$ since there are 6 i–j bonds. There are three locations I, II, and III, and one molecule of i is affixed on each location to identify it. The number of locations is the same as Y. There is one molecule of i next to I which can be placed next to II or III without changing the value of Y and this is the number of distributable molecules which is equal to $N_i - Y$. Likewise in Fig. 4.4(b) there are two locations I and II; Y is therefore 2, and there are 2 molecules ($N_i - Y = 2$) which can be placed next to II. The number of locations to which the molecules are distributed is $Y - 1$ in both figures. We emphasize that the initial configuration, or the configuration we have to start with is such that all the i beads are placed together in I except the individual ones identifying II, III, ... We assume at first that the locations I, II, . . . are at fixed positions, but we shall remove this restriction later. The number of ways of arranging $N_i - Y$ distributable molecules to the sites next to $Y - 1$ locations is

$$[(N_i - Y) + (Y - 1)]!/(N_i - Y)!(Y - 1)! \qquad (4.21)$$

The same formula is also applicable to the j molecules; hence, for both permutations we have

$$(N_i - 1)!(N_j - 1)!/(N_i - Y)!(Y - 1)!(N_j - Y)!(Y - 1)! \qquad (4.22)$$

We remove the restriction on the system that I, II, and III have fixed positions by rotating each configuration by one molecular position at one time and repeat this procedure N times. When the rotations are constructed on paper for all the configurations for a given value of Y, it is seen that each configuration is unnecessarily repeated by a factor of Y. The number of ways of arranging the molecules for a given value of Y is therefore N/Y times the preceding expression, i.e.,

$$D = (N_i - 1)!(N_j - 1)! N/Y!(Y - 1)!(N_i - Y)!(N_j - Y)!$$
$$\approx N_i! N_j!/(Y!)^2 (N_i - Y)!(N_j - Y)! \qquad (4.23)$$

This equation is similar to but not identical with the Ising equation for ferromagnetism. A stringent test for equation (4.23) is that the sum of D for all the permissible values of Y is exactly equal to the sum of all the possible configurations given by the random distribution, D_{rm}:

$$D_{rm} = \sum_{Y=m}^{Y^\circ} D = N!/N_i!N_j! \qquad (4.24)$$

where m is the lowest geometrically possible value of Y, which is 1 for linear crystals when the number of unlike bonds is 2, and the upper limit is $Y^\circ = N_j$ for $N_i > N_j$, i.e., all the j molecules have unlike neighbors. This equation can readily be verified for the system $N_i = N_j = 4$ of Fig. 4.4 as follows: $Y = 1$, $D = 8$; $Y = 2$, $D = 36$; $Y = 3$, $D = 24$; $Y = 4$, $D = 2$; the sum of all D's is 70, which is equal to $N!/(N_i!N_j!) = 8!/(4!4!)$.

Two-Dimensional Crystals

A two-dimensional crystal lattice with a coordination number of 4 is shown in Fig. 4.5. The molecules in the crystal occupy the points of intersection of the lattice. It is also possible to construct a close-packed hexagonal two-dimensional crystal with $Z = 6$ but we shall not be concerned with it because of geometrical difficulties in enumerating various configurations. The dangling bonds in a crystal having a limited number of molecules cause a relatively large error in our enumerations but for a very large number of molecules, again the error becomes quite negligible. We can eliminate this error by tying the dangling bonds 1, 2, 3, and 4 with the lattice points

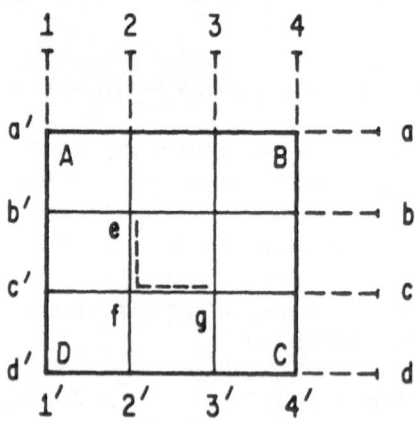

Figure 4.5. Two-dimensional crystal lattice with $Z = 4$.

1′, 2′, 3′, and 4′, respectively, away from the reader on the reverse side of the page and in such a manner that we form three squares, e.g., 1, 2, 1′, 2′ form a square. This process is best visualized if the lattice is stretched and wrapped over a sphere so that 1, 2, 1′, 2′ become quadrangular in shape. Similarly, a, b, c, and d are joined to the lattice points a′, b′, c′, and d′, respectively, to form three more squares. The four corners A, B, C, and D form their square in this process so that there are altogether 16 squares for 16 lattice sites. (There are, likewise, N cubes for N lattice sites in a simple cubic system for which $Z = 6$.) The corners A and C, as well as B and D are diagonally opposite to each other. This arrangement shows that any configuration, such as the dashed L located on efg, returns to its original position after four diagonal jumps along BD.

It is convenient to formulate a procedure which can make the enumeration of configurations clear and systematic. For this purpose, we take the number of one of the component molecules either equal to, or larger than the other without losing generality, e.g., $N_i \geq N_j$, and write a bond balance for N_j molecules from equation (4.16), i.e.,

$$j\text{-}j \text{ bonds} + (0.5)(ZY) = (0.5)(ZN_j) \qquad (4.25)$$

This equation states that there are altogether ZN_j half-bonds joined to the j molecules, and two j-j of these are the j-j half-bonds and the remainder, ZY, are the i-j half-bonds. Equation (4.25) is useful in computing Y from the known numbers of j-j bonds. As an example, we consider $N_i = N_j = 8$ in some detail. The number of configurations for all the permissible values of Y has been counted and listed in Table 4.1. A stringent check on the results is that the total number of configurations for all the possible values of Y be equal to that given by equation (4.24).

When j-j is zero, Y is 8 and j molecules occupy diagonally opposite positions, leaving the remaining positions for the i molecules. The sets of positions for i and j molecules may be all interchanged once; therefore, the total number of configurations is 2. The configurations for the values of 1 and 2 for j-j are geometrically impossible as can be verified by construction. The next configuration contains the cluster for which j-$j = 3$, and there is only one such cluster as shown in Fig. 4.6(a). The four remaining j molecules are surrounded entirely by the i molecules and they can be arranged only one way. The entire configuration can be rotated four times in 90° steps and placed in 16 different ways in the lattice, or stated differently, Fig. 4.6(a) can be rotated in four ways and placed in 16 different positions. The resulting number of configurations is $4 \times 16 = 64$. The configurations containing the clusters of j-$j = 4$ are shown in Fig. 4.6(b)-(e). A selected molecule in the cluster shown in Fig. 4.6(b) can be placed in 16 different

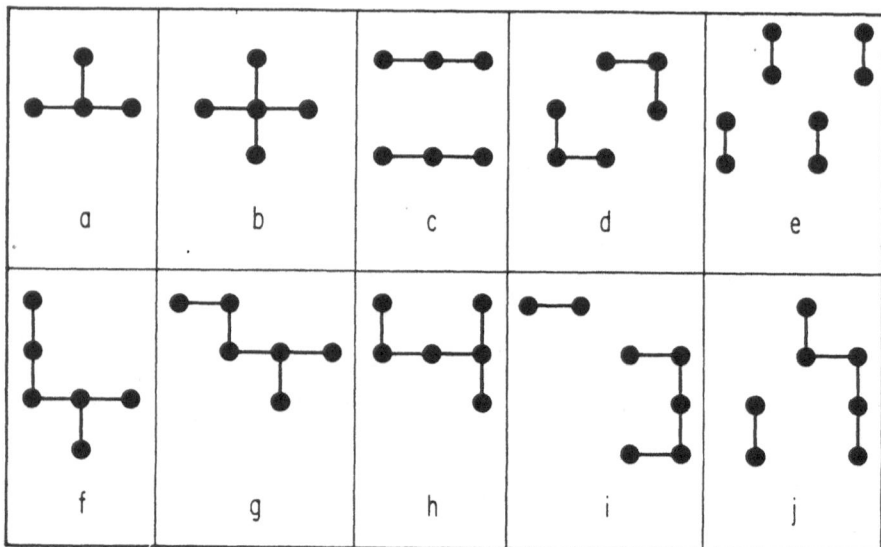

Figure 4.6. Arrangements of molecules for three, four, and five $j-j$ bonds for $N_i = N_j = 8$ and $Z = 4$.

lattice points and for each position the remaining three j molecules may be arranged in four different positions and thus yield $4 \times 16 = 64$ configurations. Figure 4.6(c) and (e) give 16 and 8 configurations, respectively, because of symmetry, and Fig. 4.6(d) gives 32 configurations. The total number of configurations for $j-j = 4$ is therefore 120. The configuration for $j-j = 5$ is shown in the remaining parts of Fig. 4.6. Each cluster and its mirror image in (f), (g), and (h) can be rotated four times and placed in 16 different sites to give $4 \times 2 \times 16 = 128$ configurations for each figure and 384 configurations for three figures. The arrangement in (i) can be rotated four times and thus yield $(4 + 4 + 4)16 = 192$ arrangements. The total for $j-j = 5$ is then 576 configurations. The process of enumerating the configurations for $j-j = 6$ is a very time-consuming task. There are three single clusters of 6 $j-j$ bonds involving squares with two attached bars that can be rotated four ways and yield $3 \times 4 \times 16 = 192$ configurations, and 15 one-piece clusters similar to (f) giving 1216 configurations. In addition, there are nine two-piece clusters similar to (i) yielding 704 configurations, all adding up to 2112 configurations. The results for $j-j = 7$ to 12 are listed in Table 4.1 and it is readily seen that the total number of configurations is equal to D_{rm}. Similar enumerations for $N_i = 9$ and $N_j = 7$ are also listed in Table 4.1. The last column for the permutation of bonds, g, will be discussed on page 107.

Table 4.1. Arrangements of Molecules for $N_i = N_j = 8$ and $N_i = 9$, $N_j = 7$ in Two-Dimensional Lattice with $Z = 4^a$

System	j-j bonds	Y	Y/Y^*	Actual arrange-ments	D	g
$N_i = 8$	12	2	0.500	8	392	257
$N_j = 8$	11	2.5	0.625	0	1,110	1,482
$Y^* = 4$	10	3	0.750	768	2,532	4,983
$D_{rm} = 12,870$	9	3.5	0.875	1,600	3,837	10,169
	8	4	1.000	4,356	4,900	12,870
	7	4.5	1.125	3,264	4,934	10,169
	6	5	1.250	2,112	3,920	4,983
	5	5.5	1.375	576	2,442	1,482
	4	6	1.500	120	1,176	2,574
	3	6.5	1.625	64	427	24.4
	2	7	1.750	0	112	1.12
	1	7.5	1.875	0	20	0.20
	0	8	2.000	2	2	7.8×10^{-5}
Total				12,870	25,624	48,995
$N_i = 9$	9	2.5	0.635	64	1,073	1,488
$N_j = 7$	8	3	0.762	624	2,240	4,838
$Y^* = 3.9375$	7	3.5	0.889	1,920	3,588	9,478
$D_{rm} = 11,440$	6	4	1.016	3,680	4,480	11,403
	5	4.5	1.143	3,136	4,386	8,446
	4	5	1.270	1,392	3,360	3,801
	3	5.5	1.397	512	1,993	1,005
	2	6	1.524	96	896	147
	1	6.5	1.651	0	293	10.41
	0	7	1.778	16	64	0.27
Total				11,440	22,373	40,617

$^a D$ is the distribution of molecules; g, the distribution of bonds; D_{rm} is given by equation (4.24).

The process of enumeration may be summarized as follows: (1) Construct all the possible clusters for a given value of j-j, (2) eliminate duplications created by rotation, (3) place each cluster in the lattice and eliminate those which do not fit or accommodate the other molecules, (4) eliminate the duplications which occur when multiple piece clusters, such as those in Fig. 4.6(c) and (e), are placed in various locations in the lattice, and (5) check the results by comparing the sum of all configurations with D_{rm} to ensure that the duplications due to symmetry have been eliminated.

The *results selected* from enumerations for 20 additional systems[4,6] with various values of N and N_j/N_i are listed in Table 4.2 with the values of

Table 4.2. Arrangements of Molecules for Various Values of N_i and N_j and for $Z = 4$

System	j–j bonds	Y	Y/Y^*	Actual arrangements	D/actual arrangements
$N_i = 11$	5	2.5	0.727	192	4.144
$N_j = 5$	4	3	0.873	688	2.093
$Y^* = 3.4375$	3	3.5	1.018	1,664	1.161
$D_{rm} = 4,368$	2	4	1.164	1,248	1.538
	1	4.5	1.309	448	3.080
	0	5	1.455	128	5.250
$N_i = 15$	15	2.5	0.417	10	745.7
$N_j = 10$	14	3	0.500	0	—
$Y^* = 6$	13	3.5	0.583	2,300	34.74
$D_{rm} = 3,268,760$	⋮	⋮	⋮	⋮	⋮
	3	8.5	1.417	21,200	8.958
	2	9	1.500	4,800	15.64
	1	9.5	1.583	400	57.32
	0	10	1.667	10	500.5
$N_i = 20$	5	2.5	0.625	210	15.79
$N_j = 5$	4	3	0.750	1,650	5.182
$Y^* = 4$	3	3.5	0.875	8,850	1.884
$D_{rm} = 53,130$	2	4	1.000	18,675	1.297
	1	4.5	1.125	17,000	1.533
	0	5	1.250	6,745	2.873
$N_i = 18$	30	3	0.333	12	18,490
$N_j = 18$	29	3.5	0.389	0	—
$Y^* = 9$	28	4	0.444	3,456	1,204
$D_{rm} = 9.0751 \times 10^0$	27	4.5	0.500	34,704	404
	⋮	⋮	⋮	⋮	⋮
	6	15	1.667	9,780	113.5
	5	15.5	1.722	1,152	206
	4	16	1.778	504	82.6
	3	16.5	1.833	144	40.3
	2	17	1.889	0	—
	1	17.5	1.944	0	—
	0	18	2.000	2	1
$N_i = 24$	18	3	0.375	12	13,920
$N_j = 12$	17	3.5	0.438	72	10,096
$Y^* = 8$	16	4	0.500	8,820	298
$D_{rm} = 1.2517 \times 10^9$					
$N_i = 84$	24	4	0.298	100	1.047×10^7
$N_j = 16$	23	4.5	0.335	218,000	3.556×10^5
$Y^* = 13.44$	22	5	0.372	498,400	1.006×10^5
$D_{rm} = 1.3459 \times 10^{18}$					

Table 4.2 (*continued*)

System	j-j bonds	Y	Y/Y^*	Actual arrangements	D/actual arrangements
$N_i = 92$	10	3	0.408	600	4,778
$N_j = 8$	9	3.5	0.476	13,400	1,438
$Y^* = 7.36$	8	4	0.544	284,450	373
$D_{rm} = 1.8608 \times 10^{11}$	7	4.5	0.611	3,368,200	144
$N_i = 95$	5	2.5	0.526	800	186
$N_j = 5$	4	3	0.632	14,300	61.1
$Y^* = 4.75$	3	3.5	0.737	260,000	15.08
$D_{rm} = 7.5288 \times 10^7$	2	4	0.842	3.7726×10^6	3.553
	1	4.5	0.947	2.2033×10^7	1.551
	0	5	1.053	4.9208×10^7	1.240
$N_i = 884$	24	4	0.255	900	1.1707×10^{13}
$N_j = 16$	23	4.5	0.286	196,200	1.4660×10^9
$Y^* = 15.7156$	22	5	0.318	8,805,600	7.0191×10^8
$D_{rm} = 7.7449 \times 10^{33}$					
$N_i = 895$	5	2.5	0.503	7,200	5,467
$N_j = 5$	4	3	0.603	848,700	847
$Y^* = 4.9722$	3	3.5	0.704	2.3940×10^7	419
$D_{rm} = 4.8663 \times 10^{12}$	2	4	0.804	3.5389×10^6	30.18
	1	4.5	0.905	2.0937×10^{11}	4.068
	0	5	1.006	4.6534×10^{13}	1.023

D from equation (4.23). A complete description of all enumerations is presented elsewhere in detail.[6] Accurate computations of D for the fractional values of factorials require the use of numerical tables for the gamma function which is defined by

$$f! = \text{gamma function of } (f + 1) \qquad (4.26)$$

where f is a fraction. For sufficiently large values of f, e.g., $f > 50$, Stirling's approximation may be used to obtain the numerical value of $f!$. The maximum value of Y/Y^* is given by $(N_i + N_j)/N_i$ and decreases with decreasing values of N_j/N_i for our convention of choosing $N_i \geq N_j$. The complexity of constructing and counting the configurations for high values of both N and Y/Y^* has limited our enumerations to relatively low values of Y/Y^* for high values of N. The values of $\log(D/\text{actual arrangements})$ are plotted versus Y/Y^* as distinct points in Fig. 4.7. The results may be

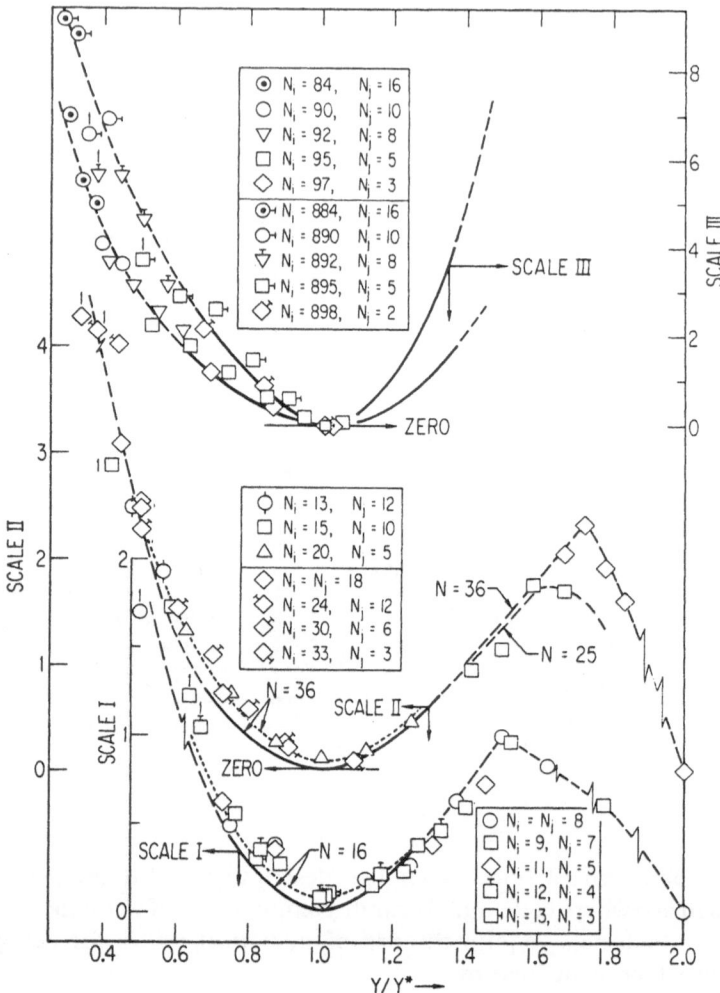

Figure 4.7. Number of arrangements of molecules in two-dimensional crystals with $Z = 4$. Vertical scales are $\log(D/\text{actual arrangements})$ and horizontal scale is Y/Y^*.

summarized as follows: (1) All the points for a given value of N may be fitted by a curve independent of the value of N_i/N_j. The dotted curves pass through the actual points for $N = 16$ and $N = 36$ and the solid curves ending in broken terminal portions on the left and on the right pass through the points for $N = 100$ and 900. The dotted line for $N = 25$ is so close to that for $N = 36$ that it has not been drawn except on the right side where there is a small systematic difference between the two sets as shown by the broken lines. The nature of the solid lines ending in broken lines for $N = 16$ and 36 will be discussed later, but in this paragraph the dotted curves for $N = 16$ and 36 and the solid curves for $N = 100$ and 900 will be considered. It is clear that D/actual arrangements is a function of N and Y/Y^* and independent of N_i/N_j. (2) The range of Y/Y^* wherein the points follow a concave-up curve becomes wider with increasing N on both sides of the vertical line passing through $Y/Y^* = 1$, or $Y = Y^*$. (3) The minimum point at $Y = Y^*$ approaches the abscissa and becomes tangent to it with increasing N. (4) The initial points on the left, marked as 1 close to the vertical axis in Fig. 4.7, scatter more than the others because of the coarseness of the clusters. A certain degree of scattering is also due to the fact that the values of N_i and N_j are rather small, and as they approach Avogadro's number the scattering would undoubtedly smooth out.

We next consider the solid/broken lines for $N = 16$ and 36 in Fig. 4.7. If we lower the dotted curves vertically so that they become tangent to the abscissa at $Y = Y^*$, we obtain very nearly the set of solid and broken curves. We observe that as N increases, the distinction between the dotted and solid curves disappears, and the dotted curve becomes tangent to the abscissa as shown by the curves for $N = 100$ and 900. The solid portions of these curves which extend from about 0.6 to 1.4 for Y/Y^* have actually been obtained in the following manner. We correct equation (4.23) to obtain D_c so that D/D_c represents the solid portions of the curves, or D_c is the actual number of arrangements for sufficiently large values of N. The correction c, and the resulting corrected distribution function D_c are as follows:

$$c = [(Z - 1)(Z - 2)/Z]Y^b[(Y/Y^*) - 1] \qquad (4.27)$$

$$b = (Z - 1)/(2Z - 1) \qquad (4.28)$$

$$D_c = \frac{(N_i - 1)!(N_j - 1)!N}{(Y + c)!(Y + c - 1)!(N_i - Y - c)!(N_j - Y - c)!}$$

$$= \frac{N_i!N_j!}{[(Y + c)!]^2(N_i - Y - c)!(N_j - Y - c)!} \qquad (4.29)$$

Both c and b were obtained by trial and error with the requirement that $c = 0$ for $Z = 2$ and for $Y/Y^* = 1$. The empirical forms of c and b are not unique because other similar formulas can be obtained for reasonably close values of D_c. The correction given by equation (4.27) is negative for $Y/Y^* < 1$, positive for $Y/Y^* > 1$, and zero for $Y = Y^*$ because equation (4.23) represents the correct number of configurations at $Y/Y^* = 1$. This

Table 4.3. Arrangements of Molecules for Simple Cubic Crystals; $N = 512$, $Z = 6$

N_i	N_j	D_{rm}	$j\text{-}j$	Y	Y/Y^*	Actual arrangements	D	D/actual arrangements
507	5	2.871×10^{11}	3	4	0.8080	32,114,688	4.396×10^7	1.37
508	4	2.830×10^0	3	3	0.7559	42,496	49,317	1.16
508	4	2.830×10^0	2	10/3	0.8399	4,943,616	3,763,200	0.76

correction is imposed on the system purely by geometry when we proceed from $Z = 2$ to 4. The functional form of c is appropriate in the range of interest, Y/Y^* from 0.6 to 1.4, and it is valid not only for $Z = 2$ and 4, but also for $Z = 6$ as will be seen later. For $Z = 2$, c becomes zero and equation (4.29) becomes identical with equation (4.23). The functional form for c has been obtained empirically by fitting a large number of enumerations listed in Tables 4.1 and 4.2. It is not important whether or not equation (4.27) is rigorously correct and unique, but it is important for our purposes

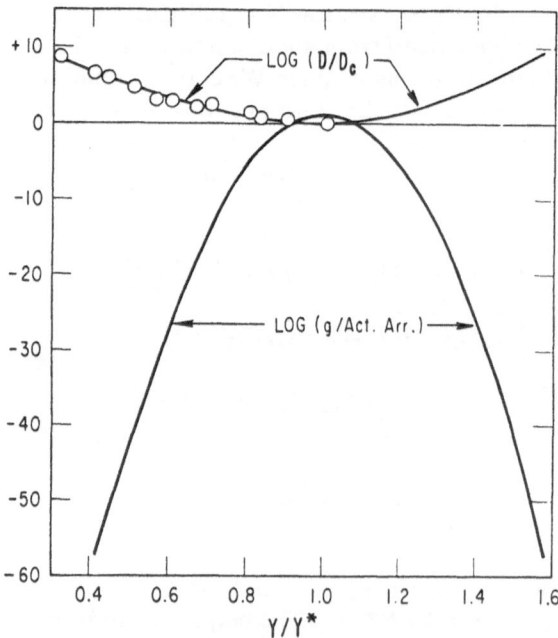

Figure 4.8. Variation of numbers of arrangements of molecules with Y/Y^* in two-dimensional crystals with $Z = 4$. Vertical scale is $\log[(D$ or $g)$/actual arrangements]; horizontal scale, Y/Y^*. Upper curve is $\log(D/D_c)$; circles are $\log (D$/actual arrangements) from Fig. 4.7, showing that $D_c \approx$ actual arrangements.

that c is a very small and negligible fraction of each set of terms in each pair of parentheses in equation (4.29) when N_i and N_j become large enough to be comparable to Avogadro's number. For example, at $Y/Y^* = 1.4$, c is about 7.3% of $Y + c$ for $N_i = N_j = 50$, but for $N_i = N_j = 450$, it is only 2.2% of $Y + c$. Likewise, at $Y/Y^* = 1.2$ the corresponding percentages are 4.1 and 1.2 for $N = 100$ and $N = 900$, respectively.

The configurations for the simple cubic system having $Z = 6$ are much more difficult to enumerate. The error from the dangling bonds is eliminated by a procedure similar to that for $Z = 2$ or 4, but for the limited number of clusters we shall consider, it is important to observe that the number of cubes is equal to the number of molecules and the total number of bonds is $ZN/2$. A limited number of enumerations are listed in Table 4.3. The results in the last column show that equations (4.27)–(4.29) are also satisfactory for $Z = 6$. A greater number of enumerations with $j-j = 2$ to 5 and $N_j = 4$ to 6 is possible but quite time-consuming.

The variation of D_c with Y/Y^* for $N_i = N_j = 450$ and $Z = 4$, or an equimolecular crystal of 30×30, is shown by the upper curve in Fig. 4.8. The circles represent the actual points from Fig. 4.7 for D/actual arrangements and show that D/D_c and D/actual arrangements are identical and therefore D_c is equal to the actual number of arrangements. The lower curve for log (g/actual arrangements) versus Y/Y^* will be discussed on page 107.

Equations

The energy of a system consisting of N_i and N_j molecules is equal to the sum of the bond energies for the sum of each type of bond. Therefore, from equation (4.18) we obtain

$$E = (N_i - Y)Ze_{ii}/2 + (N_j - Y)Ze_{jj}/2 + YZe_{ij}; \qquad (H \approx E) \quad (4.30)$$

where e_{ii}, e_{jj}, and e_{ij} are the energies for i-i, j-j, and i-j types of bonds, respectively, and further, $H \approx E$. Substitution of equation (4.6) for $W = W_{ij}$ in equation (4.30) gives

$$E = N_i Ze_{ii}/2 + N_j Ze_{jj}/2 + YW \qquad (4.31)$$

The configurational partition function corresponding to equation (3.51) on page 72 is

$$Q = \sum_Y D_c \, e^{-E/kT} \qquad (4.32)$$

where D_c is number of possible configurations or arrangements, having the same energy E as given by equation (4.31) for a given value of Y. For each value of E, the corresponding value of Y is fixed, and the summation is carried out over all values of Y corresponding to all the possible values of E. According to Chapter 3, the sum in equation (4.32) may be replaced by its maximum term, i.e.,

$$Q(\text{max}) = D_c \, e^{-E/kT} = D_c \exp[(-N_i Z e_{ii}/2 - N_j Z e_{jj}/2 - YW)/kT] \tag{4.33}$$

The value of Y for which this equation is the maximum term in equation (4.33) is obtained by differentiating $\ln Q(\text{max})$ with respect to Y and setting the result to zero:

$$\partial \ln Q(\text{max})/\partial Y = (1 + c')[-2 \ln(Y + c) + \ln(N_i - Y - c)$$
$$+ \ln(N_j - Y - c)] - (W/kT) = 0 \tag{4.34}$$

In this equation, c' is given by

$$c' = \partial c/\partial Y \approx (1 + b)[(z - 1)(z - 2)/z](Y^b/Y^*) \tag{4.35}$$

where the approximate equality holds for large values of Y; therefore, the term bY^{b-1} is not included in this equation. As $N = N_i + N_j$ approaches Avogadro's number, Y^b/Y approaches a value close to zero in the range of $0.5 < Y/Y^* < 1.5$. In addition, c in each of the logarithmic terms is negligible in comparison with Y, $N_i - Y$, and $N_j - Y$; therefore, we could have obtained an equation equivalent to (4.31) by using D of equation (4.23) instead of D_c in equations (4.29) and (4.33). We rearrange equation (4.34) and neglect c to derive

$$\frac{Y^2}{(N_i - Y)(N_j - Y)} = e^{-W/kT} = e^{-NW/RT} \tag{4.36}$$

where $R = Nk$ is the gas constant. The left side of this equation is the law of mass action or the pseudo-equilibrium constant for the following reaction:

$$-i- + -j- = \cdots i \cdots + \cdots j \cdots \tag{4.37}$$

This reaction states that one molecule of i with Z half-bonds of i–i type and one molecule of j with Z half-bonds of j–j type, both on the left and shown with solid half-bonds, react to form one molecule of i with Z half-bonds of $i \cdots j$ type and one molecule of j with Z half-bonds of $j \cdots i$ type shown with dotted lines. The number of molecules of the reactants are $(N_i - Y)$ and $(N_j - Y)$ and the products are Y and Y for the respective species in reaction (4.37). The corresponding energy effect is clearly $W = (Z/2)(2e_{ij} - e_{ii} - e_{jj})$. Reaction (4.37) is our short-range order–disorder reaction.

First Approximation

The correct first approximation to the regular solutions may now be obtained by starting with equation (4.36). We set, for brevity in notation,

$$n = e^{W/kT} \qquad (4.38)$$

where n is always a positive quantity. We substitute n into equation (4.36) to obtain

$$(1 - n)Y^2 - NY + N_iN_j = 0 \qquad (4.39)$$

There are two sets of values of Y satisfying this equation:

$$Y = \{N \pm [N^2 - 4N_iN_j(1 - n)]^{0.5}\}/2(1 - n); \qquad (+\text{ sign unacceptable})$$
$$(4.40)$$

The set with the plus sign after N is not acceptable for two reasons: (1) The maximum value of Y for $N_i = N_j$ is $Y = N_i$ for a perfectly ordered crystal in which the number of j–j bonds is zero [(see equation (4.25)]. However, for $n < 1$, the plus sign would give $Y > N_i$. For example, if $n = 0.64$ and $N_i = N_j = N/2$, then $Y = 2.5N$. (2) The value of Y cannot be negative but for $n > 1$ it becomes negative for the plus sign in equation (4.40) and positive for the minus sign. Therefore, we accept equation (4.40) with the minus sign after N.

The square root in equation (4.40) is inconvenient; hence, for simplicity we introduce

$$\beta = [1 - 4x_ix_j(1 - n)]^{0.5} \qquad (4.41)$$

which transforms equation (4.40) into

$$Y = N[(1 - \beta)/2(1 - n)] = 2Nx_ix_j/(\beta + 1) \qquad (4.42)$$

where the last equality is obtained by substituting for $(1 - n)$ from equation (4.41). It is seen that for $W = 0$, β is unity and $Y = Nx_ix_j = Y^*$ as in a random solution. We subtract $E_{ii} = N_iZe_{ii}/2$ and $E_{jj} = N_jZe_{jj}/2$ from equation (4.31) to obtain $\Delta E \approx \Delta H$ in the same way that we obtained equation (4.5); thus,

$$\Delta H = H^e = YW = 2Nx_ix_jW/(\beta + 1) \qquad (4.43)$$

The equations for $\bar{H}_i^e$ and $\bar{H}_j^e$ can be obtained by substituting equation (4.43) into (1.60); the results are

$$\bar{H}_i^e = NWx_j\frac{\beta + x_j - x_i}{\beta(\beta + 1)}; \qquad \bar{H}_j^e = NWx_i\frac{\beta + x_i - x_j}{\beta(\beta + 1)} \qquad (4.44)$$

It is evident that $\bar{H}_j^e$ can be obtained from $\bar{H}_i^e$ by interchanging the subscripts i and j. For $\beta = 1$, these equations give the zeroth approximation form of $\bar{H}_i^e = NWx_j^2$, but $\beta = 1$ signifies that $n = 1$ or $W = 0$. Therefore, it is not possible to have a random solution when W is not zero. The zeroth approximation is consequently inferior to the first approximation from a theoretical point of view. The excess molar Gibbs energy of mixing, G^e, is obtained by integrating $d(G^e/T) = H^e d(1/T)$, i.e.,

$$\frac{G^e}{T} = \int_{T^{-1}=0}^{T^{-1}} \frac{2Nx_ix_jW}{\beta + 1} d(T^{-1}) \qquad (4.45)$$

The lower integration limit for the left side is zero because at very high temperatures or for $1/T = 0$, G^e/T is zero. We wish to express $d(T^{-1})$ in terms of β and for this purpose we differentiate equation (4.38) as follows:

$$dn = n(W/k)d(T^{-1}) \qquad (4.46)$$

We solve for n from equation (4.41) and then eliminate n from equation (4.46) to obtain

$$d(T^{-1}) = (2k/W)\beta \, d\beta/[\beta^2 - (1 - 4x_ix_j)] \qquad (4.47)$$

where $1 - 4x_ix_j$ is equal to $(x_i - x_j)^2$ after we use $(x_i + x_j)^2 = 1$ for a binary system. The substitution of equation (4.47) into equation (4.45) gives

$$\frac{G^e}{T} = Nk \int_{\beta=1}^{\beta} \frac{4x_ix_j\beta \, d\beta}{(\beta + 1)(\beta + x_i - x_j)(\beta + x_j - x_i)} \tag{4.48}$$

where $T^{-1} = 0$ makes $n = 1$ in equation (4.38); hence, $\beta = 1$ from equation (4.41). The quantity after the integral sign *without* $d\beta$ may be expanded into

$$\frac{J}{\beta + 1} + \frac{L}{\beta + x_i - x_j} + \frac{M}{\beta - x_i + x_j} \equiv \frac{-1}{\beta + 1} + \frac{x_i}{\beta + x_i - x_j} + \frac{x_j}{\beta - x_i + x_j} \tag{4.49}$$

where the numerators on the left side were determined by a well-known algebraic method. Each term may now be integrated to obtain

$$G^e = NkTx_i \ln\left[\frac{\beta + x_i - x_j}{x_i(\beta + 1)}\right] + NkTx_j \ln\left[\frac{\beta + x_j - x_i}{x_j(\beta + 1)}\right] \tag{4.50}$$

This equation constitutes the first approximation to the regular solutions. All the remaining equations in this section are derived from this equation. The partial molar excess Gibbs energy of component i, $\bar{G}_i^e$, can be obtained by substituting G^e in equation (1.60) and carrying out the necessary differentiation but this procedure is long and tedious. A simple and concise procedure is to compare the terms of equation (4.50) with the right side of $G^e = x_i\bar{G}_i^e + x_j\bar{G}_i^e$ and immediately obtain

$$\bar{G}_i^e = NkT \ln\left[\frac{\beta + x_i - x_j}{x_i(\beta + 1)}\right]; \qquad \bar{G}_j^e = NkT \ln\left[\frac{\beta + x_j - x_i}{x_j(\beta + 1)}\right] \tag{4.51}$$

Equation (4.50) is symmetrical in x_i and x_j, i.e., the interchange of subscripts leaves G^e unchanged; hence, G^e versus x_i is a symmetrical curve about the line perpendicular to the x_i axis at $x_i = x_j = 0.5$.

The excess entropy S^e is expressed by

$$S^e = \frac{2Nx_ix_jW}{T(\beta + 1)} - Nkx_i \ln\left[\frac{\beta + x_i - x_j}{x_i(\beta + 1)}\right] - Nkx_j \ln\left[\frac{\beta + x_j - x_i}{x_j(\beta + 1)}\right] \tag{4.52}$$

For an ideal solution $W = 0$, $\beta = n = 1$, and equation (4.52) is zero as expected, because S^e for a random solution is zero by definition. The equations for $\bar{S}_i^e$ and $\bar{S}_j^e$ are

$$\bar{S}_i^e = \frac{NWx_j}{T}\left[\frac{\beta + x_j - x_i}{\beta(\beta + 1)}\right] - Nk \ln\left[\frac{\beta + x_i - x_j}{x_i(\beta + 1)}\right] \tag{4.53}$$

$$\bar{S}_j^e = \frac{NWx_i}{T}\left[\frac{\beta + x_i - x_j}{\beta(\beta + 1)}\right] - Nk \ln\left[\frac{\beta + x_j - x_i}{x_j(\beta + 1)}\right] \qquad (4.54)$$

Again, equation (4.54) can be obtained from (4.53) by interchanging i and j.

We proceed to test the success of the first approximation by setting $x_i = x_j = 0.5$ and by using the available data on mixtures. The resulting form of equation (4.50) is then

$$G^e = RT \ln[2\beta/(1 + \beta)] \qquad (4.55)$$

The corresponding form of equation (4.43) is

$$H^e = 0.5NW/(\beta + 1) \qquad (4.56)$$

At $x_i = x_j = 0.5$, equations (4.38) and (4.41) give

$$2 \ln \beta = \ln n = W/kT = NW/RT \qquad (4.57)$$

As a convenient test for equations (4.55) and (4.56), we take the experimental value of G^e and solve for β and then compute W from equation (4.57). Next we substitute β and W into equation (4.56) to calculate H^e. The reason for using this procedure is the simplicity of computation. The calculated and the experimental values of H^e are compared in Table 4.4 for a number of randomly selected alloys.[4,7] It is evident that the first approximation is good when G^e and H^e do not differ greatly, i.e., when the entropy term in $G^e = H^e - TS^e$ is small. There is a substantial degree

Table 4.4. Experimental and Calculated Values of H^e from First Approximation to Binary Regular Solutions[a]

Alloy	$T(°K)$	G^e	H^e(exptl)	H^e(calc)
Ag–Au	800	-837 ± 150	-1111 ± 50	-974
Bi + Pb	700	-300 ± 75	-265 ± 15	-326
Cd + Hg	600	-490 ± 60	-645 ± 30	-559
Cd + Mg	923	-1100 ± 100	-1341 ± 100	-1306
Cd + Zn	800	478 ± 50	500 ± 20	378
Hg + In	433	-495 ± 40	-540 ± 30	-580
In + Pb	673	140 ± 50	230 ± 50	132
K + Na	298.15	160 ± 30	125 ± 50	131
Mg + Pb	833	-1630 ± 200	-2100 ± 200	-1980
Sb + Zn	823	-750 ± 100	-830 ± 200	-863

[a] G^e and H^e are in calories for one gram atom (or mole) of solution at $x_1 = x_2 = 0.5$ (1 cal = 4.184 J).

of success in the results and in fact, for about half of the alloy systems for which the data are reliable,[7] the calculated and the experimental values of H^e agree fairly well.

Approximate Equations

Simple approximate equations can be derived in the narrow range of $-0.5 < W/kT < 0.5$ by using $\beta \approx (1 + 4ax_ix_j)^{0.5} \approx 1 + 2ax_ix_j$. Next, β is substituted in $2/(\beta + 1) \approx 1 - ax_ix_j$ for use in equations (4.42) and (4.43); then H^e is substituted in equation (4.45) to derive G^e. The results corresponding to equations (4.42), (4.43), (4.50), (4.51), and (4.52), in that sequence, are as follows:

$$Y \approx Nx_ix_j[1 - x_ix_j(W/kT)]$$

$$H^e = YW \approx NWx_ix_j[1 - x_ix_j(W/kT)]$$

$$G^e \approx NWx_ix_j[1 - x_ix_j(W/2kT)] \tag{4.58}$$

$$\bar{G}_i^e = NWx_j^2[1 - x_ix_j(W/kT) + x_i^2(W/2kT)]$$

$$S^e = -\frac{N}{2k}\left(\frac{Wx_ix_j}{T}\right)^2$$

The remaining partial molar properties can be derived from the molar properties as shown in Chapter 1. It is important to remember that these equations exclude the majority of alloys for which W/kT is larger than 0.5, i.e., $|NW| > 0.5RT \approx 1000$ cal/mole at 1000 K. However, it is likely that for the alloys for which $|NW/RT| > 0.5$, the criteria for regular behavior cannot generally be met; hence, equations (4.58) may be useful for systems in which the deviations from ideality are not great.

Short-Range Order

The relationship expressing Y^* for the random distribution can be obtained by setting the exchange energy W to zero in equation (4.38) to obtain $n = 1$, leading to $\beta = 1$, from equation (4.41) and $Y^* = Nx_ix_j$ from equation (4.42). The ratio Y/Y^* is therefore

$$Y/Y^* = 2/(\beta + 1)$$

We define a short-range order parameter α_s by

$$\alpha_s = 1 - \frac{Y}{Y^*} = \frac{\beta - 1}{\beta + 1} \tag{4.59}$$

For a random solution $\beta = 1$ and $\alpha_s = 0$, signifying that there is no order, or the solution is random. Positive values of W yield $n > 1$ and also $\beta > 1$, $\alpha_s > 0$, indicating that the net number of atoms surrounded by the unlike atoms is less than the random number, or there is a tendency for the like atoms to cluster together. Negative values of W lead to the opposite conclusion, i.e., when $W < 0$, then $n < 1$ and $\beta < 1$ and $\alpha_s < 0$, or there is a tendency for unlike atoms to surround each other. Very large negative values of W lead to β approaching zero and α_s approaching -1. There are other definitions of the short-range order which will not be discussed here. Actually, Y/Y^* itself may be considered as an order parameter and reaction (4.37) may be regarded as the order–disorder reaction. It will be seen later that in ordered alloys, a different parameter, r, will be more convenient than α_s.

Dilute Binary Regular Solutions

The equations for dilute binary regular solutions according to the first approximation can be obtained by setting one of the mole fractions close to zero. Since the equations derived in the previous sections are symmetric with respect to composition, it is sufficient to consider the case for $x_j \to 0$. The equations for H^e and $\bar{H}_i^e$ are zero as expected, and the equation for $\bar{H}_j^e$ is

$$\bar{H}_j^e = \partial H^e / \partial x_j = NW; \qquad (x_j \to 0) \tag{4.60}$$

This result can be obtained by substituting $x_j \to 0$ in equation (4.44) and noting that β becomes 1 for $x_j \to 0$. Likewise, G^e and $\bar{G}_i^e$ are zero, and $\bar{G}_j^e$ can be obtained from equation (4.51); however, for $x_j \to 0$ the resulting equation contains $\ln(0/0)$ and the indeterminate ratio can be obtained by applying the L'Hopital theorem, i.e., by taking the derivative of the numerator and then the denominator with respect to x_j and then setting $x_j \to 0$ in the resulting ratio. The derivative of the numerator, $\beta + 2x_j - 1$, is $2n$, and that of the denominator, $x(\beta + 1)$, is $\beta + 1 = 2$; therefore,

$$\bar{G}_j^e = \partial G^e / \partial x_j = NkT \ln n = NW; \qquad (x_j \to 0) \tag{4.61}$$

where $n = e^{W/kT}$ has been used to obtain the last equality. Since equations (4.60) and (4.61) are equal, it is evident that

$$\bar{S}_j^e = 0; \qquad (x_j \to 0) \tag{4.62}$$

Consequently, in dilute solutions, the solute can have only one type of configuration in which the j molecules are entirely surrounded by the solvent molecules, and this arrangement corresponds to the random dispersion of the j molecules in the solvent i molecules. The foregoing relationships can also be derived by using equations (4.58).

Critique of Treatments Based on Permutation of Bonds

The first approximation that existed prior to 1970 was based on the permutation of bonds as developed mainly by Bethe[5] and by Guggenheim.[1] It was shown earlier in this chapter that such a permutation leads to physically impossible spliced molecules. While the method has been the subject of numerous papers, some based on abstruse mathematical arguments, the most elegant and rigorous method is due to Guggenheim.[1]

The numbers of i-i, j-j, and i-j bonds used by Guggenheim are also given by equation (4.18). The corresponding bonds for the random distribution involve Y^* of equation (4.16). The energy E is also given by equation (4.30). However, D_c used in equations (4.29) and (4.32) are quite different, since the bonds are used in permutation. Guggenheim assumed that an observer is capable of distinguishing an i-j bond from a j-i bond and therefore there are $ZY/2$ of i-j and $ZY/2$ of j-i bonds. There are altogether $ZN/2$ bonds consisting of $Z(N_i - Y)/2$, $ZY/2$, $ZY/2$, and $Z(N_j - Y)/2$ bonds of i-i, i-j, j-i, and j-j bonds, respectively. The permutation of these bonds is given by

$$g' = \frac{(ZN/2)!}{[Z(N_i - Y)/2]!(ZY/2)!(ZY/2)![Z(N_j - Y)/2]!} \tag{4.63}$$

This equation is inexact as admitted by Guggenheim because the sum of g' for all the possible values of Y exceeds all the possible configurations $N!/(N_i! N_j!)$ by a very large factor for Z greater than 2. To remove this objection, Guggenheim introduces an arbitrary normalizing factor $h = h(N_i, N_j)$ to obtain a corrected expression $g = hg'$. The assumption that h is a function of N_i and N_j and independent of Y is incorrect as will be seen later. The sum of g for all values of Y must be equal to all the possible configurations, i.e.,

$$\sum_Y hg' = N!/(N_i!\, N_j!) \tag{4.64}$$

The sum in equation (4.64) is then replaced by its maximum term $g(\text{max}) = hg'(\text{max})$. To accomplish this, hg' is differentiated with respect to Y and the result is set to zero to solve for the corresponding value of Y. This solution gives $Y = Y^* = Nx_ix_j$, and when Y^* is substituted in $g(\text{max}) = N!/(N_i!\, N_j!) = hg'$, an equation is obtained for h. The substitution of h into $g = hg'$ gives

$$g = \frac{N!}{N_i!\, N_j!} \times \frac{[Z(N_i - Y^*)/2]!(ZY^*/2)!(ZY^*/2)![Z(N_j - Y^*)/2]!}{[Z(N_i - Y)/2]!(ZY/2)!(ZY/2)![Z(N_j - Y)/2]!}$$

$$\tag{4.65}$$

It is very important to observe that equation (4.65) differs from equation (4.63) by the factor h, which is assumed to be independent of Y. The partition function Q is then expressed by

$$Q = \sum_Y g\, e^{-E/kT} \tag{4.66}$$

where E is given by equation (4.30). We wish to replace g by g' to show that the same final equations of Guggenheim can be obtained without h and to prove that the use of the correction factor h is an exercise in futility. Therefore, we write

$$Q' = \sum_Y g'e^{-E/kT} \tag{4.67}$$

The sum in equation (4.67) can be replaced by its maximum term to obtain the equation for Y. For this purpose, $\partial \ln Q'/\partial Y$ is set to zero in the same way that equation (4.34) was set to zero. The solution of the resulting equation for $\partial \ln Q'/\partial Y = 0$ yields

$$Y^2/(N_i - Y)(N_j - Y) = \exp(-2W/ZkT)$$

$$= \exp[-(e_{ij} + e_{ji} - e_{ii} - e_{jj})/kT] \tag{4.68}$$

It is now evident that whether we had used g' or $g = hg'$, we would still obtain the same preceding equation because Guggenheim assumed that h was independent of Y. The *left side* of equation (4.68) is the equilibrium constant for the following reaction among the bonds:

$$i\text{-}i + j\text{-}j = i \cdots j + j \cdots i \tag{4.69}$$

In contrast, the *right side* of equation (4.68) may be regarded as the pseudo-equilibrium constant for reaction (4.69) where the exponential terms give the energies corresponding to the same bond reaction. Therefore, equations (4.68) and (4.69) reemphasize the independence of bonds as permutable entities despite the attempt to normalize g' by using h. Permutation of bonds as independent entities leads to enormously large values of g'; hence, g' cannot represent the order of magnitude of the correct distribution, D_c, for the molecules in solution. In addition, h is so small that $g = hg'$ quickly becomes smaller with increasing values of Y. Therefore, g cannot represent D_c as shown in Table 4.1, and by the curve for g/actual arrangements in Fig. 4.8. The values of g decrease sharply when Y/Y^* becomes either larger or smaller than unity as shown in Fig. 4.8. This is also evident in the last column of Table 4.1 at high values of Y/Y^* but the permissible values of Y/Y^* are not small enough for such a small number of molecules to show a similar trend at low values of Y/Y^*. The curves for g/actual arrangements at various values of N are not shown in Fig. 4.8, but they shift lower relative to D/actual arrangements with increasing N. It is evident that g does not represent the actual number of arrangements with a realistic degree of approximation and therefore it cannot be used in deriving meaningful statistical thermodynamic equations.

The ratio g/D_c for $N_i = N_j = L$, $Y^* = N_i/2$, $Y = N_i/4$, and for any value of Z is given by

$$\log(g/D_c) = -0.057(Z - 2)L + 0.316L^b + 0.5 \log L\pi \tag{4.70}$$

where Stirling's approximation for factorials has been used. The value of D_{rm} is expressed by

$$\log D_{rm} = 0.602L - 0.5 \log L\pi \tag{4.71}$$

Therefore, g and D_c differ by a factor of e^L or by about the same factor as D_{rm}. On the other hand, D/D_c is given by

$$\log(D/D_c) \approx 2 \log[Y/(L - Y)] = 0.316L^b \tag{4.72}$$

where the expression in the center is for any value of Y obtained by using $\ln(1 + \alpha) \approx \alpha$ and the last equality is for $Y = L/4$. It is seen that D and D_c differ by a much smaller factor than g and D_c. For $Z = 2$, equation (4.70) becomes comparable to equation (4.72) as $\log L$ is much smaller than L^b. This is evidently due to the fact that L and the number of permutable bonds become identical for $Z = 2$. However, the permutation of bonds again creates spliced molecules.

The thermodynamic equations resulting from equation (4.68) require that n of equation (4.38) must now be written as

$$n_g = e^{2W/ZkT} \tag{4.73}$$

When the procedure followed by equations (4.39)–(4.48) is repeated by using the preceding n_g, the result for G^e is obtained, i.e.,

$$G^e(\text{bond permutation}) = \frac{ZNkT}{2}\left\{x_i \ln\left[\frac{\underline{b} + x_i - x_j}{x_i(\underline{b} + 1)}\right] + x_j \ln\left[\frac{\underline{b} + x_j - x_i}{x_j(\underline{b} + 1)}\right]\right\} \tag{4.74}$$

where $\underline{b}$, *used only in this equation*, is defined by $\underline{b}^2 = 1 - 4x_ix_j(1 - n_g)$. The success of equation (4.74) in correlating G^e and H^e is considerably less than equation (4.50) based on permutation of molecules. It should be reemphasized that Guggenheim's method is the most rigorous and the simplest treatment based on the permutations of the bonds. Therefore, all the remaining treatments based on permutation of bonds add nothing significantly new to the resulting thermodynamic equations; hence, they are subject to the same criticism that the bonds are not permutable entities.

We next proceed to comment on a treatment called the "surrounded atom" or the "central atom" model for binary alloys.[8,9] The results can be interpreted in terms of the pseudo-equilibrium constant as in equation (4.36) and its reaction similar to (4.37). These relationships in terms of the notation in this book are

$$\frac{Y^2}{(N_i - Y)(N_j - Y)} = e^{-W/ZkT}; \qquad \left[\frac{W}{Z} = \frac{1}{2}(e_{ij} + e_{ji} - e_{ii} - e_{jj})\right] \tag{4.75}$$

$$\frac{1}{2}(i\text{-}i) + \frac{1}{2}(j\text{-}j) = \frac{1}{2}(i\cdots j) + \frac{1}{2}(j\cdots i) \tag{4.76}$$

where the coefficient $1/2$ in this reaction comes from the fact that W/Z contains $1/2$ as shown in equation (4.75). The difference between equation (4.75) and Guggenheim's equation (4.68) is that the latter contains $2W/Z =$

$(e_{ij} + e_{ji} - e_{ii} - e_{jj})$ in its exponent, or a quantity twice as much as in equation (4.75). The mathematical procedure and some of the postulates in the surrounded atom model are interesting; however, the resulting thermodynamic relations, starting with equation (4.75), are subject to the same criticism as equation (4.74) based on the permutation of bonds.

Ternary Regular Solutions

The zeroth approximation to multicomponent solutions was given by equation (4.7) in which $\Delta H = H^e = G^e$, and by equation (4.11). The first approximation requires the additional pairwise terms.[10] For example, a ternary system consisting of i-j-k components requires labeling β as β_{ij}, β_{ik}, and β_{jk}, and then writing equation (4.50) for the pairwise terms as follows:

$$\frac{G^e}{NkT} = x_i \ln\left[\frac{\beta_{ij} + x_i - x_j}{x_i(\beta_{ij} + 1)}\right] + x_j \ln\left[\frac{\beta_{ij} + x_j - x_i}{x_j(\beta_{ij} + 1)}\right]$$

$$+ x_i \ln\left[\frac{\beta_{ik} + x_i - x_k}{x_i(\beta_{ik} + 1)}\right] + x_k \ln\left[\frac{\beta_{ik} + x_{ik} - x_k}{x_k(\beta_{ik} + 1)}\right]$$

$$+ x_j \ln[\cdots] + x_k \ln[\cdots] \tag{4.77}$$

where the last two terms are for the j-k system and identical in form with the preceding terms. The equations for β_{ij} require writing $W_{ij}\ldots$ for pairwise interactions of net numbers of atoms. This relatively complicated equation can be simplified by using equation (4.58), i.e.,

$$\frac{G^e}{NkT} \approx W_{ij}x_ix_j\left[1 - x_ix_j\left(\frac{W_{ij}}{kT}\right)\right] + W_{ik}x_ix_k\left[1 - x_ix_k\left(\frac{W_{ik}}{kT}\right)\right]$$

$$+ W_{jk}x_jx_k\left[1 - x_jx_k\left(\frac{W_{jk}}{kT}\right)\right] \tag{4.78}$$

Liquid ternary alloys of some elements such as Mn, Fe, Co, and Ni are expected to obey qualitatively this equation.

Regular Associated Solutions

The phase diagrams in which a compound semiconductor is a solid phase, generally have a congruently melting solid, e.g., CuTe, SnTe, GaAs. Such 1 : 1 or other intermetallic and metal–metalloid compounds in solid state do not dissolve the component elements in significant concentrations.

The liquidus for these binary systems at the melting point in the vicinity of the compound has a temperature peak, the viscosity of liquid is a maximum, and the conductivity is a minimum. Further, the enthalpy and entropy of mixing show sharp minima in the vicinity of the compound. These properties indicate the existence of a compound species such as AB in the liquid phase of the binary system A–B; hence, the system may be considered as consisting of three species, A, B, and AB in the liquid phase. Other compounds, AB_n, may also exist, but the treatment presented here for the equiatomic compound AB would require only minor modifications to extend it to other intermetallic compounds, AB_n. The existence of compound species along with A and B forms the basis of associated solutions.

The solid phase, AB(s), the simplest compound to be considered here, is in equilibrium with the liquid phase at the liquidus composition in which the activities of A(l) and B(l) are denoted by a_1 and a_2, respectively. The accompanying equilibrium among A(l), B(l), and AB(s) is

$$AB(s) = A(l) + B(l); \qquad \Delta G_s^\circ = -RT \ln \frac{[a_1 \cdot a_2][\text{at liquidus}]}{a_3^s} \qquad (4.79)$$

where the activity of solid AB(s), denoted by a_3^s, will be taken as unity, because AB(s) is assumed to be a pure compound. Liquid AB(l) is formed by A and B in the liquid phase at the equiatomic composition, $x_1 = x_2 = 0.5$, according to

$$A(l) + B(l) = AB(l); \qquad \Delta G_1^\circ = -RT \ln \frac{a_3}{a_1(0.5) \cdot a_2(0.5)} \qquad (4.80)$$

We emphasize that a_1, a_2, and a_3 without superscripts always refer to the liquid phase. If the compound formation is also very strong in the liquid phase so that the activity of AB(l) can be taken as unity only in equation (4.80) at equiatomic composition, then the sum of the two preceding reactions and their $\Delta G°$ is

$$AB(s) = AB(l); \qquad \Delta G_m^\circ = RT \ln \frac{a_1(0.5) \cdot a_2(0.5)}{[a_1 \cdot a_2][\text{at liquidus}]} \qquad (4.81)$$

The resulting ΔG_m° refers to the melting of 1 mole of AB(s) into AB(l); therefore,

$$0.5\Delta G_m^\circ = \Delta H_m^\circ - T\Delta S_m^\circ \qquad (4.82)$$

where $\Delta H_m^\circ = T_m \Delta S_m^\circ$ refers to 1 g-atom of $(A+B)$ or 0.5 mole of AB in accord with the convention in thermochemistry of alloy phases, and T_m is the melting point. If it is assumed that the activity coefficients in the liquid phase obey the regular zeroth behavior, i.e., $\ln \gamma_1 = \alpha x_2^2$ and $\ln \gamma_2 = \alpha x_1^2$, then with $a_1 = x_1 \gamma_1$ and $a_2 = x_2 \gamma_2$, the logarithmic terms in equation (4.81) become

$$0.5\alpha + \ln 0.25 - \alpha x_2^2 - \ln x_1 - \alpha x_1^2 - \ln x_2$$

$$= -\ln 4x_1 x_2 - \alpha(x_1^2 + x_2^2 - 0.5); \qquad (x_i \text{ at liquidus}) \qquad (4.83)$$

The elimination of x_1 by using $1 - x_2$ in the last term gives $-2\alpha(x_2 - 0.5)^2$, and substitution of this result and equation (4.82) into equation (4.81) yields

$$\alpha = \frac{0.5RT \ln 4x_1 x_2 - \Delta H_m^\circ + T\Delta S_m^\circ}{(x_2 - 0.5)^2} \qquad (4.84)$$

A single point on the liquidus yields one value of $x_1 x_2$, from which α can be calculated if ΔH_m° and ΔS_m° are known from measurements, e.g., from calorimetry. After the evaluation of α, the activity coefficients can be calculated from $\ln \gamma_1 = \alpha x_2$ and $\ln \gamma_2 = \alpha x_1^2$. Calculations show that α varies slowly with temperature along the liquidus of a number of binary systems. This equation, first derived by Wagner,[11] was later used by Vieland,[12] Thurmond,[13] and other investigators. A modified treatment has also been presented by Jordan.[14]

The regular associated solution model does not comply with the requirements for regularity because the compound AB is so much larger in size than A and B, and the distinction between the A–B bond within the compound AB and A–B bond in AB–A between the compound and A are difficult to reconcile with the equality of bond energies for the same type of bonds. The term regular in this case originates from G^e expressed as a function of composition corresponding to the zeroth approximation to the regular solutions, often modified by various investigators to enhance the success of representation of experimental data.

A more accurate and elaborate treatment is presented by Hsieh et al.[15] correlating thermodynamic data[15-19] with the phase diagram. The results for ΔG and ΔH of alloy formation per gram atom of Sn–Te alloy are shown in Fig. 4.9, and the phase diagram, in Fig. 4.10.

The regular associated solution models can be simplified by dividing the system into two pseudobinary systems because the solid phase is assumed to be nearly stoichiometric with or without the assumption that the compound AB exists in the liquid phase. If the solid phase is indeed stoichiometric, then it can be shown that the discontinuity in the liquidus

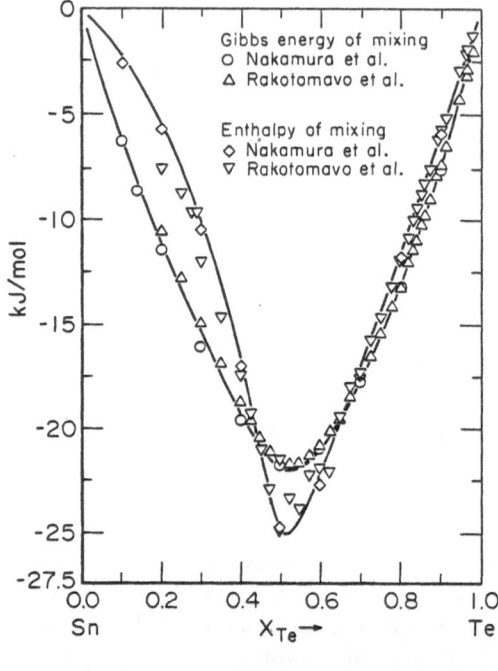

Figure 4.9. Gibbs energy and enthalpy of formation of Sn-Te alloys at 1100 K. (Courtesy of Y. A. Chang.[15] Data of Nakamura *et al.*[16] and Rakotomavo *et al.*[17])

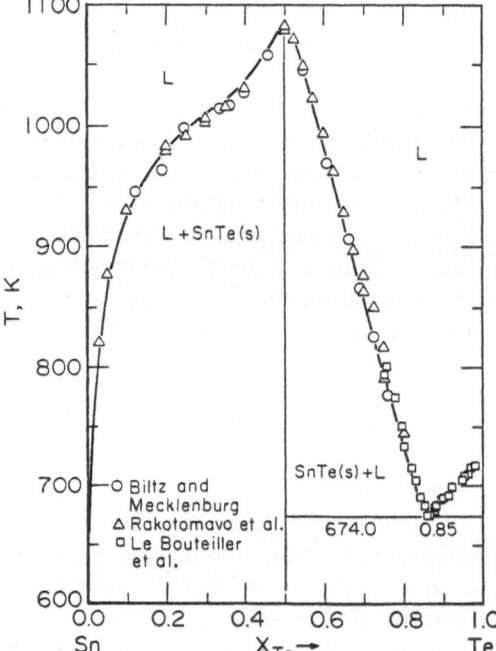

Figure 4.10. Sn-Te phase diagram according to Hsieh *et al.*[15] (Courtesy of Y. A. Chang.[15] Data of Biltz and Mecklenburg,[18] Rakotomavo *et al.*,[17] and Le Bouteiller *et al.*[19])

at AB is required both by the Gibbs-Konovalow theorem, and by Raoult's law because the depression of freezing temperature has to be proportional to the mole fraction of AB. Actually, even the solid compound AB deviates slightly from stoichiometry and the liquidus shows smoother changes in the slope with increasing deviations from the stoichiometry of AB(s).

Theories of interstitial solutions will be presented in Chapter 6 where modified forms of the regular solution equations and the Wagner equations[20] will be discussed in detail.

References

1. E. A. Guggenheim, *Mixtures*, Oxford University Press, London (1952).
2. C. Wagner, *Thermodynamics of Alloys*, translated by S. Mellgren and J. H. Westbrook, Addison-Wesley, Reading, Massachusetts (1952).
3. J. H. Hildebrand and R. L. Scott, *The Solubility of Nonelectrolytes*, Third Edition, Reinhold, New York (1950).
4. N. A. Gokcen and E. T. Chang, *J. Chem. Phys.* **55**, 2279 (1971).
5. H. A. Bethe, *Proc. R. Soc. London Ser. A* **150**, 552 (1935).
6. N. A. Gokcen and E. T. Chang, A New Method for Enumerating Molecular Configurations in Propellant Mixtures, Aerospace Report No. TR-0172 (2210-10)-1, The Aerospace Corp., El Segundo, California (1971).
7. R. Hultgren, P. D. Desai, D. T. Hawkins, M. Gleiser, and K. K. Kelley, *Selected Values of the Thermodynamic Properties of Binary Alloys*, ASM, Metals Park, Ohio (1973).
8. J.-C. Mathieu, F. Durand, and E. Bonnier, *J. Chim. Phys.* **62**, 1289, 1297 (1965); B. Brion, J.-C. Mathieu, P. Hicter, and P. Desré, *J. Chem. Phys.* **66**, 1238, 1745 (1970).
9. C. H. P. Lupis and J. F. Elliott, *Acta Metall.* **15**, 265 (1967).
10. N. A. Gokcen, *Scr. Metall.* **16**, 723 (1982).
11. C. Wagner, *Acta Metall.* **6**, 309 (1958).
12. L. J. Vieland, *Acta Metall.* **11**, 137 (1963).
13. C. D. Thurmond, *J. Phys. Chem. Solids* **26**, 785 (1965).
14. A. S. Jordan, *Metall. Trans. AIME* **1**, 239 (1970).
15. K.-C. Hsieh, M. S. Wei, and Y. A. Chang, *Z. Metallkd.* **74**, 330 (1983).
16. Y. Nakamura, S. Himuro, and M. Shimoji, *Ber. Bunsenges. Phys. Chem.* **84**, 240 (1980).
17. J. Rakotomavo, M.-C. Baron, and C. Petot, *Metall. Trans. AIME* **12B**, 461 (1981).
18. W. Biltz and W. Mecklenburg, *Z. Anorg. Chem.* **64**, 226 (1909).
19. M. Le Bouteiller, A. M. Martre, R. Farhi, and C. Petot, *Metall. Trans. AIME* **8B**, 339 (1977).
20. C. Wagner, *Acta Metall.* **21**, 1297 (1973).

5

Long-Range Order

Chapter 2 dealt with the phase equilibria in alloys involving only the first-order phase transitions. The Gibbs energy of a given multicomponent system is a function of its variables of state P, T, n_1, n_2, ..., n_c, i.e., $\mathscr{G} = \mathscr{G}(P, T, n_1, n_2, ..., n_c)$. This function is continuous for the first-order phase transitions but its derivative, with respect to one of its variables, becomes discontinuous upon a first-order transition. The variable of the greatest importance is the temperature; therefore, we limit our discussion to the derivatives of $\mathscr{G}$ with respect to T. A first-order transition is accompanied with a discontinuity in the first derivatives of $\mathscr{G}$; thus,

$$\frac{\partial \mathscr{G}}{\partial T} = -S; \quad \text{or} \quad \frac{\partial(\mathscr{G}/T)}{\partial(1/T)} = H \tag{5.1}$$

would show a discontinuity in the entropy or enthalpy when these properties are measured from a reference temperature such as $T = 0$, or often more conveniently, $T = 298.15$ K. Condensation and freezing of pure components provide some of the most elementary examples of first-order transitions. At a second-order transition, the second derivatives of $\mathscr{G}$ exhibit a discontinuity; i.e.,

$$\frac{\partial^2 \mathscr{G}}{\partial T^2} = -\frac{C_p}{T}; \quad \text{or} \quad \frac{\partial^2(\mathscr{G}/T)}{\partial(1/T)\partial T} = C_p \tag{5.2}$$

In summary, S, or H, is discontinuous for the first-order phase transitions and C_p is discontinuous for the second-order phase transitions.

The order–disorder phenomena are second-order phase transitions in which discontinuities are observed in C_p of metals and alloys. In pure elements, such as iron, order–disorder in the vicinity of 1043 K produces magnetic order–disorder, but in alloys such as Cu–Zn at 742 K, the arrangement in the crystal lattice causes order–disorder.

Theoretical treatments of order–disorder are basically applicable to any type of second-order transition. However, we limit our presentation to the order–disorder in binary alloys, with the tacit assumption that the treatment of multicomponent systems do not present unusual difficulties.

Numerous statistical thermodynamic attempts have been made[1-10] to formulate the behavior of long-range order in alloys after the significant initial work of Gorsky,[1] and Bragg and Williams[2] (GBW). The quasi-chemical method, largely due to Bethe[3] and to Guggenheim[4] and based on the permutation of bonds, is often erroneously claimed to be an improvement over the GBW method. The permutation of bonds in the quasi-chemical method is made to become the permutation of atoms for zero exchange energy, or when the temperature is sufficiently high, through a dubious normalization process. It has been shown conclusively by Gokcen and Chang[11] that (1) the normalization process of the quasi-chemical method yields the same equations for the excess Gibbs energy of solution as without normalization, and (2) the actual enumerations of configurations[11] prove that the permutable entities are the atoms, not the bonds, as discussed in the preceding chapter.

Ordering and Clustering

The number of AB bonds in a substitutional binary solution is more than the random number $(ZN/2)(2x_A x_B)$ when the exchange energy $W = (Z/2)(2e_{AB} - e_{AA} - e_{BB})$ is negative, and less when W is positive. Very large negative values of W/kT may cause formation of one or more intermetallic compounds in the same system. For positive values of W/kT, clustering, or the association of like atoms, occurs, and in fact, when W/kT is sufficiently large, the clustering causes separation of a phase into two phases. On the other hand, if W/kT is negative to some optimal degree that cannot always be quantified, A atoms may form nearly entirely A–B bonds with virtually no A–A bonds if $x_A \leq x_B$. This is possible when certain sites in the crystal lattice are occupied by A atoms and others by B atoms. In such substitutional solid solutions, we shall see that the A atoms may be regarded as having formed their own lattice, interpenetrating the lattice formed by the B atoms. This phenomenon is known as the *long-range* order. Since W itself may not often be sufficiently negative, the long-range order is expected to occur at sufficiently low temperatures; however, the kinetic energy of atoms may not often be sufficient to move the atoms into their ordered state below ambient temperatures. The clustering may occur in liquid and solid alloys but the long-range order may occur only in solid solutions; therefore, this chapter is largely concerned with solid solutions.

Order–Disorder in Binary Alloys

We consider order–disorder phenomena in binary substitutional alloys wherein the lattice structure of solid solution is retained upon transition from the ordered state to the disordered state. It will be assumed that minor changes in crystal structure upon transformation are not of primary importance. This type of transition is therefore a second-order transition, which may be called the Curie-type transition, or the coherent transition, but we retain the term order–disorder transition as the more descriptive term for our purposes. The treatment presented in this chapter can, however, be readily extended to noncoherent transitions. For a body-centered cubic (bcc) structure, shown in Fig. 5.1(a), the random distribution of A and B atoms for an equimolar or equiatomic solution exhibits no particular preference in the locations of A and B atoms. Each lattice site in such a solid solution may therefore be occupied by either A or B whether the solution is ideal or nonideal. However, if all the A atoms occupy only the corners of the cube as shown in Fig. 5.1(b), and all the B atoms, the body-centered sites, then we have a perfectly ordered structure as in beta brass, Cu–Zn. We designate the ordered sites by a for A atoms, and by b for B atoms when the ordering is perfect, but when the ordering is partial, the sites still retain their identities, even if they are occupied by a number of wrong atoms. The identities of sites disappear when nearly half of the atoms of each component occupy their right places, and the remaining half, the wrong places for an equimolecular solution. We note that 1/8 of each corner atom belongs to each cube, and that the nearest neighbors lie along the diagonals of the cube by 0.866 of the cube edge. All the A atoms may be regarded as forming their cubic lattice by interpenetrating a similar cubic lattice formed by the B atoms, and for this reason an ordered structure is said to form a superlattice. Tetrahedral crystals of the type exemplified by ZnS, may be projected on two dimensions as in Fig. 5.1(c) to illustrate this type of order in which each atom has four unlike nearest neighbors. The picture is somewhat different for a close-packed two-dimensional ordered structure for AB_2-type phase as shown in Fig. 5.1(d). In this case, each A atom is surrounded by six B atoms, but each B atom contributes one-third to each hexagon with A in the center; therefore, the composition is AB_2.

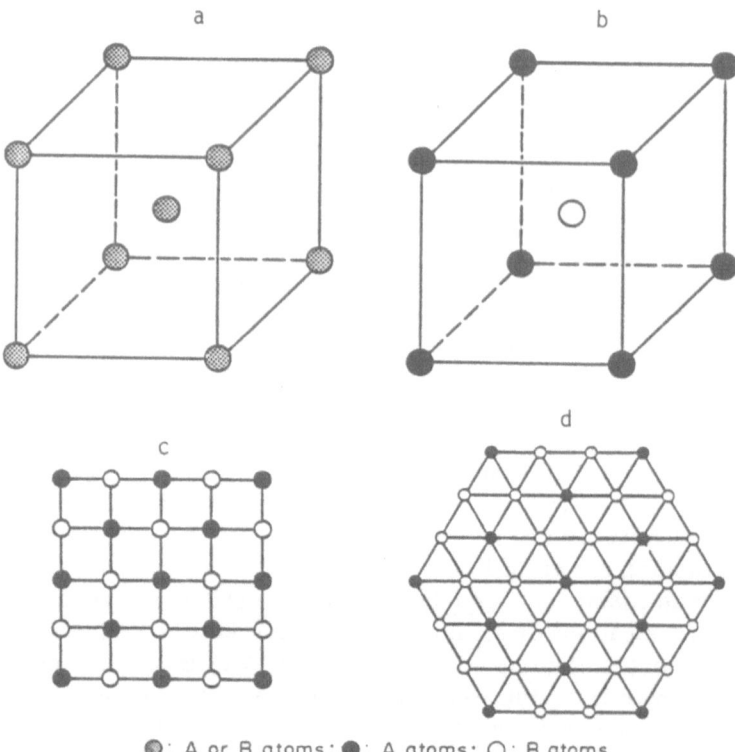

⊚: A or B atoms; ●: A atoms; ○: B atoms

Figure 5.1. (a) Disordered structure in body-centered cubic crystals; each site is occupied randomly by either type of atom. (b) Ordered AB type of alloy; corner atoms occupied by A atoms form cubes interpenetrating cubes formed by B atoms occupying body-centered positions. (c) ZnS-type tetrahedral ordered structure projected on two-dimensional coordinates; A = Zn, B = S. (d) Two-dimensional AB_2-type hexagonal close-packed structure. Each A atom is entirely surrounded by B atoms, and every third atom in any direction is an A atom.

A face-centered cubic structure of AB_3 type, such as $AuCu_3$, has Au at the cube corners and Cu at the face centers; corner atoms contribute one atom and face-center atoms three atoms per cube.

The existence of order can be determined semiquantitatively, or in favorable cases quantitatively, by X-ray diffraction. New diffraction lines are observed when ordering initiates in a random alloy. The intensities of the new lines are than a measure of the degree of ordering. Electron and neutron diffraction methods, and some of the physical properties also determine the extent of order with various degrees of accuracy. Selected reviews, discussions, and summaries on various aspects of ordering are to be found in Refs. 5–10.

The atomic ratio of A to B may not always be a ratio of whole numbers, and the numbers of a- and b-lattice sites may or may not be determined by the crystal structure. As the composition of a perfectly ordered phase is changed, the degree of order also changes because the excess number of atoms of one of the components must be accommodated in the wrong sites. Atomic vibrations increase with increasing temperature to such an extent that, finally at a critical temperature, the degree of order disappears and the solution becomes disordered in the sense that no measurable degree of long-range order can be observed. We shall be concerned mainly with the statistical thermodynamic treatment of long-range order, leaving physics and physical metallurgy of this interesting type of phenomena to the appropriate monographs and treatises.

Long-Range Order Parameter

Statistical thermodynamic treatment of ordering in alloys requires a convenient definition of the long-range order parameter. There are several definitions[1-6] but the most frequently used parameter, r, is defined by

$$\text{A atoms on a-sites} = N_A^a = 0.5 \times \text{all A atoms} \times (1 + r) \qquad (5.3)$$

where it is assumed that the number of A atoms is equal to or smaller than the number of a-sites. For equal numbers of a- and b-sites, each designated by L, and for equal numbers of A and B atoms, or $x_A = x_B$, this equation is written as

$$N_A^a = x_A(1 + r)L = 0.5(1 + r)L \qquad (5.4)$$

Thus, if all A atoms are distributed randomly, r is zero and the probability of finding A atoms on a-sites is equal to its mole fraction. The parameter r is unity for a perfectly ordered alloy and all the A atoms are on the a-sites, i.e., $N_A^a = N_A = L$. The recognition of two or more different types of sites is essential in formulating the properties of ordered solutions. The sets of a- and b-sites are sometimes called the a- and b-sublattice respectively. Each site has Z neighbors, Z being the coordination number. The more complicated cases of unequal numbers of A and B atoms and a- and b-sites will be discussed later.

The parameter r for the preceding simple case with equation (5.4) and $x_A = 0.5$ leads to

$$\text{A atoms on a-sites} = N_A^a = 0.5(1 + r)L$$

$$\text{A atoms on b-sites} = N_A^b = 0.5(1 - r)L$$

$$\text{B atoms on b-sites} = N_B^b = 0.5(1 + r)L$$

$$\text{(5.5)}$$

$$\text{B atoms on a-sites} = N_B^a = 0.5(1 - r)L$$

The second of these equations is obtained by subtracting the first equation from the total number of A atoms, $N_A = L$, because the remaining A atoms must be accommodated on the b-sites. The third equation is the same as the first, and the fourth equation is the same as the second as can be shown by similar arguments. The ordering parameter r varies between zero and unity, i.e.,

$$0 \leqslant r \leqslant 1 \tag{5.6}$$

When r is zero, the solution is called "long-range disordered" or briefly "disordered" but not necessarily random or ideal. The distinction between "disordered" and "random" is therefore important in this chapter.

Gorsky, and Bragg and Williams (GBW) Approximation

The simplest treatment of order–disorder in alloys is the zeroth approximation due to GBW.[1-5] This model is similar to the zeroth approximation to the regular solutions. For simplicity, we first consider the body-centered cubic (bcc) lattice in our discussions with $N_A = N_B = L$ as in the Cu–Zn system. The nearest neighbors to an atom on the body-centered site are the corner sites, and since $Z = 8$, we take into account the eight bonds emanating from an a-site to the neighboring b-sites. The fraction of L sites occupied by A atoms in N_A^a/L and this is also equal to the probability of finding an A atom on the a-sites. The probability of finding an A atom on the b-sites is N_A^b/L so that the probability of finding an AA pair from equation (5.5) is

$$\text{Probabilty of AA} = \frac{N_A^a}{L} \frac{N_A^b}{L} = (1 - r^2)/4 \tag{5.7}$$

Likewise,

$$\text{Probability of BB} = (1 - r^2)/4 \tag{5.8}$$

The probability of AB pairs is the sum of two probabilities, i.e., the probability of A on a-sites times B on b-sites, i.e., $0.25(1 + r)^2$, plus the probability of A on b-sites times B on a-sites, i.e., $0.25(1 - r)^2$; hence,

$$\text{Probabilty of AB} = (1 + r)^2/4 + (1 - r)^2/4 = (1 + r^2)/2 \qquad (5.9)$$

Multiplication of each probability term by the total number of bonds, ZL, gives the total number of each type of bond, and multiplication of each set of bonds by its bond energy e_{ij} gives the total energy of the alloy as in the zeroth approximation to the regular solutions, i.e.,

$$E \approx H(r) \approx (ZL/4)(1 - r^2)e_{AA} + (ZL/4)(1 - r^2)e_{BB} + (ZL/2)(1 + r^2)e_{AB}$$

$$= (ZL/4)(e_{AA} + e_{BB} + 2e_{AB}) + (ZLr^2/4)(2e_{AB} - e_{AA} - e_{BB}) \qquad (5.10)$$

For the disordered solution, r is zero and equation (5.10) becomes

$$H(r = 0) = (ZL/4)(e_{AA} + e_{BB} + 2e_{AB}) \qquad (5.11)$$

For convenience, we rewrite the exchange energy of equation (5.10) as follows:

$$W = (Z/2)(2e_{AB} - e_{AA} - e_{BB}) \qquad (5.12)$$

Equations (5.10)–(5.12) give

$$H(r) - H(r = 0) = Lr^2W/2 \qquad (5.13)$$

It is assumed that the atoms in each group as given in each one of equations (5.5) can be rearranged randomly and this is the reason that the model is called the zeroth approximation. The resulting distribution for all the atoms on the a-sites is

$$D_a = \frac{L!}{[(1 + r)L/2]![(1 - r)L/2]!} \qquad (5.14)$$

The distribution D_b for the atoms on the b-sites is identical, i.e., $D_a = D_b$; hence, the overall distribution D_r is

$$D_r = D_a D_b = \left[\frac{L!}{[(1 + r)L/2]![(1 - r)L/2]!} \right]^2 \qquad (5.15)$$

The entropy of solution is $S = k \ln D_r$; substitution of this entropy and H of equation (5.10) in $G = H - TS$ gives

$$G = H - Tk \ln D_r \tag{5.16}$$

The term $\ln D_r$ can be obtained by using the Stirling approximation,

$$S(r) = 2k \ln D_r = Lk[2 \ln 2 - (1 + r) \ln(1 + r) - (1 - r) \ln(1 - r)] \tag{5.17}$$

where $S(r)$ is the entropy of the solution. For $r = 0$, $S(r = 0) = 2Lk \ln 2$ as expected since the distribution of atoms is random; therefore,

$$S(r) - S(r = 0) = -Lk[(1 + r) + (1 - r) \ln(1 - r)] \tag{5.18}$$

Equation (5.16) now becomes

$$G(r) = \frac{ZL}{4}(e_{AA} + e_{BB} + 2e_{AB}) + \frac{Lr^2 W}{2}$$
$$- LkT[2 \ln 2 - (1 + r) \ln(1 + r) - (1 - r) \ln(1 - r)] \tag{5.19}$$

Likewise, from equations (5.13) and (5.18), or directly from equation (5.19), we obtain

$$G(r) - G(r = 0) = \frac{Lr^2 W}{2} + LkT[(1 + r) \ln(1 + r) + (1 - r) \ln(1 - r)] \tag{5.20}$$

The equilibrium state of the alloy corresponds to the value of r which makes $\partial G / \partial r$ zero:

$$\frac{\partial G}{\partial r} = LrW + LkT[\ln(1 + r) - \ln(1 - r)] = 0 \tag{5.21}$$

At the critical temperature, T_c, the second derivative of $G(r)$ with respect to r is zero as r approaches zero, i.e., the order disappears; thus,

$$\frac{\partial^2 G(r)}{\partial r^2} = LW + LkT \left(\frac{1}{1 + r} + \frac{1}{1 - r} \right) = 0; \qquad \text{for } r = 0 \text{ and } T = T_c \tag{5.22}$$

Substitution of $r \to 0$ yields

$$T_c = -W/2k; \qquad (W < 0) \qquad (5.23)$$

The critical temperature must be positive; therefore, W must be negative, i.e., the energy of 2AB bonds must be lower than the sum of the energies of AA and BB bonds. The derivation of equation (5.23) from (5.22) is the rigorous method of determination for T_c, because, for a second-order transition, the second derivative of G with respect to a variable of state must be zero at the critical point.

Substitution of $W = -2kT_c$, obtained from equation (5.23), in equation (5.21) gives

$$\frac{1+r}{1-r} = \exp\left(\frac{-rW}{kT}\right) = \exp\left(\frac{2rT_c}{T}\right) \qquad (5.24)$$

A relationship equivalent to equation (5.24) was first derived by Gorsky.[1] This equation can be rearranged to obtain

$$r = \frac{\exp\left(\dfrac{-rW}{kT}\right) - 1}{\exp\left(\dfrac{-rW}{kT}\right) + 1} \equiv \tanh\left(\frac{-rW}{2kT}\right) \equiv \tanh\left(\frac{rT_c}{T}\right) \qquad (5.25)$$

where the identity sign defines tanh. The numerical computation of T_c/T from this equation is very simple. For this purpose, it is necessary to select an arbitrary value of $\alpha = rT_c/T$ and read the value of $\tanh \alpha$ from an appropriate calculator or obtain it from the ratio immediately after the first equal sign in equation (5.25). Then, $\tanh \alpha$ is the value of r, and this value substituted in $\alpha = rT_c/T$ yields the value of T from the experimentally known value of the critical temperature. The reader can verify that for the equiatomic Cu–Zn alloy having $T_c = 742$ K, with $\alpha = 0.7$, $\tanh 0.7 = 0.6044 = r$, and then $T = rT_c/\alpha = 640.66$ K.

Equation (5.25) has a solution that is $r = 0$ for any value of T_c/T or W/k. When $T > T_c$, $r = 0$ is the only solution because above the critical temperature, disorder prevails. At $T < T_c$, there is another solution in the range of $0 < r < 1$, and this root corresponds to the minimum in $G(r)$ as can be shown by using equation (5.22) with (5.23) as follows:

$$\frac{\partial^2 G(r)}{\partial r^2} = -2LkT_c + LkT\left(\frac{1}{1+r} + \frac{1}{1-r}\right) \qquad (5.26)$$

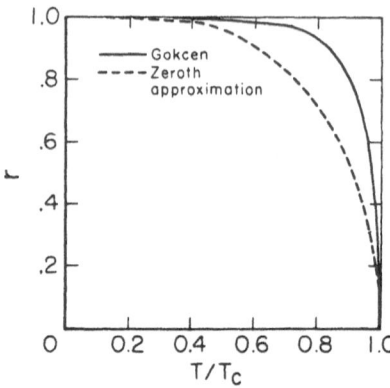

Figure 5.2. Variation of long-range order parameter with temperature. T_c is 742 K. Curve for quasi-chemical approximation is not shown because it follows closely the zeroth approximation curve. (From Gokcen.[12])

The substitution of $T_c = 742$, $T \doteq 640.66$, and $r = 0.6044$ yields $535Lk$, which is a positive quantity, and this value of r corresponds to a minimum in $G(r) - G(0)$. At 640.66 K, however, $r = 0$ corresponds to a maximum in $G(r) - G(0)$ because equation (5.26) becomes negative in value. The equilibrium values of r and the corresponding temperature are plotted in Fig. 5.2, as the reduced temperature T/T_c versus r.

Above the critical temperature, the alloy is disordered, and therefore the equations derived for the zeroth approximation to the regular solution become applicable if the GBW approximation is assumed to apply for the ordered solution. The available data indicate that this approximation does not predict the properties of ordered solutions satisfactorily,[6,9] and for ordered and disordered solutions of the same components, different values of W have to be used to express $G(r) - G(0)$ and the Gibbs energy of formation of the disordered alloy, ΔG. Nevertheless, because of its simplicity, the GBW approximation has been used in a number of recent publications.

Heat Capacity

The heat capacity $C (= C_p)$ can be obtained by differentiation of equation (5.13) with respect to temperature:

$$\frac{\partial H(r)}{\partial T} - \frac{\partial H(r = 0)}{\partial T} = C(r) - C(r = 0) = LrW\frac{\partial r}{\partial T} \tag{5.27}$$

where $LW = -2LkT_c = -RT_c$ from equation (5.23). The expression for $\partial r/\partial T = dr/dT$ can be obtained by taking the logarithm of equation (5.24) and then differentiating it; the result, substituted in equation (5.27), is

$$\Delta C = C(r) - C(r = 0) = \frac{Rr^2 T_c^2(1-r^2)}{T^2 - T_c T(1-r^2)} \tag{5.28}$$

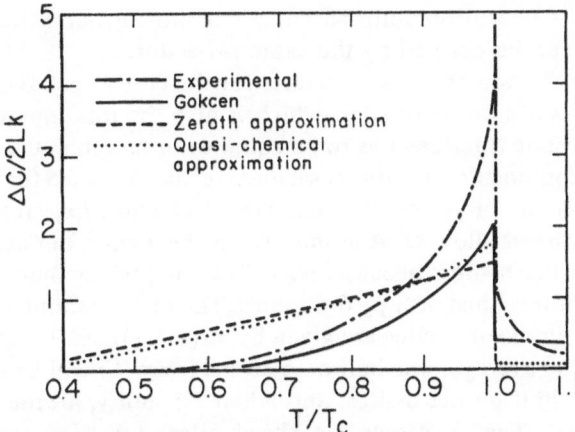

Figure 5.3. Variation of heat capacity with temperature for equiatomic Cu–Zn alloy. T_c is 742 K, and measured heat capacities of pure components were subtracted from that of alloy to obtain experimental curve. ΔC is the same as the excess heat capacity C^e as shown on page 139. (From Gokcen.[12])

where $C(r = 0)$ is the heat capacity of disordered alloy. The heat capacity $C(r = 0)$ of a random disordered solution is the same as the heat capacities of its component elements.* The values of r and T are related by equation (5.25) and shown in Fig. 5.2; therefore, for each value of r, T must be computed and then the result must be substituted in equation (5.28) to obtain ΔC. The result for the previous example, for which $T = 640.66$ K, $T_c = 742$ K, or $T/T_c = 0.8634$, and $r = 0.6044$, is simply $\Delta C/R = 1.174$. Above the critical temperature, $T > T_c$, r is zero and ΔC rapidly approaches zero because of r^2 in the numerator of equation (5.28). The values of ΔC calculated from equation (5.28) are plotted in Fig. 5.3. The results will be discussed in conjunction with those obtained by the refined methods to be considered later in this chapter.

More General Cases

We have thus far considered the equal numbers of A and B atoms distributed on equal numbers of a- and b-sites. The unequal numbers of A and B atoms on equal numbers of a- and b-sites will be considered first and then the unequal numbers of a- and b-sites with N_A = a-sites = L_a, and

---

*ΔH in equation (4.5) for a random disordered solution is independent of temperature; hence, $\partial \Delta H/\partial T = C(r = 0) - C$ (component elements) $= 0$.

N_B = b-sites = L_b will be outlined later. The most general treatment is not given, but it can be derived by the same procedure.

Let L be the number of a- or b-sites for the case of equal a- and b-sites, and $2L$, the total number of sites, which is also the total number of atoms. One of the atomic fractions has to be equal to or less than 0.5 to generalize the range of composition for this case, and we take $x_A \leq 0.5$ for this purpose. The total numbers of A and B atoms are $2Lx_A$ and $2Lx_B$, respectively. In a perfectly ordered alloy, all A atoms are on the a-sites, but all the B atoms cannot be on the b-sites, because $N_B = 2Lx_B$ is greater than L; therefore, the excess B atoms must occupy the a-sites. The order parameter r is defined so that the A atoms on a-sites are given by $x_A(1 + r)L$. Consequently, when r is zero, on the average, one half of A atoms (or $x_A L$) will be on the a-sites and the other half on the b-sites, and when r is unity, all the A atoms are on the a-sites. The A atoms on the b-sites are $2Lx_A - x_A(1 + r)L = x_A(1 - r)L$. The B atoms on the a-sites occupy the remaining a-sites, i.e., $L - x_A(1 + r)L = (x_B - x_Ar)L$. The remaining B atoms, $2Lx_B - (x_B - x_Ar)L = (x_B + x_Ar)L$, must then be on the b-sites. In summary,

$$\text{A atoms on a-sites} = N_A^a = x_A(1 + r)L$$

$$\text{A atoms on b-sites} = N_A^b = x_A(1 - r)L$$

$$\text{B atoms on b-sites} = N_B^b = (x_B + x_Ar)L$$

$$\text{B atoms on a-sites} = N_B^a = (x_B - x_Ar)L$$

(5.5a)

Here the equation number (5.5a) is used to indicate its correspondence with equation (5.5).

The probability of A–A bonds is now $N_A^a N_A^b / L^2 = x_A^2(1 - r^2)$, that of B–B is $(x_B^2 - x_A^2 r^2)$, and that of A–B is $x_A(1 + r)(x_B + x_Ar)$ plus $x_A(1 - r) \times (x_B - x_Ar)$ or a total of $2x_Ax_B + 2x_A^2 r^2$. Multiplication of each of these probabilities with the total number of all bonds, ZL, yields the number of each type of bond, and multiplication with the corresponding bond energy yields the total energy of each type of bond; the sum of all energies is then

$$E \approx H \approx LZ(x_A^2 e_{AA} + x_B^2 e_{BB} + 2x_Ax_B e_{AB}) + 2Lx_A^2 r^2 W$$
$$= H(r = 0) + 2Lx_A r^2 W$$

The distribution function corresponding to equation (5.15) is

$$D_r = \frac{(L!)^2}{[x_A(1 + r)L]![x_A(1 - r)L]![(x_B + x_Ar)L]![(x_B - x_Ar)L]!}$$

(5.15a)

The Gibbs energy difference $G(r) - G(r = 0)$ corresponding to equation (5.20) is

$$G(r) - G(r = 0) = 2Lx_A^2 r^2 W + LkT[(x_A + x_A r) \ln(x_A + x_A r)$$
$$+ (x_A - x_A r) \ln(x_A - x_A r) + (x_B + x_A r) \ln(x_B + x_A r)$$
$$+ (x_B - x_A r) \ln(x_B - x_A r) - 2x_A \ln x_A - 2x_B \ln x_B] \qquad (5.20a)$$

Likewise, the relationships corresponding to equations (5.21) for any value of r and the corresponding T, and (5.22) for $r = 0$ and $T = T_c$, are as follows:

$$\frac{\partial G}{\partial r} = 4Lx_A^2 r W + LkTx_A[\ln(x_A + x_A r) + \ln(x_B + x_A r)$$
$$- \ln(x_A - x_A r) - \ln(x_B - x_A r)] = 0 \qquad (5.21a)$$

$$\frac{\partial^2 G}{\partial r^2} = 4Lx_A^2 W + 2LkT_c x_A \left(1 + \frac{x_A}{x_B}\right) = 0; \qquad \text{for } r = 0 \qquad (5.22a)$$

The last equation gives

$$T_c = -2x_A x_B W/k; \qquad (W < 0) \qquad (5.23a)$$

The next case to be considered is the unequal numbers of A and B atoms, as well as a- and b-sites. However, we limit the treatment to $N_A =$ a-sites $= L_a$, and $N_B =$ b-sites $= L_b$, so that when the alloy is perfectly ordered, all A atoms are on a-sites and all B on b-sites. The number of a-sites is taken as $L_a = x_a N$, and that of b-sites as $L_b = x_B N$. Next, we take $N_A < N_B$ or $x_A < x_B$ since one of the components has to be larger than the other; this also signifies that $L_a < L_b$. The order parameter is defined so that $N_A^a = (x_A + x_B r)L_a$, and when $r = 1$, or for perfect order, $N_A^a = L_a$, or all A atoms are on the a-sites, and for $r = 0$, $x_A L_a$ atoms are on the a-sites. The total number of A atoms, i.e., L_a, minus N_A^a gives A atoms on b-sites, which are $N_A^b = x_B(1 - r)L_a$, and for $r = 1$, there are no A atoms on b-sites. The distributions of atoms on a- and b-sites are therefore

$$\text{A atoms on a-sites} = N_A^a = (x_A + x_B r)L_a$$

$$\text{A atoms on b-sites} = N_A^b = x_B(1 - r)L_a$$

$$(5.5b)$$

$$\text{B atoms on b-sites} = N_B^b = (x_B + x_A r)L_b$$

$$\text{B atoms on a-sites} = N_B^a = x_A(1 - r)L_b$$

The resulting equation for $G(r) - G(r = 0)$ can be derived by using the foregoing procedure; the result is

$$G(r) - G(r = 0) = NkTx_A^2 r^2 W + NkT[(x_A^2 + x_A x_B r) \ln(x_A^2 + x_A x_B r)$$
$$+ 2x_A x_B(1 - r) \ln(x_A x_B - x_A x_B r) + x_B(x_B + x_A r)$$
$$\times \ln(x_B^2 + x_A x_B r) - 2x_A \ln x_A - 2x_B \ln x_B] \quad (5.20b)$$

An ordered phase such as AuCu$_3$ has $x_A = 0.25$ and $x_B = 0.75$ and the equilibrium values of r are given by

$$\frac{\partial G}{\partial r} = \frac{NkT}{16}\left\{3\ln(1 + 3r)(3 + r) - 3\ln\left[3(1 - r)^2\right] + \frac{2rW}{kT}\right\} = 0 \quad (5.21b)$$

The critical temperature is related to W as required by

$$\frac{\partial^2 G}{\partial r^2} = \frac{NkT_c}{8}\left(8 + \frac{W}{kT_c}\right) = 0; \qquad \text{for } r = 0 \quad (5.22b)$$

from which $T_c = -W/(8k)$. We give these relations for the sake of completeness without making use of them.

First Approximation

The zeroth approximation to the long-range order is the GBW approximation which is analogous to the zeroth approximation to the regular solutions. The first approximation to the long-range order attempts to correct the random distribution assumption used in the GBW method, and to resolve the discrepancy between the exchange energy W for the ordered solutions and that for the disordered solutions.

We consider a binary alloy of equiatomic composition for simplicity in our formulation because the results can easily be extended to non-equiatomic compositions. Again, $N_A = N_B = L$ so that $2L$ is 1 g-atom of an alloy, and in a perfectly ordered alloy, all N_A are on the a-sublattice, and all N_B are on the b-sublattice. Let R be the actual number of A atoms on a-sites and Q the remaining A atoms on b-sites, and let the number of nearest neighbors to an atom be Z. The occupancy of B atoms is similar, i.e., R is the number of B atoms on b-sites and Q is the remaining B atoms on a-sites. The long-range order parameter r from equation (5.3) gives

$$N_A^a = N_B^b \equiv R = \frac{L}{2}(1 + r); \qquad N_A^b = N_B^a = Q = \frac{L}{2}(1 - r)$$

$$\left(\frac{L}{2} \leqslant R \leqslant L; 0 \leqslant Q \leqslant \frac{L}{2}\right) \quad (5.29)$$

If X is the net number of A atoms on a-sites whose neighbors are A atoms, then $R - X$ is the net number of A atoms surrounded by B atoms. In summary, for all the atoms, we have

    (i)   A on a-sites with A neighbors $= X$

    (ii)   A on a-sites with B neighbors $= R - X$

    (iii)   A on b-sites with A neighbors $= X$

    (iv)   A on b-sites with B neighbors $= Q - X$

          Total number of A atoms $= R + Q = L$

    (v)   B on b-sites with B neighbors $= X$ $(5.30)^*$

    (vi)   B on b-sites with A neighbors $= R - X$

    (vii)   B on a-sites with B neighbors $= X$

    (viii)   B on a-sites with A neighbors $= Q - X$

          Total number of B atoms $= R + Q = L$

The net numbers X, $R - X$, and $Q - X$ represent the numbers of atoms such that when each of these quantities is multiplied by $Z/2$, which is the number of bonds per atom, we obtain the number of corresponding types of bonds. This concept, based on previous publications,[11] ensures that the permutable entities are the net numbers of each type of atom, conceived to have only one type of neighbor. The correction required to equate the actual permutations to the permutations based on the net numbers of atoms is small and will be discussed later. This concept, fully justified earlier, is essential in pursuing the ensuing argument. The types of atoms shown in equations (5.30) are illustrated in Fig. 5.4 for a unidimensioal crystal having 16 atoms and 16 bonds, with $N_A = N_B = L = 8$. The net number of atoms in part (i) is illustrated in this figure. The A atoms on a-sites with A neighbors generate the bonds numbered 1 through 4. The bonds belonging to all of such atoms alone is half of four, because the other half belongs to the neighboring atoms; therefore, $X = 2$. The remaining types of atoms in equation (5.30) can be obtained by the same procedure.

    The number of A atoms on a-sites is R and it consists of the sum of (i) and (ii), the latter being X and $R - X$, respectively. Likewise, the number of B atoms on the same a-sites is Q from (vii) and (viii), the latter being

---

$^*R$ is used here and in writing the equations for permutations $(D_a, D_b,$ and $D_r)$; avoid confusion with the gas constant; see also pages 143 and 145.

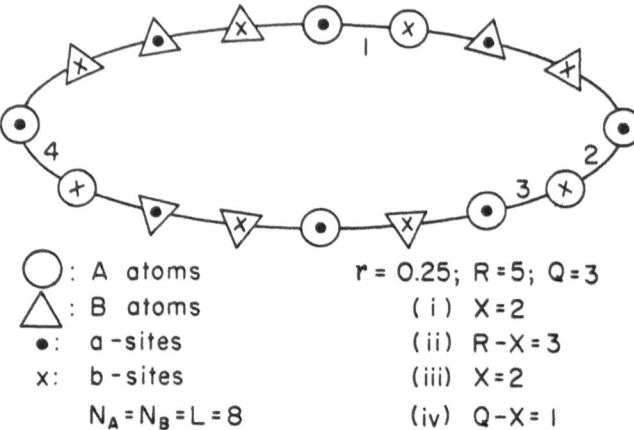

: A atoms $r = 0.25$; $R = 5$; $Q = 3$

: B atoms   (i) $X = 2$

●: a-sites   (ii) $R - X = 3$

x: b-sites   (iii) $X = 2$

$N_A = N_B = L = 8$ (iv) $Q - X = 1$

Figure 5.4. Net numbers of different types of atoms on a- and b-sites for $N_A = N_B = L = 8$. Four numbered bonds involve the A atoms on a-sites with A neighbors; half of these bonds belong to A on a-sites; therefore, $X = 2$ atoms for part (i). Remaining parts in equation (5.30) can be obtained similarly.

X and $Q - X$, respectively. The permutation of these atoms as in equations (5.14) and (5.15) is

$$D_a = \frac{R!}{(R - X)!X!} \frac{Q!}{(Q - X)!X!} \tag{5.31}$$

where D_a is the distribution for the a-sites. The distribution, D_b, for the b-sites from [(iii), (iv)] and [(v), (vi)] is identical with D_a; consequently, the overall distribution D_r is

$$D_r = D_a D_b = \left[\frac{R!Q!}{(R - X)!(X!)^2(Q - X)!} \right]^2 \tag{5.32}$$

The following Stirling approximations are useful to convert the second order in equation (5.32) into the first order:

$$[(R - X)!]^2 = (2R - 2X)!2^{2X - 2R}; \qquad (X!)^2 = (2X)!2^{-2X}$$
$$[(Q - X)!]^2 = (2Q - 2X)!2^{2X - 2Q} \tag{5.33}$$

As a result, equation (5.32) becomes

$$D_r = \frac{(2R)!(2Q)!}{(2R - 2X)![(2X)!]^2(2Q - 2X)!} \tag{5.34}$$

The elimination of R and Q by using equation (5.29) in (5.34) gives

$$D_r = \frac{[L(1 + r)]![L(1 - r)]!}{(L + Lr - 2X)![(2X)!]^2(L - Lr - 2X)!} \tag{5.35}$$

Comparison of D_r from this equation and D_r from equation (5.15) is now appropriate. The use of relationships similar to those given by equation (5.33), converts equation (5.15) into

$$D_r = \frac{(2L)!}{(L + Lr)!)(L - Lr)!} \tag{5.15a}$$

This equation states that $R = 0.5(L + Lr)$ and $Q = 0.5(L - Lr)$ can be permutated for any value of R, Q, and r, irrespective of the value of X. When $r = 1$, then $R = L$; hence, both X and Q must be zero since there are no A–A bonds, and $D_r = 1$ as expected from equation (5.35). Conversely, if $X = 0$ or if there are no A–A bonds, R has to be equal to L by geometrical requirements because when R is less than L, it can be easily shown by one- and two-dimensional crystal constructions that X cannot remain zero if the permutation given by equation (5.15a) were carried out for other values of R than $R = L$. The GBW method assumes that R and Q may have values unrestricted by the values of X so that equation (5.15a) is valid for any value of X, and this assumption is analogous to the random distribution permitted in the zeroth approximation to the long-range disordered regular solutions.

All A atoms have $ZN_A/2$ bonds (or ZN_A half-bonds) belonging to A atoms. *If for convenience we define Y such that $ZY/2$ are the A–B bonds emanating from A atoms, then the remaining bonds are the A–A bonds;* hence, (A–A bonds) $+ ZY/2 = ZN_A/2$, but A–A bonds are equal to $(2X)Z/2$ from (i) and (iii); consequently,

$$2X + Y = N_A = L \tag{5.36}$$

For the disordered alloy, r is zero but the solution is not necessarily random, and substitution of $r = 0$ in equation (5.35) gives

$$D_r = \frac{L!L!}{[(L - Y)!Y!]^2} \tag{5.37}$$

which is exactly equation (4.23) for equiatomic solutions for the first approximation to the regular solutions.

Equation (5.35) is not yet useful for our purposes when the solution is highly ordered or Y is considerably high as was shown in equation (4.29). The correction suggested there may be simplified as $F^L \equiv [F(r)]^L$ in the numerator of equation (5.35):

$$D_r = D_r(\text{corrected}) = \frac{[L(1 + r)]![L(1 - r)]!F^L}{(L + Lr - 2X)![(2X)!]^2(L - Lr - 2X)!}$$

(5.38)

The requirement for $F(r)$ is that $F(r) = 1$ for both $r = 0$ and $r = 1$, or $\ln F = 0$ in both cases. The exponent L over $F = F(r)$ is for convenience in deriving the succeeding equations from (5.38). The form of $\ln F$ obeying these requirements, and with one adjustable parameter,[11,12] g, is

$$\ln F = g(1 + r)\ln(1 + r) + g(1 - r)\ln(1 - r)$$

(5.39)

It will be seen later that without g, it is impossible to make the exchange energy, W, the same in the ordered solution and the disordered solution of the same alloy. Further, g is sufficiently small to make F a correction equivalent to c in equation (4.29).

The energy E of the alloy is $Ze_{hl}/2$ times each term in (i)–(viii) in equation (5.30) with the subscripts hl representing AA and AB, or BB, and e_{hl}, the corresponding bond energy; the result is

$$E = LZe_{AB} - 2XW$$

(5.40)

where W, the exchange energy, was defined by $W = (Z/2)(2e_{AB} - e_{AA} - e_{BB})$. The partition function (P.F.) can be written by using equations (5.38) and (5.40):

$$\text{P.F.} = \sum_{X=0}^{X} De^{-E/kT}$$

(5.41)

The maximum term in this equation is obtained by setting $\partial \ln(\text{P.F.})/\partial X$ to zero; the result is

$$(L + Lr - 2X)(L - Lr - 2X) = 4X^2 e^{-W/kT} = 4X^2 n;\ (n = e^{-W/kT})$$

(5.42)*

The value of X from this equation yields the maximum term in equation (5.41), the term that replaces the summation in P.F. The solution for X from this quadratic equation in X is

$$\frac{2X}{L} = \frac{1 \pm \sqrt{n + r^2(1 - n)}}{1 - n}$$

*Mathematical procedure requires that n here be $1/n$ of Chapter 4; see page 101.

The minimum value of X is zero according to equations (5.30) when all the A molecules are on the a-sites, i.e., $r = 1$; consequently, the positive sign before the square root would yield an unacceptable nonzero value of X. Therefore, we take the root with the minus sign preceding the square root, and multiply the numerator and the denominator of the right side with its complement, $1 + \sqrt{n + r^2(1 - n)}$, and then simplify it to derive

$$\frac{2X}{L} = \frac{1 - r^2}{1 + [n + r^2(1 - n)]^{0.5}} = \frac{1 - r^2}{1 + b} \tag{5.43}$$

where b is used for brevity to denote

$$[n + r^2(1 - n)]^{0.5} = b \tag{5.44}$$

We now proceed to derive the Gibbs energy $G(r, T)$ at constant composition and constant r by following the same procedure as that in Chapter 4 (see also Refs. 4 and 11). We assume that the enthalpy H is nearly identical with E of equation (5.40) for the alloys, and write

$$\int_{T^{-1}=0}^{T^{-1}} d\left[\frac{G(r, T)}{T}\right] = \int_{T^{-1}=0}^{T^{-1}} H \, d(T^{-1}) \tag{5.45}$$

where the integration is carried out at constant composition and constant r, and for simplicity $T \to \infty$ is written as $T^{-1} = 0$. The upper integration limit for the left side is simply $G(r, T)/T$. The lower integration limit for the left side is

$$\left[\frac{G(r, T^{-1})}{T}\right]_{T^{-1}=0} = \left(\frac{H}{T}\right)_{T^{-1}=0} - S(r, T^{-1} = 0) \tag{5.46}$$

Since $H \approx E$ is finite according to equation (5.40), H/T is zero for $T^{-1} = 0$, and the entropy in this equation can be obtained from equation (5.38) by writing $S(r, T^{-1} = 0) = k \ln D_r$. The value of X for $T^{-1} = 0$ can be obtained from equation (5.42) by setting $n = e^{-W/kT} = 1$ for $T^{-1} = 0$. The result is

$$X(T^{-1} = 0) = (1 - r^2)L/4 \tag{5.47}$$

Substitution of equation (5.47) in equation (5.38) and in $S(r, T^{-1} = 0) = k \ln D_r$, results in

$$S(r, T^{-1} = 0) = k \ln D_r$$
$$= Lk \ln F - Lk(1 + r) \ln(1 + r)$$
$$- Lk(1 - r) \ln(1 - r) + 2Lk \ln 2 \tag{5.48}$$

For the right side of equation (5.45), the integration procedure is similar to that used for the long-range disordered regular solutions with $r = 0$ as in Chapter 4. The integration of the right side of equation (5.45) requires expressing $d(T^{-1})$ by using b of equation (5.44) as follows:

$$d\left(-\frac{W}{2kT}\right) = \frac{bdb}{b^2 - r^2} \tag{5.49}$$

Next, equations (5.40), (5.43), and (5.49) are substituted in equation (5.45) to obtain

$$\int_{T^{-1}=0}^{T^{-1}} H \, dT^{-1} = \int_{T^{-1}=0}^{T^{-1}} \left[ZLe_{AB} \, dT^{-1} + \frac{2Lk(1-r^2)bdb}{(1+b)(b+r)(b-r)} \right] \tag{5.50}$$

The left side of equation (5.45) is equal to $[G(r,T)/T] + S(r, T^{-1} = 0)$ from equation (5.46). The first term on the right side of equation (5.50) is ZLe_{AB}/T, and the last term can be integrated by parts to derive

$$\frac{G(r, T)}{T} + S(r, T^{-1} = 0)$$

$$= Lk\left[\frac{Ze_{AB}}{kT} + (1+r)\ln\frac{b+r}{1+r} + (1-r)\ln\frac{b-r}{1-r} - 2\ln\frac{1+b}{2} \right] \tag{5.51}$$

Substitution of equation (5.48) for the entropy on the left side and rearrangement of the result gives

$$\frac{G(r, T)}{LkT} = \frac{Ze_{AB}}{kT} - \ln F + (1+r)\ln(b+r)$$

$$+ (1-r)\ln(b-r) - 2\ln(1+b) \tag{5.52}$$

The equilibrium value of r is obtained by setting $\partial G/\partial r = 0$:

$$\frac{1}{LkT}\frac{\partial G}{\partial r} = -\frac{\partial \ln F}{\partial r} + \ln\left(\frac{b+r}{b-r}\right) = 0 \tag{5.53}$$

where all the remaining terms including those containing $\partial b/\partial r$ cancel out. For $r = 0$, this equation requires that $\partial F/\partial r$ be zero since $F(r = 0) = 1$. Differentiation of equation (5.53) and thereafter substitution of $r = 0$ at the critical temperature gives

$$\frac{1}{LkT_c}\frac{\partial^2 G}{\partial r^2} = -\frac{d^2 \ln F(r)}{dr^2} + \frac{2}{b_c} = 0; \quad (r = 0, T = T_c, b_c = b \text{ at } T_c) \tag{5.54}$$

For $r = 0$, equation (5.44) gives $b_c = n^{0.5} = e^{-W/2kT_c}$, and the use of equation (5.53) and the rearrangement of equation (5.54) yields

$$0.5 \frac{d^2 \ln F}{dr^2} = \exp\left(\frac{W}{2kT_c}\right); \qquad (r = 0, \ T = T_c) \tag{5.55}$$

Equation (5.52) is also applicable to the regular disordered solutions for which $r = 0$, because D_r of equation (5.38) becomes identical with equation (4.29) for $r = 0$. When r is set to zero, equation (5.52) assumes the simple form given by

$$\frac{G(r = 0, T)}{LkT} = \frac{Ze_{AB}}{kT} - \frac{W}{kT} - 2 \ln(1 + b) \tag{5.56}$$

For an ideal equiatomic solution the enthalpy H is equal to the enthalpy of pure components, $H = (e_{AA} + e_{BB})ZL/2$, and the configurational entropy is the ideal entropy of the alloy so that $S(\text{ideal}) = 2Lk \ln 2$; hence, the ideal Gibbs energy of solution is

$$\frac{G(\text{ideal})}{LkT} = (e_{AA} + e_{BB})\frac{Z}{2kT} - 2 \ln 2; \qquad (x_A = x_B) \tag{5.57}$$

Subtraction of this equation from equation (5.56) and substitution for W from equation (5.12) yields the following excess Gibbs energy of solution, G^e for an equiatomic solution:

$$\frac{G^e}{LkT} = 2 \ln 2 - 2 \ln(1 + b) = 2 \ln\left(\frac{2e^{W/2kT}}{1 + e^{W/2kT}}\right) \tag{5.58}$$

This equation, where $2Lk$ is the gas constant, is identical[13] with equation (4.55).

The correction term, $\ln F$, in equation (5.39) may now be substituted in equation (5.55) to obtain the value of g for any system. We take, as an example, the Cu–Zn system in the vicinity of equiatomic composition, known as beta brass. The value of W/k is the only parameter permitted to be determined from experimental data in any approximation to the regular disordered solutions. The compilation of Hultgren *et al.*[14] extrapolated a short distance to the equiatomic composition yields $G^e/2LkT = -1.202$ at 773 K, which substituted in equation (5.58), gives $W/k = -2678$. The regular solution model assumes that W/k is independent of temperature. The critical temperature T_c is 742 K. The explicit functional form of $F(r)$ is not known and its determination would require extensive and laborious enumeration of configurations for one- and two-dimensional crystals.[11] Such a task is expected to be very difficult and time-consuming. However, the function $F(r)$ suggested in equation (5.39) is a very small correction meeting the conditions that it vanish for $r = 1$ and $r = 0$, and further, it is a much smaller factor in D_r than any factorials that yield terms of the order of $\ln M! \approx M \ln M$ for any factorial term $M!$. We shall soon see that g is also a number of considerably smaller than unity.

The successive derivatives of $\ln F$ are

$$\frac{\partial \ln F}{\partial r} = g \ln \frac{1+r}{1-r}; \qquad \frac{\partial^2 \ln F}{\partial r^2} = \frac{2g}{1-r^2} \qquad (5.59)$$

The last of these equations gives $2g$ at $r = 0$, and substitution in equation (5.55) with $W/k = -2678$, and $T_c = 742$ K yields

$$g = \exp\left(\frac{W}{2kT_c}\right) = \exp(-1.8046) = 0.16454 \qquad (5.60)$$

The first derivative of $\ln F$, substituted in equation (5.53), yields the equilibrium values of r at various temperatures:

$$0.16454 \ln \frac{1+r}{1-r} = \ln \frac{b+r}{b-r} \qquad (5.61)$$

A numerical example for this equation is useful. Let $r = 0.80$; the left side of this equation is then 0.36153, so that

$$1.4355 = \frac{b+0.8}{b-0.8} \qquad (5.62)$$

from which $b = 4.4739$, or $b^2 = 20.0158 = 0.64 + 0.36n$, with $\ln n = 2678/T$. Equation (5.62) therefore yields $n = 53.8217$, and $T = 671.91$, or $T/T_c = 0.9055$. The points, such as this one, are plotted in Fig. 5.2. The values of g play an important role in the result for W/k. For example, if we set $g = 0.1$, instead of 0.1645, then equation (5.55) would yield $W/k = -3417$ instead of -2678.

The curve for r versus T/T_c from the quasi-chemical treatment of Guggenheim[4] is close to that for the GBW method; therefore, it is not shown in Fig. 5.2. The X-ray data of Chipman and Warren,[15] not plotted in Fig. 5.2, are fairly close to the curve obtained from equation (5.61). However, a more stringent test for any theory is the success in representing $\Delta C = C^e$ versus T/T_c shown in Fig. 5.3 and discussed in the next section.

Enthalpy and Heat Capacity

The enthalpy of an ordered phase AB is given by equation (5.40), in which equation (5.43) can be substituted to obtain

$$E \approx H \approx LZe_{AB} - \frac{LW(1 - r^2)}{1 + b} \qquad (5.63)$$

The total differential of $H = H(r, T)$ is

$$dH = \left(\frac{\partial H}{\partial T}\right)_r dT + \left(\frac{\partial H}{\partial r}\right)_T dr \qquad (5.64)$$

which yields, after division by dT, the following useful equation:

$$\frac{dH}{dT} = C = C^e = \left(\frac{\partial H}{\partial T}\right)_r + \left(\frac{\partial H}{\partial r}\right)_T \frac{dr}{dT} \qquad (5.65)$$

Since H(ideal), as obtained by setting $W = 0$ in equation (5.63), is independent of r and T, $H^e = H - H$(ideal) can be substituted in equation (5.65) to derive the same equation for C^e, which is the excess heat capacity. Substitution of equation (5.63) in $(\partial H/\partial T)_r$ gives

$$\left(\frac{\partial H}{\partial T}\right)_r = \frac{LW(1 - r^2)}{(1 + b)^2}\left(\frac{\partial b}{\partial T}\right)_r \qquad (5.66)$$

Equation (5.44) can be rewritten as $b^2 = r^2 + (1 - r^2)n$, and at constant r, with $n = \exp(-W/kT)$, its differential is $2b\,db = (1 - r^2)nW\,dT/kT^2$; hence,

$$\left(\frac{\partial b}{\partial T}\right)_r = \frac{(1 - r^2)nW}{2bkT^2} \qquad (5.67)$$

Substitution of this equation in equation (5.66) and division by $2Lk = R$ gives

$$\frac{1}{R}\left(\frac{\partial H}{\partial T}\right)_r = \frac{(1 - r^2)^2nW^2}{4(1 + b)^2bk^2T^2} \qquad (5.68)$$

Similarly, the substitution of equation (5.63) in $(\partial H/\partial r)_T$, and $(\partial b/\partial r)_T = (r - rn)/b$, obtained by differentiation of b^2, yields

$$\frac{1}{R}\left(\frac{\partial H}{\partial r}\right)_T = \left(\frac{W}{2k}\right)\left[\frac{2rb + 2rb^2 + (r - r^3)(1 - n)}{b(1 + b)^2}\right] \qquad (5.69)$$

Equation (5.53), in which $\partial \ln F/\partial r$ is replaced by its equivalent in equation (5.59), is

$$-g \ln\left(\frac{1+r}{1-r}\right) + \ln\left(\frac{b+r}{b-r}\right) = 0 \tag{5.70}$$

The derivative dr/dT is obtained from this equation, after using $b\,db = r\,dr - rn\,dr - (1 - r^2)nW\,dT/2kT^2$ to eliminate db; the result is

$$\frac{dr}{dT} = \left(\frac{W}{2k}\right)\frac{(r^3 - r)}{T^2(bg - 1)} \tag{5.71}$$

Substitution of equations (5.68), (5.69), and (5.71) in (5.65) gives the final equation for C^e/R:

$$\frac{C^e}{R} = \frac{(1 - r^2)^2 n}{4(1 + b)^2 bT^2}\left(\frac{W}{k}\right)^2$$
$$+ \left(\frac{W}{2k}\right)^2 \left[\frac{2rb + 2rb^2 + (r - r^3)(1 - n)}{b(1 + b)^2}\right]\frac{(r^3 - r)}{T^2(bg - 1)} \tag{5.72}$$

At 1000 K, $r = 0$, and $C^e/R = 0.295$, which is the value for the regular long-range disordered solution of Cu–Zn. For $r = 0.404$, $b = 5.7406$, $n = 39.1873$ from equation (5.61), and using $-W/k = 2678$ in $\ln n = 2678/T$, we obtain $T = 730.03$ or $T/T_c = 0.9839$ with $T_c = 742$ K. Substitution of these values with $g = 0.16454$ in equation (5.72) yields

$$C^e/R = 1.798 \tag{5.73}$$

This value, and others obtained similarly, are plotted in Fig. 5.3. It is evident that the results are in fair agreement with the closely concordant and independent experimental values of Moser,[16] and Sykes and Wilkinson[17] plotted as $C^e/R = [C(\text{alloy}) - C(\text{component elements})]/R$. The values calculated from the quasi-chemical method are also plotted in Fig. 5.3, indicating that these results are not significantly different from the zeroth approximation of GBW.

We now justify the functional form of $F(r)$ by the way of summary of the foregoing procedure. Equation (5.39) for $F = F(r)$ must satisfy (I) the boundary conditions that $\ln F$ be zero for $r = 0$ and $r = 1$, (II) equation (5.53) for all the equilibrium values of r, and (III) equation (5.54) for $r = 0$ and $T = T_c$. For example, $\ln F = g(1 - r)\ln(1 + r)$ would satisfy all the requirements but not (II) for $r = 0$. It may be stated that equation (5.39) is not unique, but it is a useful and relatively small correction meeting the foregoing requirements.

Further justification for equation (5.39) can be made by comparing D_r calculated from equations (5.38) and (5.39) and from equation (4.29). The latter can be written with its correction term c as follows:

$$D_r = \frac{[L(1+r)]![L(1-r)]!}{(Y+Lr+c)![(L-Y-c)!]^2(Y-Lr+c)!} \tag{5.74}$$

For $r = 0$, this equation reduces to equation (4.29) corresponding to equiatomic solutions. For example, when $L = 500,000$, $r = 0.4$, and $b \approx 5.74$, then, from equation (5.43), $2X = 500,000 \times 0.84/6.74 = 62,130$, $Y = L - 2X = 437,690$. In addition, $Y^* = 10^6 \times 0.25 = 250,000$, and $Y/Y^* = 1.75$ so that $c = 1690$. The substitution of these values in equation (5.74) yields $\ln D_r(5.74) = 357,200$ whereas equations (5.38) and (5.39) yield $\ln D_r = 376,980$, which is within 5% of the preceding value for $\ln D_r(5.74)$. These computations show that F in equation (5.38) and c in equation (5.74) or (4.29) are of the correct order of magnitude, but the actual enumerations are not yet available even for unidimensional crystals with various values of the ordering parameter r. Therefore, equations (5.38) and (5.39) are to be preferred for the supporting evidence that equations (5.53) and (5.54) are satisfied, and that for $r = 0$ and $r = 1$, equation (5.38) reduces to equation (4.29) based on actual enumeration of configurations. It is also necessary to emphasize that, in essence, $F(r)$ is not only a correction factor for the ordered solutions, *but also for the existence of two sublattices for the component atoms.*

It is essential to reiterate here that unless equation (5.39) is used with g adjusted to satisfy b_c at the critical point, as required by equation (5.55), then it would be impossible to reconcile the value of the exchange energy W from the regular disordered solutions, with W from equation (5.55) with the correction required by equation (4.29) or (5.74) or any other similar equation. Determination of c in equation (5.74) for values of Y/Y^* considerably farther from unity, as in the preceding example, would require enumeration of configurations for highly ordered systems. This is a very difficult and time-consuming task, yet to be undertaken.

Comments on Previous Approximations

The quasi-chemical method is an early method that attempted to improve on the GBW method. The main assumption of the quasi-chemical method requires permutation of bonds that can be obtained from (i)–(viii) of equation (5.30). This permutation requires that each factorial such as $X!$ be written as $(ZX/2)$, a procedure that was criticized in Chapter 4. The resulting D_r for the permutation of bonds is then normalized by an incorrect procedure as in the quasi-chemical method in long-range disordered regular solutions in Chapter 4. In addition, it has been shown elsewhere conclusively,[11,12] and in Chapter 4, that the permutation of bonds is incorrect since it creates impossible spliced molecules, and contradicts the numbers of configurations obtained by actual enumerations. Further, it is a geometrical fact that for $Z = 2$, or for unidimensional crystals, as beads on a necklace, the long-range order exists; for example, when $r = 1$ and $N_A = N_B$, then N_A and N_B occupy alternate succeeding sites as ABABA.... The quasi-chemical method, however, predicts that there can be no order when $Z = 2$. For these reasons, the quasi-chemical method, despite the claims by numerous investigators, is not am improvement over the GBW method.

The value of W/k for an ordered solution is calculated from the critical temperature T_c in the GBW and quasi-chemical approximations. For the same solution in the disordered state, W/k is determined from experimental results, and we now proceed to show that each of these methods leads to inconsistent results within itself. For Cu–Zn, the GBW method requires $2kT_c = -W(\text{GBW})$ from which $W(\text{GBW})/k = -1484$, and the zeroth approximation to the regular long-range disordered solutions, the analog of the GBW approximation, yields $W(\text{zeroth})/k = -3717$ from $G^e/2LkT = -1.202$ with $T = 773$ and $G^e = 2Lx_A x_B W = 0.5LW$. The quasi-chemical approximation for ordered Cu–Zn gives $W/k = -1708$ from $W/k = T_c Z \ln[(Z - 2)/Z]$, and the same approximation for the disordered solution gives $W/k = -3285$ from $G^e/2LkT = -1.202 = (Z/2) \ln\{2 \exp(W/ZkT)/[1 + \exp(W/ZkT)]\}$, with $T = 773$ and $Z = 8$. Similar calculations for other binary systems show similar discrepancies. The results for r versus T/T_c and for $C^e/2Lk$ versus T/T_c, obtained by these methods, are presented in Figs. 5.2 and 5.3 for comparison.

A significant improvement over the quasi-chemical method is the cluster variation method developed and perfected by Kikuchi.[18] A simplified version of the earlier development by Kikuchi has been presented by Guggenheim and McGlashan.[19] The simplified version for the square as the cluster would require writing quasi-chemical equilibria and their pseudoequilibrium constants for pseudoreactions such as

$$
\begin{array}{ccc}
\text{A}\!-\!\!-\!\!-\!\!-\text{A} & \text{B}\!-\!\!-\!\!-\!\!-\text{B} & \text{A}\!-\!\!-\!\!-\!\!-\text{B} \\
\left|\qquad\right| \;\; + \;\; \left|\qquad\right| \;\; = \; 2 \; \left|\qquad\right| \\
\text{A}\!-\!\!-\!\!-\!\!-\text{A} & \text{B}\!-\!\!-\!\!-\!\!-\text{B} & \text{B}\!-\!\!-\!\!-\!\!-\text{A}
\end{array}
$$

and accounting for all the possible lower hierarchy of configurations down to single bonds. While with increasingly complex clusters the method would minimize the error originating from the permutation of single bonds and formation of unrealistic spliced atoms, clearly a great number of different clusters must be used as shown in Ref. 12. Further, much larger and varied clusters than those used by Kikuchi pose vastly increasing mathematical complexities.[20]

The mean field theory, as presented by Burley,[21] shows that the curve for r versus T/T_c is farther the left of the curves in Fig. 5.2, or somewhat closer to the origins of coordinates. The calculations based on such curves show decreasing degrees of success for representation of C^e/R versus T/T_c.

Unequal Numbers of Atoms and Equal Numbers of Sites

The first approximation can be extended to the case of unequal numbers of A and B atoms with equal numbers of a- and b-sites. Again, we take one of the numbers of atoms smaller than the other, i.e., $N_A < N_B$. The order parameter r is defined as in equations (5.5a) and (5.29), and we repeat them here for convenience:

$$N_A^a = R = x_A(1 + r)L; \qquad N_A^b = Q = x_A(1 - r)L \qquad (5.5c)$$

$$N_B^b = R' = (x_B + x_A r)L; \qquad N_B^a = Q' = (x_B - x_A r)L \qquad (5.5d)$$

Various types of atoms on a- and b-sites, corresponding to equations (5.30), are

(i)   A on a-sites with A neighbors $= X$

(ii)   A on a-sites with B neighbors $= R - X$

(iii)   A on b-sites with A neighbors $= X$

(iv)   A on b-sites with B neighbors $= Q - X$

Total number of A atoms $= R + Q = N_A$

(5.30a)

(v)   B on b-sites with B neighbors $= X'$

(vi)   B on b-sites with A neighbors $= R' - X'$

(vii)   B on a-sites with B neighbors $= X'$

(viii)   B on a-sites with A neighbors $= Q' - X'$

Total number of B atoms $= R' + Q' = N_B$

These atoms are shown and summarized in Fig. 5.5 for $N_A = 8$, $N_B = 12$, $x_1 = 0.4$, and $L = 10$, for sufficiently large numbers of atoms to avoid objectionable fractional values for items other than r, x_A, and x_B in equations (5.5c), (5.5d), and (5.30a). The determination of X is given in the figure caption, and that of X' will now be presented as an additional example. Atoms of B on b-sites have 8 bonds with B atoms, and the latter are on the a-sites. Eight bonds are shared by B on b-sites, and B on a-sites; hence, $8/2 = 4 = X'$. The remaining parts of equations (5.30a) are determined by the same procedure, and the results are listed in Fig. 5.5.

The a-sites contain R atoms of A, and Q' atoms of B, and the distribution of these atoms on the a-sites is similar to equation (5.31), i.e.,

$$D_a = \frac{R! Q'}{(R - X)! X! (Q' - X')! X'!}$$

(5.31a)

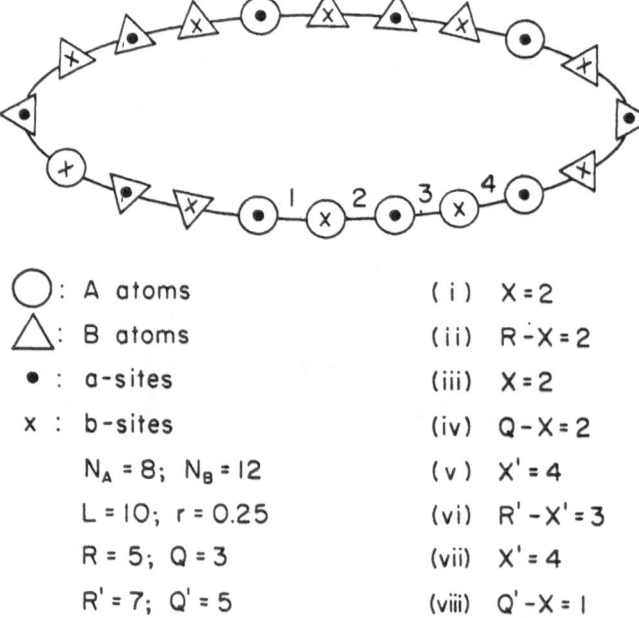

$\bigcirc$: A atoms   (i) $X = 2$

$\triangle$: B atoms   (ii) $R - X = 2$

$\bullet$: a-sites   (iii) $X = 2$

$\times$: b-sites   (iv) $Q - X = 2$

$N_A = 8$; $N_B = 12$ (v) $X' = 4$

$L = 10$; $r = 0.25$ (vi) $R' - X' = 3$

$R = 5$; $Q = 3$ (vii) $X' = 4$

$R' = 7$; $Q' = 5$ (viii) $Q' - X = 1$

Figure 5.5. Net numbers of different types of atoms on a- and b-sites for $N_A = 8$, $N_B = 12$, $L = 10$, and $r = 0.25$. Four numbered bonds involve the A atoms on a-sites with A neighbors; half of these bonds belong to A on a-sites; therefore, $X = 2$ atoms for part (i): the other four half-bonds belong to A on b-sites with A as neighbors, and this constitutes part (iii). Remaining parts in equation (5.30a) are summarized above and can be obtained similarly. All the numbers are selected to avoid fractional atoms on a- and b-sites and fractional neighbors.

Likewise, D_b is given by

$$D_b = \frac{R'!Q!}{(R' - X')!X'!(Q - X)!X!} \qquad (5.31b)$$

The overall distribution function D_r is the product of D_a and D_b, i.e., $D_r = D_a D_b$.

The energy of the system consists of $Ze_{ij}/2$ times the number of bonds of each type, summed up for all types of bonds. The procedure described by equations (5.33)–(5.74) can then be pursued to obtain all the thermodynamic properties of order–disorder in these types of alloys.

The preceding method can be generalized to any numbers of N_A, N_B, a-sites, and b-sites. For brevity, however, we limit our treatment to the case described by equations (5.5b), i.e., $N_A < N_B$, $L_a = N_A$, $L_b = N_b$. Various types of atoms are then summarized as follows:

$$N_A^a = R = (x_A + x_B r)L_a; \qquad N_A^b = Q = x_B(1 - r)L_a \qquad (5.5e)$$

$$N_B^b = R' = (x_B + x_A r)L_b; \qquad N_B^a = Q' = x_A(1 - r)L_b \qquad (5.5f)$$

The types of atoms on a- and b-sites corresponding to equations (5.5e), (5.5f), and (5.30a) are

    (i)   A on a-sites with A neighbors $= X$

    (ii)   A on a-sites with B neighbors $= R - X$

    (iii)   A on b-sites with A neighbors $= X$

    (iv)   A on b-sites with B neighbors $= Q - X$

        Total number of A atoms $= R + Q = L_a$

$$(5.30b)$$

    (v)   B on b-sites with B neighbors $= X'$

    (vi)   B on b-sites with A neighbors $= R' - X'$

    (vii)   B on a-sites with B neighbors $= X''$

    (viii)   B on a-sites with A neighbors $= Q' - X''$

        Total number of B atoms $= R' + Q' = L_b$

These atoms are shown and summarized in Fig. 5.6 for $N_A = 6 = L_a$, $N_B = 18 = L_b$, $x_A = 0.25$, and $r = 5/9$. Again, these numbers are selected to avoid fractional values for all the items in equations (5.5e), (5.5f), and (5.30b). The enumeration $R - X$ is illustrated in the caption for Fig. 5.6; and this value also yields X because R is given by equation (5.5e).

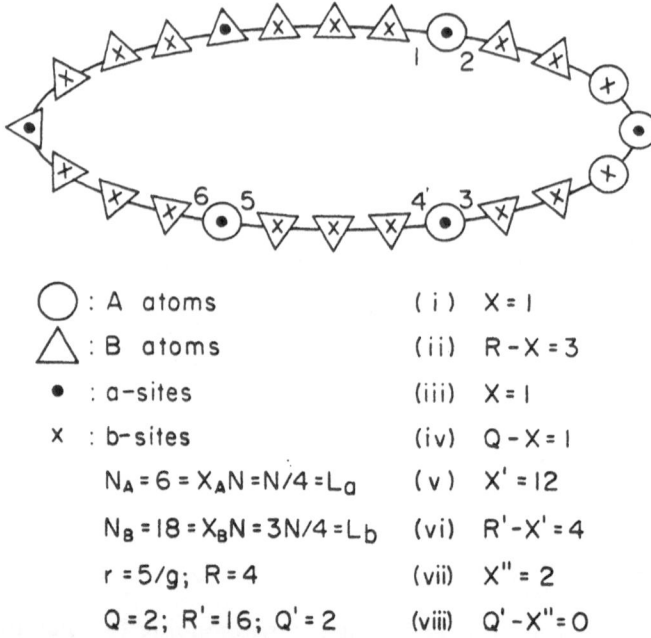

⃝ : A atoms   (i) $X = 1$

△ : B atoms   (ii) $R - X = 3$

● : a-sites   (iii) $X = 1$

x : b-sites   (iv) $Q - X = 1$

$N_A = 6 = X_A N = N/4 = L_a$ (v) $X' = 12$

$N_B = 18 = X_B N = 3N/4 = L_b$ (vi) $R' - X' = 4$

$r = 5/9; \ R = 4$ (vii) $X'' = 2$

$Q = 2; \ R' = 16; \ Q' = 2$ (viii) $Q' - X'' = 0$

Figure 5.6. Net numbers of A and B atoms on a- and b-sites for $N_A = 6 = L_a$, $N_B = 18 = L_b$, and $r = 5/9$. Six numbered bonds involve A atoms on a-sites with B neighbors; therefore, $R - X = 3$ for part (ii). Remaining parts in equation (5.30b) are obtained similarly and summarized above. All the numbers are selected to avoid fractional atoms on a- and b-sites and fractional neighbors.

The distribution functions are similar to equations (5.31a) and (5.31b); thus,

$$D_a = \frac{R!Q'}{(R - X)!X!(Q' - X'')!X''!} \tag{5.31c}$$

$$D_b = \frac{R'!Q!}{(R' - X')!X'!(Q - X)!X!} \tag{5.31d}$$

Again, the overall distribution function, D_r, is given by $D_r = D_a D_b$.

The energy of the system can be obtained by multiplying Ze_{ij} and the net numbers of each type of atom in equations (5.30b) and adding up the results; e.g., for the A atoms on a-sites with A neighbors, the energy term is $Ze_{AA}X/2$. Again, the procedure described by equations (5.33)–(5.74) can be pursued to obtain all the thermodynamic properties of order–disorder in these types of alloys. Some of the most interesting alloys of this category are Ag_2Al, $AlFe_3$, Au_3Cu, $AuCu_3$, $CoPt_3$, Cu_3Pd, $FePd_3$, Fe_3Pt, $FePt_3$, Ni_4Mo, and Ni_2Mo, to name a few. The preceding equations for $N_A < N_B$, $L_a = L_b = L$, and for $(N_A = L_a) < (N_B = L_b)$ have not yet been applied to such alloy systems.

References

1. W. S. Gorsky, *Z. Phys.* **50**, 64 (1928).
2. W. L. Bragg and E. J. Williams, *Proc. R. Soc. London Ser. A* **145**, 699 (1934); **151**, 540 (1935).
3. H. A. Bethe, *Proc. R. Soc. London Ser. A* **150**, 552 (1935).
4. E. A. Guggenheim, *Mixtures*, Oxford University Press, London, Chapters IV and VII (1952).
5. R. Fowler and E. A. Guggenheim, *Statistical Thermodynamics*, Cambridge University Press, London, Chapter XII (1956).
6. M. A. Krivoglaz and A. A. Smirnov, *The Theory of Order-Disorder in Alloys*, Macdonald, London (1965); see also J. M. Cowley, *Phys. Rev.* **120**, 1648 (1960).
7. R. M. White and T. H. Geballe, *Long Range Order in Solids*, Supplement 15, Solid State Physics, Student Edition, Academic Press, New York (1983).
8. I. Prigogine, *The Molecular Theory of Solutions*, North-Holland, Amsterdam (1957).
9. D. de Fontaine, *Solid State Phys.* **34**, 73 (1979); *Acta Metall.* **23**, 553 (1975).
10. H. Sato, in *Physical Chemistry: An Advanced Treatise*, Volume X, edited by W. Jost, Academic Press, New York, p. 579 (1970).
11. N. A. Gokcen and E. T. Chang, *J. Chem. Phys.* **55**, 2279 (1971); *Scr. Metall.* **4**, 941 (1970); A New Method for Enumerating Molecular Configurations in Propellant Mixtures, Aerospace Report No. TR-0172 (2210-10)-1, The Aerospace Corp., El Segundo, California (1971).
12. N. A. Gokcen, *Scr. Metall.* **17**, 53 (1983). (The treatment presented in this reference contains minor initial statistical errors that have been corrected in this book. However, final equations and conclusions are correct in this reference.)
13. N. A. Gokcen, *Thermodynamics*, Techscience, Hawthorne, California, Chapter XI (1975).
14. R. Hultgren, P. D. Desai, D. T. Hawkins, M. Gleiser, and K. K. Kelley, *Selected Values of the Thermodynamic Properties of Binary Alloys*, ASM, Metals Park, Ohio (1973).
15. D. Chipman and B. E. Warren, *J. Appl. Phys.* **21**, 696 (1950).
16. H. Moser, *Phys. Z.* **37**, 737 (1936).
17. C. Sykes and H. Wilkinson, *J. Inst. Met.* **61**, 223 (1937).
18. R. Kikuchi, *Phys. Rev.* **81**, 988 (1951); *J. Chem. Phys.* **60**, 1071 (1974); and the intervening papers.
19. E. A. Guggenheim and M. L. McGlashan, *Mol. Phys.* **5**, 433 (1962).
20. R. Kikuchi, *J. Chem. Phys.* **60**, 1071 (1974); R. Kikuchi and C. M. van Baal, *Scr. Metall.* **8**, 425 (1974).
21. D. M. Burley, in *Phase Transitions and Critical Phenomena*, edited by C. Domb and M. S. Green, Volume 2, Academic Press, New York, p. 329 (1972).

6

Interstitial Solutions

Introduction

It was stated in Chapter 2 that when the atomic diameter of a metalloid is about 59% or less than that of a solvent metal, then the metalloid may form an interstitial solid solution. Such metalloids are hydrogen, boron, carbon, nitrogen, and oxygen, but silicon, phosphorus, and sulfur may also form interstitial solid solutions in certain favorable cases. The interstitial solid solutions of hydrogen and carbon in metals have received particular attention because they form some of the most interesting alloy systems. Metal–hydrogen systems are very useful in hydrogen storage, hydrogen purification, and isotope separation, and metal–carbon systems have unusual structural and mechanical properties. We shall present and discuss the Pd–H and Fe–C systems and then the Wagner theory on the solutions of interstitials in binary metal solutions.

The interstitial lattice sites available for metalloids depend on the crystal structure of the solvent metal. The number of such sites, c, is considered to be 3 per metal atom for body-centered cubic structure and 1 for face-centered cubic, close-packed structures, and liquid metals. Let n be the number of interstitial atoms, and N, that of the metal atoms; the total number of interstitial sites is cN, and the fraction of sites occupied by the interstitial, y, and its atomic fraction x are given by

$$y = \frac{n}{cN} = \frac{r}{c}; \qquad x = \frac{n}{n+N} = \frac{y}{y+1/c}; \qquad y = \frac{x/c}{1-x} \qquad (6.1)$$

where $r = n/N$ is a convenient variable as will be seen later. The Henrian

149

activity of the interstitial, $\dot{a}$, may be written as

$$\dot{a} = \dot{\gamma}x = \frac{fy}{1 - y} \qquad (6.2)$$

where $\dot{\gamma}$ and f are the activity coefficients for their respective concentrations, and the activity can be measured in terms of a gas phase potential or pressure in equilibrium with the solution, such as $H_2(g)$ for dissolved hydrogen [H], CH_4/H_2, or CO/CO_2 gas mixtures for dissolved carbon [C] as will be discussed later. *We emphasize here that the symbols without subscripts refer to the interstitial element* unless they are qualified by immediately succeeding words in parentheses.

Diatomic gas molecules dissociate to dissolve as monatomic solutes in metals. Small solubilities of a diatomic gas such as $H_2(g)$ in metals of Group VIA (e.g., Cr, Mo) and those to the right side in the periodic chart on page 318 can be assumed to take place in two steps: (1) by dissociation of $H_2(g)$ into monatomic gas, $H(g)$ (or as superficially adsorbed H), requiring a large positive value for ΔH_I° of dissociation, and (2) by dissolution of $H(g)$ in a metal, generally requiring a relatively small and usually negative value for ΔH_{II} of solution. The value of ΔH for the overall process is $\Delta H_I^\circ + \Delta H_{II}$, and this sum is a positive quantity because ΔH_I° is generally much greater than $|\Delta H_{II}|$; consequently, the solubility for a given pressure increases with increasing temperature in accord with Le Chatelier's principle. The elements in Groups IIA–VA, as well as Pd, La, and Ta, dissolve up to 10^5 times larger amounts of hydrogen than the remaining poor metal solvents. For most of these good solvents for hydrogen, again $H_2(g)$ dissolves as monatomic H, but $\Delta H = \Delta H_I^\circ + \Delta H_{II}$ is negative so that the solubilities decrease with increasing temperature. [An italic H refers to enthalpy; a roman H refers to dissolved hydrogen. Monatomic gaseous hydrogen is always followed by (g), i.e., H(g).] Palladium dissolves large amounts of hydrogen, and at certain pressures and temperatures up to 292°C, two phases coexist with gaseous hydrogen. The phase low in hydrogen is the α-phase, and that high in hydrogen, the β-phase. Above 292°C, known as the critical temperature, Pd and H form a single solid solution. Hydrogen in palladium is released without much difficulty even below ambient temperatures when the pressure is decreased. For this and many other interesting reasons, and for many technical applications, the solubilities of hydrogen and its isotopes in palladium have been thoroughly investigated.[1-5] It is therefore appropriate to devote the next section to the thermodynamics of the Pd–H system with more details than usual for other systems considered in this book.

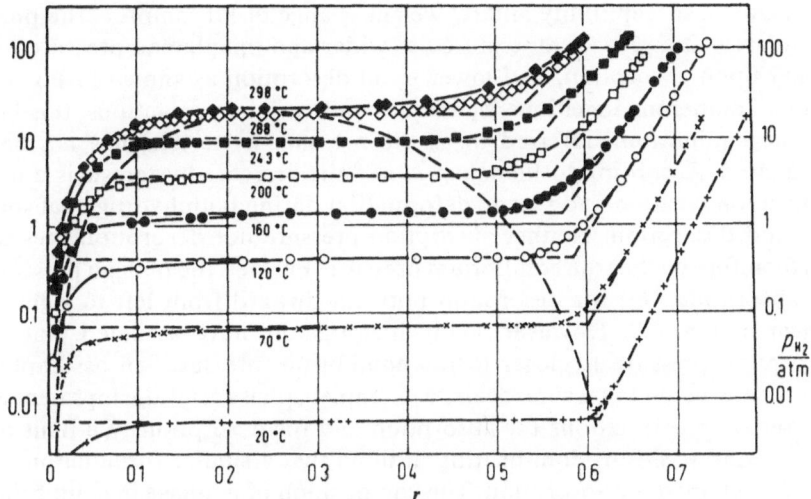

Figure 6.1. Dissolved hydrogen as $r = H/Pd$ atomic ratio versus gaseous hydrogen (H_2) pressure P at various fixed temperatures. For phase boundaries, see Table 6.1. (From Frieske and Wicke[4] with permission.)

Pd–H System

Hydrogen dissolves in palladium as shown in Fig. 6.1, where dissolved hydrogen as $r = H/Pd$ atomic ratio is plotted[5] versus gaseous diatomic hydrogen pressure P in atmospheres. The α-phase on the left coexists with the β-phase on the right at each temperature below 292°C and at each pressure, which is known as the plateau pressure. Neutron diffraction and other studies indicate that hydrogen occupies the octahedral sites in both α- and β-phases. The number of octahedral sites is the same as the number of Pd atoms; therefore, N is also the number of octahedral sites. Hydrogen must also occupy other sites when n/N is greater than 1 at very high pressures.

The α-phase has a face-centered cubic structure, slightly expanded by dissolved hydrogen, and the β-phase has a highly distorted face-centered cubic structure. The increase in volume of Pd upon transition from α to β is about 1.6 cm³/g-atom H. (The volume of pure Pd is close to 8.87 cm³/g-atom.)

The most extensive recent study of the Pd–H system has been carried out by Frieske[4,5]* in the range of 20 to 300°C at hydrogen pressures from 0.01 to 140 atm. The solubilities and the phase boundaries were determined

*The results quoted in References 4 and 5 are from H. Frieske, Dissertation, University of Münster (1972).

by magnetic susceptibility and by weight change of Pd samples. The phase boundaries at a given temperature are wider and the plateau pressures are higher upon absorption, and lower upon desorption as shown in Fig. 6.2. Despite numerous ingenious techniques and various precautions, this lack of true equilibrium, referred to as the pressure hysteresis, has not been eliminated. According to Flanagan et al.,[6] the pressure hysteresis is due to dislocation creation and plastic deformation during both hydrogen absorption and desorption. Neither absorption pressure nor desorption pressure can therefore be the true equilibrium pressure. Further, the plateau pressures for absorption and for desorption both tilt upward from left to right, as shown in Fig. 6.2. However, many investigators have assumed that the desorption pressure is closer to the equilibrium pressure, an assumption that is considered to be controversial. Complete phase relationships necessitate separate correlations for absorption and for desorption. We limit our formulation to desorption, bearing in mind that a similar formulation can also be obtained for absorption. The composition of α-phase in equilibrium with β-phase was obtained by extending the desorption plateau to intersect the absorption curve as shown at the left in Fig. 6.2. The composition of β-phase in equilibrium with α-phase was obtained by extending the straight portions of the magnetic susceptibility–composition curve and taking their point of intersection as the equilibrium composition (not shown in Fig.

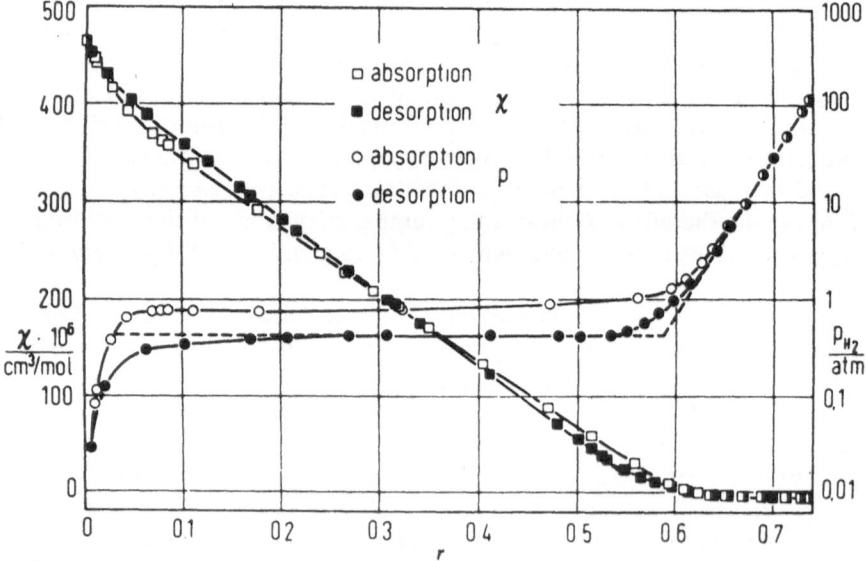

Figure 6.2. Magnetic susceptibility χ and diatomic hydrogen pressure P, versus $r = H/Pd$ atomic ratio at 120°C. (From Frieske and Wicke[4] with permission.)

6.2). No particular significance of interest will be attached in this chapter to the intersection of the dashed lines on the right side in Fig. 6.2. The compositions of coexisting phases determined by the foregoing procedure are shown by the dashed concave-down curve in Fig. 6.1, and listed in Table 6.1.

The plateau pressures, P, for 20 to 300°C in Fig. 6.1 are closely represented by

$$4PdH_{0.5}(\beta\text{-phase}) \rightarrow 4Pd(\alpha\text{-phase}) + H_2(g) \qquad (6.3)$$

$$\ln P(\text{atm}) \approx -\frac{\Delta H°}{RT} + \frac{\Delta S°}{R} = -\frac{4932}{T} + 11.73;$$

$$(\Delta H° = 9800 \text{ cal/mole } H_2; \Delta S° = 23.3 \text{ cal/K-mole } H_2) \qquad (6.4)$$

The activities of both Pd and $PdH_{0.5}$ in reaction (6.3) decrease with increasing temperature because both phases dissolve hydrogen; hence, P is nearly equal to the equilibrium constant of reaction (6.3); therefore, equation (6.4) is simply an empirical equation without the activity corrections for $PdH_{0.5}$ and Pd. Other equations for different ranges of temperature are available; e.g., $\Delta H° = 9325 \text{ cal/mole } H_2$, and $\Delta S° = 21.8 \text{ cal/K-mole } H_2$ have been reported for −80 to +50°C by Wicke and Nernst.[7] (For lower temperatures, see, e.g., Lynch and Flanagan.[8]) The values of $\Delta H°$ agree well with the $\Delta H° = 9440 \text{ cal}/4$ moles $PdH_{0.5}$ based on the calorimetric data of Nace and Aston.[9] Equilibria between α-phase and H_2, or D_2 have been investigated in the range of approximately 0 to 100°C by Wicke and Nernst[7] whose results agree very well with those of Clewley et al.[10]:

Table 6.1. Pressure-Temperature-Composition in Two-Phase Field in Pd–H System[a]

°C	r, α-phase	r, β-phase	P, atm $H_2(g)$[b]
20	0.008 ± 0.002	0.607 ± 0.002	0.0061
70	0.017 ± 0.002	0.575 ± 0.002	0.071
120	0.030 ± 0.002	0.540 ± 0.003	0.44
160	0.046 ± 0.002	0.504 ± 0.003	1.41
200	0.075 ± 0.002	0.459 ± 0.004	3.7
243	0.117 ± 0.002	0.399 ± 0.005	8.8
288	0.21 ± 0.01	0.29 ± 0.01	18.9
292 ± 2[c]	0.250 ± 0.005[c]	0.250 ± 0.005[c]	19.7 ± 0.2[c]

[a]Composition is in r = H/Pd atomic ratio, and pressure of $H_2(g)$, P, in atmospheres. From Refs. 4 and 5.
[b]Calculated from equation (6.4) using the temperatures in the first column.
[c]Critical point; values based on experiment.

Table 6.2. Values of $\Delta H(r)$ and ΔS for Equation (6.7)

Gas	$\Delta H(r)$, cal/mole H_2	ΔS, cal/K-mole H_2
H_2	23,970–21,500 r	25.6
D_2	22,820–21,500 r	25.4
T_2	22,050–21,500 r	24.6

$$0.5 H_2 (\text{or } D_2) \rightarrow [\text{H or D in } \alpha\text{-phase}]; \quad K_p(\text{H or D}) = \frac{r}{P^{0.5}}$$

$$\ln K_p(\text{H}) = \frac{1162}{T} - 6.42; \qquad \ln K_p(\text{D}) = \frac{949}{T} - 6.40 \qquad (6.5)$$

Similar results for tritium in the range of 25 to 100°C, obtained by Schmidt,[11] are represented by

$$\ln K_p(\text{T}) = \frac{832}{T} - 6.25 \qquad (6.6)$$

These equations refer to the α-phase within a limited range of temperature and concentration.

The corresponding empirical equations based on the data of Sicking[12] and Wicke and Nernst[7] for β-phase desorption in a temperature range of roughly −34 to +83°C are summarized by Wicke and Brodowsky[5] by using $\Delta H(r)$ and ΔS in the following equation:

$$\ln P = -\frac{\Delta H(r)}{RT} + 2 \ln \frac{r}{1-r} + \frac{\Delta S}{R} \qquad (6.7)$$

The results are listed in Table 6.2. for $\Delta H(r)$ and ΔS.

The gas-phase equilibria, e.g., $H_2 + D_2 = 2HD$, can be combined with these equations for use in isotope separation. The critical temperature for the Pd–D system is 276°C (292°C for H_2), and the critical pressure is 35 atm (19.7 atm for H_2).

The Pd–H and related equilibria and other properties have been discussed elsewhere in full detail.[1-8]

Statistical Treatment, Zeroth Approximation

The simplest statistical treatment of interstitial alloys, such as H in a metal M, is based on the zeroth approximation to the regular solutions. We start the alloy formation process with pure monatomic gaseous hydrogen H(g) at 1 atm, in the same way as we start with pure metals at their standard states in forming all metallic alloys. The basis of treatment is n atoms of H and N atoms of metals providing cN interstitial sites. Let $Z_{11}/2$ and Z_{12} be the respective numbers of H-H and H-M bonds per atom of dissolved H; e_{11} and e_{12}, the corresponding energies per bond; and e_1, the energy of gaseous atomic hydrogen. The probability of finding a dissolved H atom at a selected site is n/cN and the probability of finding another H atom next to the first atom is $n^2/(cN)^2$. The total maximum number of bonds, if all the interstices were occupied, is $Z_{11}cN/2$, and multiplication with $n^2/(cN)^2$ yields the actual number of existing H-H bonds, i.e., $Z_{11}n^2/(2cN)$. The number of H-H bonds times e_{11}, the H-H bond energy, yields $Z_{11}e_{11}n^2/2cN$, which is the energy due to the H-H bonds. The number of H-M bonds per atom of H is Z_{12}, and the total number of H-M bonds is $Z_{12}n$, which, multiplied by the bond energy e_{12}, yields $Z_{12}e_{12}n$. The energy of the metal is $Z_{22}Ne_{22}/2 = Nh_2^\circ$, and the enthalpy of the alloy is

$$H(\text{alloy}) \approx Z_{11}e_{11}n^2/(2cN) + Z_{12}e_{12}n + nh_1^\circ + Nh_2^\circ \qquad (6.8)$$

where $H(\text{alloy})$ is the enthalpy of alloy, and $h_1^\circ = e_1 + pv$ is the standard enthalpy of one atom of hydrogen, H(g). The energy of metal M, $Z_{22}Ne_{22}/2$, is written as its enthalpy Nh_2°, which is assumed to be unaffected by the dissolved interstitial. The entropy of hydrogen in solution is obtained in three steps. Let s_1° be the entropy per atom of gaseous hydrogen, H(g), at 1 atm; for the overall process of compressing and freezing monatomic hydrogen to a hypothetical solid having the same volume as metal M, the entropy decreases by s^* to $s_1^\circ - s^*$. The next process of random mixing by the diffusion of solid hydrogen to the interstices increases the entropy of solid hydrogen by distribution of n atoms of hydrogen in cN interstices. This distribution, D, is given by

$$D = \frac{(cN)!}{(cN - n)n!} \qquad (6.9)$$

The resulting overall entropy of the alloy, $S(\text{alloy})$, is then

$$S(\text{alloy}) = n(s_1^\circ - s^*) + Ns_2^\circ + k \ln D \qquad (6.10)$$

where Ns_2° is the entropy of metal, which is assumed to be unaffected by interstitial alloying. The Gibbs energy of alloy having n atoms of H, $(cN - n)$

interstices, and N atoms of metal is G(alloy), which is obtained by substituting equations (6.8) and (6.10) in G(alloy) $= H$(alloy) $- TS$(alloy); thus,

$$
\begin{aligned}
G(\text{alloy}) = {}& Z_{11}e_{11}n^2/(2cN) + Z_{12}e_{12}n + nh_1^\circ - nT[s_1^\circ(g) - s^*] \\
& + N(h_2^\circ - Ts_2^\circ) \\
& - kT[cN \ln cN - (cN - n)\ln(cN - n) - n \ln n]
\end{aligned}
$$
$$(6.11)$$

Here, $h_2^\circ - Ts_2^\circ = G_2^\circ/N$ is the standard Gibbs energy per atom of metal. The last set of terms represent $-kT \ln D$, expanded by using the Stirling approximation for factorials.

Equation (6.11) can be simplified by using the following notation:

$$
\begin{aligned}
&\mathcal{G} = N_0 G(\text{alloy}); \qquad N_0 k = R; \qquad (N_0 = \text{Avogadro's number}) \\
&2A = -cN_0 Z_{11} e_{11}; \qquad G_2^\circ = N_0(h_2^\circ - Ts_2^\circ); \qquad\qquad (6.12) \\
&G^\bullet = N_0[Z_{12}e_{12} + h_1^\circ - T(s_1^\circ - s^*)]
\end{aligned}
$$

Substitution of these equations in equation (6.11) results in

$$
\begin{aligned}
\mathcal{G} = {}& -A\frac{n^2}{c^2 N} + nG^\bullet + NG_2^\circ \\
& - RT[cN \ln cN - (cN - n)\ln(cN - n) - n \ln n] \qquad (6.13)
\end{aligned}
$$

The partial molar Gibbs energy of dissolved hydrogen, $\bar{G}$, is obtained by partial differentiation of $\mathcal{G}$ with respect to n:

$$
\bar{G} = \left[\frac{\partial \mathcal{G}}{\partial n}\right]_{P,T,N} = G^\bullet - \frac{2A}{c}\frac{n}{cN} + RT \ln \frac{n}{cN - n} \qquad (6.14)
$$

Substitution of $r = n/N$ gives

$$
\bar{G} = G^\bullet - \frac{2A}{c^2}r + RT \ln \frac{r}{c - r} \qquad (6.15)
$$

Dissolved hydrogen is in equilibrium with $H_2(g)$ *at its pressure P* according to

$$
0.5H_2(g) = [\text{H in metal}] \qquad (6.16)
$$

Therefore, $\bar{G}$ is equal to $0.5G(H_2)$ so that

$$\bar{G} = 0.5G(H_2) \equiv 0.5G°(H_2) + 0.5RT \ln P \qquad (6.17)$$

The use of this equation to eliminate $\bar{G}$ in equation (6.15), and a simple rearrangement of the resulting relationship leads to

$$0.5RT \ln P = G^{\bullet} - 0.5G°(H_2) - \frac{2A}{c^2}r + RT \ln \frac{r}{c-r} \qquad (6.18)$$

This is the equation frequently used by researchers in correlating P with r. The unknown terms $G^{\bullet} - 0.5G°(H_2)$ and A must be determined by using appropriate experimental data.

It is useful at this point to attach a thermodynamic significance to $G^{\bullet}$ in equation (6.18). As $y = n/cN$ approaches zero, the activity coefficient f in equation (6.2) approaches unity, $2Ar/c^2$ in equation (6.18) becomes negligible, and $y/(1-y) = r/(c-r)$ approaches $y = r/c$; therefore, rearrangement of equation (6.18) results in

$$G^{\bullet} - 0.5G°(H_2) = -RT \ln \frac{r}{cP^{0.5}} \equiv -RT \ln K_p \qquad (6.19)$$

where $K_p = r/(cP^{0.5})$ is the equilibrium constant for reaction (6.16) in the dilute range where $f = 1$. The values of K_p at various temperatures are used in equation (6.19) to determine $\Delta H^{\bullet}$ and $\Delta S^{\bullet}$ in $G^{\bullet} - 0.5G°(H_2) = \Delta H^{\bullet} - T\Delta S^{\bullet}$, and substitute the result in equation (6.18) to obtain the following alternative relationship:

$$0.5RT \ln P = \Delta H^{\bullet} - T\Delta S^{\bullet} - \frac{2A}{c^2}r + RT \ln \frac{r}{c-r} \qquad (6.20)$$

The remaining unknown parameter A can be determined from significantly large values of r at the correspondingly large values of P of $H_2(g)$. A relationship similar to equation (6.18) or (6.20) was first derived by Lacher.[13]

Experimental results for M–H systems have been interpreted in terms of three-dimensional oscillating H or H^+ in metals, by using equation (6.18). These interpretations are by no means universally accepted due to various reasons, but chiefly because (1) the assumption that the zeroth approximation is valid, i.e., that hydrogen is dissolved randomly, cannot be fully justified, particularly at high values of r, as discussed in Chapter 3, and (2) the lack of true equilibrium, or the presence of the pressure–concentration hysteresis as the two-phase region is approached, either from high- or low-r

region. The distribution of H atoms in the octahedral sites is indeed nonrandom as shown by extensive investigations.[14]

Activity and Activity Coefficients

The value of the equilibrium constant for reaction (6.16) does not vary with r. For sufficiently large values of r, we write K_p in terms of the activity $\dot{a}$ and the activity coefficient f to obtain their dependence on concentration and temperature:

$$K_p = \frac{\dot{a}}{P^{0.5}} = \frac{fy}{(1-y)P^{0.5}} \tag{6.21}$$

Recall that the symbols without subscripts refer to the solute, and P refers to $H_2(g)$. Substitution of K_p and $cy = r$ in equation (6.18) and elimination of $G^{\bullet} - 0.5G^{\circ}(H_2)$ yields

$$\bar{G} - G^{\bullet} = RT \ln \dot{a} = -\frac{2A}{c}y + RT \ln \frac{y}{1-y} \tag{6.22}$$

Partial differentiation of equation (6.13) with respect to N and elimination of r by using $cy = r$ yields $\bar{G}_2$ for the solvent metal:

$$\frac{\partial \cancel{G}}{\partial N} = \bar{G}_2 = G_2^{\circ} + Ay^2 + cRT \ln(1-y) \tag{6.23}$$

where $\bar{G}_2 - G_2^{\circ} = RT \ln a_2$, and the standard state for a_2 is the pure metal.

Solid Interstitial Solutes

We extend the derivation of the foregoing equations to solid interstitials. Let a solid interstitial be C which may be the elements such as boron, carbon, silicon, or phosphorus, for which s^* is zero, and $nh_1^{\circ} - nTs_1^{\circ}$ refers to pure C. Consequently, equations (6.13)-(6.15) are valid and equation (6.15) is the important relationship for $\bar{G}$ and for the activity of C denoted by $\dot{a}$. We rewrite equation (6.15) by using $y = r/c$ and $\dot{a} = fy/(1-y)$ as follows:

$$\bar{G}^{\bullet} + RT \ln \dot{a} = G^{\bullet} - \frac{2A}{c}y + RT \ln \frac{y}{1-y} \tag{6.24}$$

On the Henrian scale for the activity, as y approaches zero, i.e., $y \to 0$, then $f \to 1$, $y/(1 - y) \to y$, and equation (6.24) becomes $\bar{G} = G^{\bullet} + RT \ln y$. Comparison with the Henrian definition of activity $\bar{G} = G^{\bullet} + RT \ln y$ at $y \to 0$ and $\dot{a} \to y$ shows that $G^{\bullet}$ in equation (6.24) is the standard Gibbs energy with the reference state as $f \to 1$ at $y \to 0$. Equation (6.24) is therefore identical with equation (6.22).

The activity on the Henrian scale in equation (6.24) is given by $\bar{G} - G^{\bullet} = RT \ln \dot{a}$ for any concentration; however, the activity $\underline{a}$ on the Raoultian scale, in which the pure solid is the standard state for the interstitial, is obtained by subtracting G° from both sides of equation (6.24):

$$\bar{G} - G^{\circ} = RT \ln \underline{a}(\text{Raoult}) = G^{\bullet} - G^{\circ} - \frac{2A}{c}y + RT \ln \frac{y}{1 - y} \quad (6.25)$$

The equation for $G^{\bullet} - G^{\circ}$, as well as for A, must then be obtained by fitting the activity data with equation (6.25) at various temperatures and concentrations. It is quite frequently possible to express $G^{\bullet} - G^{\circ}$ as a linear function of temperature, i.e.,

$$G^{\bullet} - G^{\circ} \equiv \Delta G^{\bullet} = \Delta H^{\bullet} - T \Delta S^{\bullet} \quad (6.26)$$

where $\Delta H^{\bullet}$ and $\Delta S^{\bullet}$ are constant within experimental errors.

Substitution of $\dot{a} = fy/(1 - y)$ in equation (6.24) and rearrangement of the result gives the equation for the activity coefficient:

$$\bar{G}^{e} \equiv RT \ln f = -\frac{2A}{c}y \quad (6.27)$$

Equations (6.24) and (6.27) justify the definition of f by equation (6.2). It can be shown that the corresponding equation for the solvent metal is identical with equation (6.23):

$$\bar{G}_2 = G_2^{\circ} + Ay^2 + cRT \ln(1 - y) \quad (6.28)$$

As y decreases, Raoult's law is approached and Ay^2 becomes negligible so that this equation becomes $\bar{G}_2 - G_2^{\circ} = RT \ln(1 - y)^c$; hence, the concentration that must be used to correlate the activity a_2 and the activity coefficient f_2 is $(1 - y)^c$, i.e., $a_2 = f_2(1 - y)^c$. Substitution of a_2 in equation (6.28) as $\bar{G}_2^{\circ} - G_2^{\circ} = RT \ln f(1 - y)^c$ and the use of definitional equation $\bar{G}_2^{\circ} = RT \ln f_2$ yields

$$\bar{G}_2^{e} = RT \ln f_2 = Ay^2 \quad (6.29)$$

This equation can also be derived by using equation (6.27) in the Gibbs–Duhem relation, $d\bar{G}_2^e = -[x/(1-x)]\,d\bar{G}^e = 2Ay\,dy$ and integrating it from $y = 0$ to y, observing that $\bar{G}_2^e$ is zero at $y = 0$ since $a_2 \to 1$ with $y \to 0$.

The molar excess Gibbs energy $G^e = x\bar{G}^e + (1-x)\bar{G}_2^e$ and the molar Gibbs energy $G = x\bar{G} + (1-x)\bar{G}_2$ can be easily obtained by substituting the partial properties in these equations but we shall have no specific use for such equations.

Alternative derivations of equations (6.24) and (6.27), e.g., that given by Hillert and Staffansson,[15] are also based on the zeroth approximation to the regular interstitial solutions. In a number of applications, it is assumed that A is a linear function of temperature, i.e., the bond energy e_{11} is no longer constant but varies linearly with temperature. This is a useful empirical concept, as will be shown later in this chapter.

Application to Pd–H System

Palladium is face-centered cubic in crystal structure, which provides one interstitial site for each Pd atom, i.e., $c = 1$. Equation (6.18) for the Pd–H system is then

$$0.5 \ln P = \frac{G^{\bullet} - 0.5G^{\circ}(\mathrm{H}_2)}{RT} - \frac{2Ar}{RT} + \ln\frac{r}{1-r} \qquad (6.30)$$

The miscibility gap between α and β phases disappears above the critical temperature, T_c, which is 565 ± 2 K in Fig. 6.1, where a horizontal inflection appears in the P versus r curve. The first and second derivatives of equation (6.30) with respect to r are zero at the critical point:

$$0.5\frac{\partial \ln P}{\partial r} = \frac{1}{r} + \frac{1}{1-r} - \frac{2A}{RT} = 0; \qquad (at\ T_c\ and\ r_c) \qquad (6.31)$$

$$0.5\frac{\partial^2 \ln P}{\partial r^2} = \frac{2r-1}{(r-r^2)^2} = 0; \qquad (r_c = 0.5) \qquad (6.32)$$

The first of these equations determines A in terms of T_c and $r = r_c$, and the second, the critical value of r_c, which is 0.5. Unfortunately, the experimental value of r_c is 0.25, far below $r_c = 0.5$ from equation (6.32). An artifice defended by a number of investigators is that there are two sets of sites available for dissolved hydrogen: the first set is $0.6N$ of the total available set of N, and the second set is $0.4N$. This artifice has been generally rejected[16] because all the N sites are equivalent. As evidence for the existence of two such sites, a distinct change in the magnetic behavior of the Pd-H system is often cited, i.e., Pd-H alloys change sharply from paramagnetic state to diamagnetic state when r exceeds 0.6 or after $0.6N$ sites are occupied by dissolved H. Substitution of $0.6N$ for N in the foregoing equations would simply replace r by $\theta = r/0.6$. Equation (6.32) with θ would then yield $\theta_c = 0.5$ and $r_c = 0.3$, which is not too far from the experimental value in Table 6.1. The sites in excess of $0.6N$, i.e., $0.4N$ sites, are said to be available for H at high pressures and low temperatures, and for this range it is proposed that an empirical equation linear in r is obeyed:

$$\ln P = A + Br \qquad (6.33)$$

where A and B are independent of r, and linearly dependent on $1/T$, e.g., $A = l + (m/T)$. An alternative empirical relationship is equation (6.7). The solubilities at high pressures for Pd-H and other systems are discussed in much greater detail elsewhere,[7,17] and need not be considered here since equations such as (6.33) are essentially empirical.

An alternative procedure to obtain an equation consistent with the value of r_c is to assume the existence of interactions beyond the closest neighbors, leading to an additional term in r^2/RT in equation (6.18). This term can be generated by writing the probability for a three-atom cluster $(n/N)^3$, multiplied by its energy per cluster E' and the number of Pd atoms N, i.e., $E'n^3/N^2$, and then substituting the result in equation (6.11). This argument is not rigorous because it does not account for the corresponding entropy effect. Subsequent differentiation as in equation (6.14) contributes a term $3N_0E'r^2/RT$ to equation (6.18); thus,

$$0.5 \ln P = \frac{G^{\bullet} - 0.5G°(H_2)}{RT} - \frac{2Ar}{RT} + \ln \frac{r}{1-r} + \frac{3N_0E'r^2}{RT} \qquad (6.34)$$

The relationships corresponding to equations (6.31) and (6.32) at the critical point are

$$-\frac{2A}{RT} + \frac{1}{r - r^2} + \frac{6N_0E'r}{RT} = 0 \qquad (6.35)$$

$$\frac{2r - 1}{(r - r^2)^2} + \frac{6N_0E'}{RT} = 0 \qquad (6.36)$$

The values of $r_c = 0.25$ and $T_c = 565$ are now required for determination of A and $6N_0E'/RT$; substitution of these values in the preceding equations yields

$$2A/R = 5022, \qquad 3N_0E'/R = 4018 \qquad (6.37)$$

The resulting relationship expressing $\ln P$ as a function of r is

$$0.5 \ln P = 6.35 - \frac{1140}{T} + \ln \frac{r}{1-r} - \frac{5022}{T} r + \frac{4018}{T} r^2 \qquad (6.38)$$

where the first two terms on the right represent the solubility of H_2 in the α-phase for very small concentrations of r when $\ln[r/(1 - r)] \approx \ln r$, and when the last two terms become negligible so that

$$\frac{G^{\bullet} - 0.5 G^{\circ}(H_2)}{RT} = \ln \frac{r}{P^{0.5}} = \frac{1140}{T} - 6.35 \qquad (6.39)$$

Observe that the solubility decreases with increasing temperature according to this equation for a fixed pressure, P.

Equation (6.39) represents the data for the α-phase in Fig. 6.1 well within the errors indicated in Table 6.1 for desorption. It also represents the data for the β-phase in a qualitative way but generally this phase requires equations similar to equation (6.7) or (6.33) with empirical parameters as mentioned earlier. Equations identical in form with those for α- and β-phases may also be derived for the absorption isotherms.

In the regions of sufficiently low values of r for the α-phase, and sufficiently high values of r for the β-phase, the absorption and desorption isotherms coincide without hysteresis, well within experimental errors; consequently, equation (6.38) is obeyed by H in α-phase, and equation (6.33) in β-phase for both absorption and desorption isotherms in the single-phase regions.

There are other additional terms recommended to be included in equation (6.30) on the basis of interpretation of experimental results.[16-18] For example, the electrons from H atoms contribute to the $4d$ band of Pt, and to the electron gas of Pd, and the H atoms in the octahedral sites are conceived to be electronically screened protons. Therefore, when a third component such as Cu is added in Pd, the solubility of hydrogen decreases due to the contribution of one electron per Cu atom to the Pd–H solution, in agreement with observations.[16] However, the number of electrons contributed by the elements is not always predictable.

A pair of empirical equations similar to equation (6.30) have been fitted by Evans and Everett[18] to their data on absorption and desorption isotherms of the α-phase, and another equation, similar to equation (6.33), has been fitted for the β-phase, and further, their treatment has been extended to deuterium in palladium.

The first approximation to the regular solutions in its correct form has not yet been applied to the interstitial solutions. If e_{11} in equation (6.11) is taken to be zero, then the random distribution given by equation (6.9) would be valid, and the zeroth and first approximations would be identical.

Other Hydrogen–Metal Systems

The pressure–composition diagrams are more complex in systems having more than one set of two phases. In such cases there are corresponding numbers of plateaus, each plateau for each pair of coexisting phases. Between two plateaus there is a single phase for which the pressure and concentration both vary until the next plateau is reached. An interesting example is the $ErCo_3$-H system,[19] which contains two plateaus at 101°C, with the first plateau terminating at about 1.7 atoms of H per mole of $ErCo_3$, and the second, at about 4.5 atoms per mole of $ErCo_3$.

The concentration–temperature diagrams are simple for some systems such as the Pd-H system which contains a solubility gap with a critical point, and can be constructed from the data in Table 6.1. However, the majority of such diagrams are very complex as shown in Fig. 6.3 for the Ta-H system.[20,21]

Hydrogen Storage

Hydrogen is a versatile fuel, readily generated by various methods and easily converted to other forms of energy. Large solar farms and thermonuclear reactors of the future may generate enormous amounts of hydrogen that can be used as a convenient and efficient source of energy. These areas,[22,23] as well as the photovoltaic cells presented in Chapter 7, are active fields of research on renewable sources of energy. It is therefore appropriate to devote this section to hydrogen storage.

The storage of large amounts of hydrogen poses numerous problems. Liquefaction wastes about 25% of the available energy of hydrogen, and compression into cylinders requires large and heavy equipment that may consume 15% of the energy of hydrogen. Certain alloys can store and release hydrogen readily without consuming large amounts of energy; their hydrogen capacities per unit volume may be in excess of that attained by liquid hydrogen. A number of interesting alloys appear to be promising and useful for practical applications; therefore, a brief presentation of this topic, based largely on extensive investigations of Wiswall,[24] Reilly,[25] and Hoffman *et al.*,[26] is given in this section.

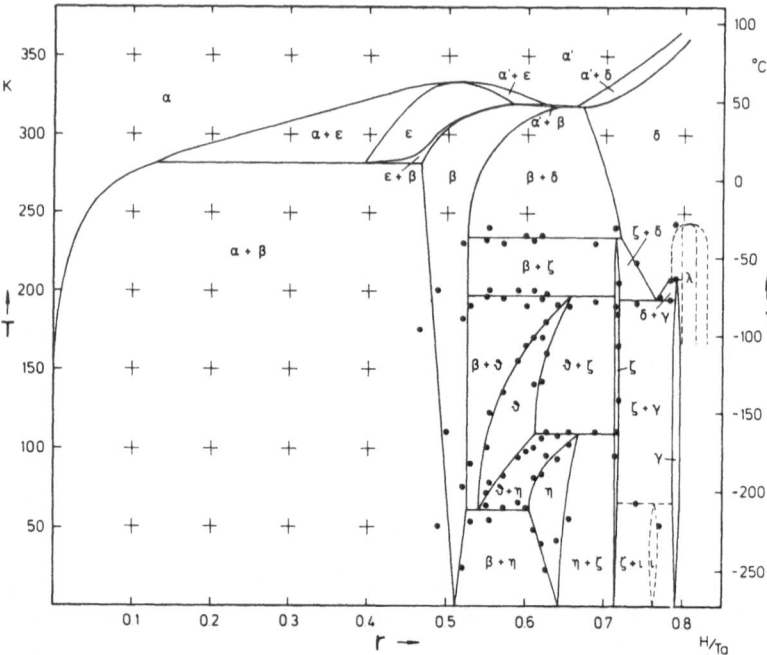

Figure 6.3. Tantalum–hydrogen system; r = H/Ta atomic ratio. (From Köbler and Welter[20] with permission.)

The plateau pressures in metal–hydrogen systems are generally linear when $\ln P$ is plotted versus $1/T$ in K^{-1} either for the absorption curve or for the desorption curve as shown by equation (6.4). Figure 6.4 shows the desorption pressures of a number of systems in which gaseous diatomic hydrogen is in equilibrium with the indicated phases.[24] All such media exhibit pressure–composition hysteresis as in the Pd–H system at temperatures below the annealing range of the metal or alloy, and further, the plateaus are seldom exactly horizontal, particularly for the alloys of two metals. Two or more phases may appear at a given temperature in many metal alloys and each pair of phases is at a different plateau pressure than another pair of phases as shown in Fig. 6.5 for the FeTi–H system.

There are two types of reactions between a hydrogen-rich phase (or hydride phase) AH_r and a metal B, capable of forming an alloy phase AB_n:

$$AH_r + nB = AB_n + \frac{r}{2}H_2(g) \tag{6.40}$$

$$AH_r + nB = AB_nH_r \tag{6.41}$$

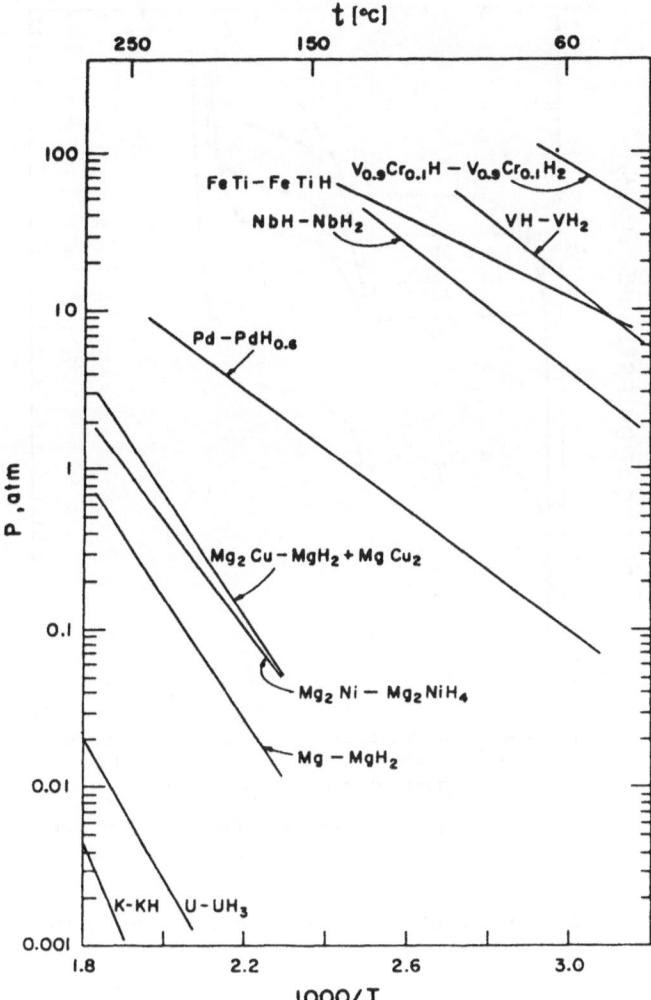

Figure 6.4. Dissociation pressures of metal–hydrogen systems. (From Wiswall[24] with permission.)

In the first case, A and B form such a stable phase that hydrogen is displaced as a gas, and in the second case, hydrogen and B share strong bonding with A. In the latter case, hydrogen from AB_nH_r is released as follows:

$$AB_nH_r = AB_nH_{r-\nu} + \frac{\nu}{2}H_2(g) \qquad (6.42)$$

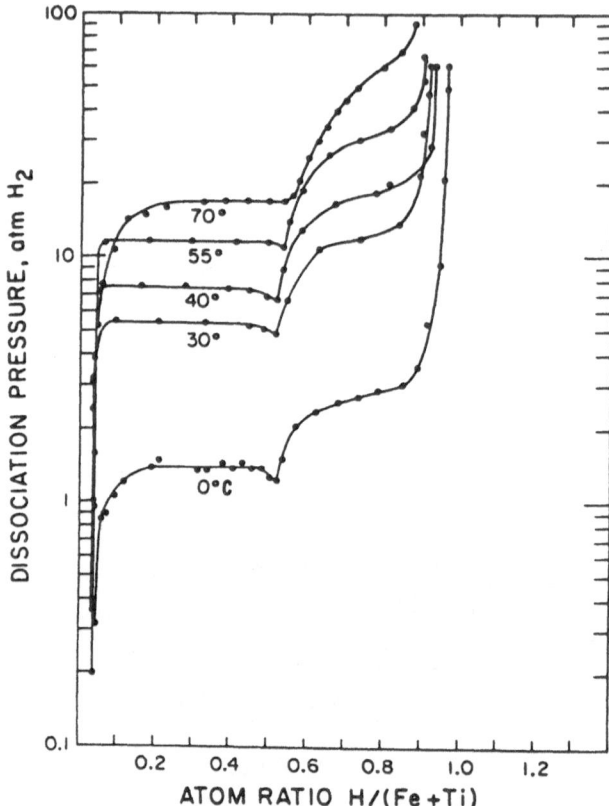

Figure 6.5. Desorption pressure in FeTi–H system. (From Wiswall[24] with permission.)

where $AB_nH_{r-\nu}$ may dissociate to release more hydrogen in the succeeding stages. If B is a non-hydride-forming element, the hydrogen pressure over AB_nH_r is often orders of magnitude larger than over AH_r. Thus, according to Libowitz *et al.*,[27] the hydrogen pressure over $ZrNiH_2$ is about 10^{10} times greater than that over ZrH_2 at 250°C because Ni does not form a stable hydride phase at moderate pressures. In some cases, other types of reactions may occur as in the reaction involving Mg_2Cu:

$$MgCu_2 + 3MgH_2 \rightarrow 2Mg_2Cu + 3H_2 \tag{6.43}$$

If, however, both A and B are stable hydride formers, more complex reactions may often occur.

Hydrogen Storage Metals and Alloys

The alkali elements and their alloys form stable hydrides, unsuitable for hydrogen storage at temperatures below 300°C. Alkaline earth elements (Ca, Br, Sr) also form stable hydrides, but an alloy phase of calcium, $CaNi_5$, is of possible interest because it stores hydrogen up to $CaNi_5H_6$ and dissociates at room temperature in the vicinity of 1 to 15 atm. Other calcium alloys also have possible uses because they are similar to $CaNi_5$.

Hydrides of magnesium and its alloys are more promising media for storage. Magnesium dihydride, MgH_2, contains 7.65% by weight hydrogen and dissociates at 287°C by adsorbing 17.8 kcal/mole H_2. The waste heat from a combustion chamber or from a hot exhaust given off by a motor can be used to supply the necessary heat for dissociation. The hydride is re-formed at higher temperatures and pressures. The intermetallic phase,

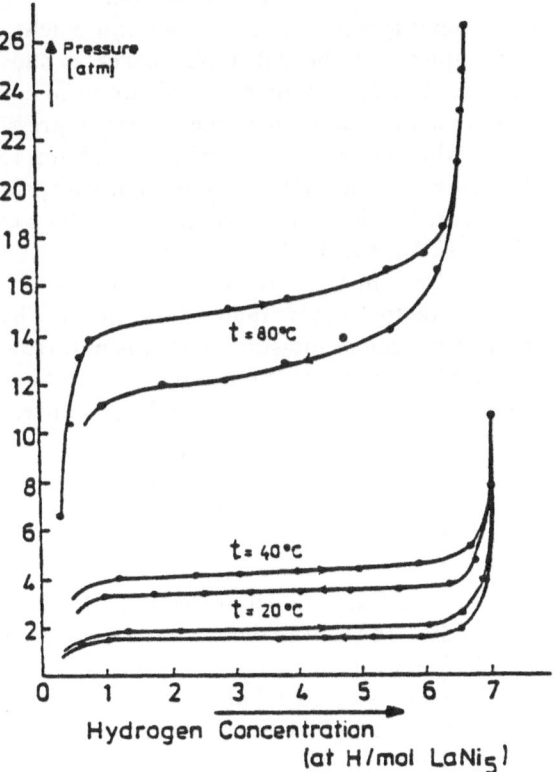

Figure 6.6. Absorption–desorption isotherms for $LaNi_5$ at various temperatures. (From Kuijpers and van Mal[31] with permission.)

Mg_2Cu, absorbs H_2 according to the reverse of reaction (6.43). The equilibrium pressure is about 1 atm at 239°C, lower than that for MgH_2. A similar phase in the Mg–Ni system, Mg_2Ni, reacts with H_2 to form Mg_2NiH_4. Both Mg_2Cu and Mg_2Ni are much more easily hydrided than Mg, but their hydrogen contents are less than half as much as that for Mg. Some alloy phases of Mg–Al might also be useful. Other alloys of magnesium appear to have much fewer potential uses.

Scandium, yttrium, and the lanthanides form very stable dihydrides, and less stable trihydrides. Dissociation pressures of dihydrides are very low; hence, reversible storage is not possible. Some alloys of AB_5 type, where A is Y, Th, or a lanthanide and B is usually Co or Ni, have interesting properties.[28-30] Results on pressure–temperature–hydrogen/metal atomic ratio, P-T-r, are available for Y, Th, La, Ce, Pr, Nd, Sm, Gd as metals A and Co and Ni as metals B. A number of ternary metal-alloys of $AB_{5-x}M$ and $A_{1-y}C_yB_5$ types, where M is Pd, Ag, Cu, Fe, Cr, or Co and C is Er, Y, Gd, Nd, Th, or Zr, have also been investigated.[24,25]

The results for lanthanum pentanickel by van Vucht, Kuijpers, and co-workers[28,31] are typical for the AB_5-type phases as shown in Fig. 6.6, where it is seen that $LaNi_5$ may contain close to 7 atoms of easily recoverable hydrogen at ambient temperatures where the hydrogen pressure, P, is about 2 atm. The concentration of H in the β-phase is about 1.4% by weight because the molecular weight of $LaNi_5$ is rather high. The pressure hysteresis is moderate and rehydriding is easily accomplished. When cobalt progressively replaces nickel in $LaNi_5$, the hydrogen pressure also decreases progressively. Further, when n in $LaNi_n$ increases from 4.9 to 5.5, the plateau pressure increases by a factor of more than 3.3. While most hydride-forming phases are sensitive to oxygen or moisture, $LaNi_5$ is not. However, if carbon monoxide is present, $LaNi_5$ becomes "poisoned" or inhibited for hydrogen absorption but a small amount of copper slightly reduces the poisoning tendency.

Ti, Zr, Hf, and Their Alloys

Group IVA elements, Ti, Zr, and Hf, form binary hydrogen metal alloys of considerable stability; therefore, they are not suitable for hydrogen storage. Some metal binary alloys of the type TiFe have potential applications because they absorb H_2 up to $TiFeH_2$ and release it as shown in Fig. 6.5. The pressure hysteresis is high and the concentration of H is only 1.91% by weight. Powdered TiFe is not pyrophoric and not very expensive, and for stationary storage it is possibly the best storage medium according to Wiswall.[24]

Partial substitution of Fe by Ni or Co in TiFe results in considerable changes in hydrogen capacity and pressure. For example, the hydrogen desorption pressure for $TiFe_{0.9}Ni_{0.1}$ is about an order of magnitude lower at 50°C than for TiFe. Both TiNi and TiCo form more stable hydride phases than TiFe. Intermetallic compounds or phases such as $TiFe_2$, $TiCo_2$, and $TiNi_3$ do not absorb significant amounts of hydrogen. The compound $TiCr_2$ absorbs hydrogen at ambient temperatures at well above 100 atm.

Group VA elements, V, Nb, and Ta, form stable monohydrides of narrow stoichiometry. Their dihydrides dissociate to monohydrides under near-ambient conditions. Thus, for VH_2, dissociation into VH occurs at 13°C and 1 atm, yielding 1.9% by weight hydrogen, without large pressure hysteresis. The enthalpy of dissociation of VH_2 to VH is 9.6 kcal/mole H_2 released. Addition of small amounts of Ti and Zr lowers the hydrogen pressure; however, addition of Ta, Mo, Si, Fe, and possibly other elements to the right side of V-A on the periodic chart, increases the hydrogen pressure. Vanadium is expensive; therefore, it can be used in applications requiring small quantities of hydrogen where the cost can be justified.

Other compounds such as TiV_4H_8, Th_4H_{15}, and VH_3 are fairly stable and expensive hydrides. The behavior of $ThFe_5$, $ThCo_5$, and $ThNi_5$ is similar to the corresponding phases of lanthanum, but the dissociation pressures of their hydrides are higher than those of LaB_5 phases where B is a nonlanthanide transition metal. The properties of selected useful hydrides are listed in Table 6.3. (See also Bambakidis.[32])

Practical Applications

The processes requiring hydrogen storage and use are in their infancy. Nevertheless, possible uses have already either been proposed or made in

Table 6.3. Thermodynamics and Hydrogen Contents of Selected Substances[a]

Reaction with H_2	$\Delta Hf°$, kcal/mole H_2	$\Delta Sf°$, cal/K-mole H_2	Density of hydride, g/cm^3	% H by wt	g-atoms H/cm^3
$Mg \rightarrow MgH_2$	−17.8	−32.3	1.4	7.6	0.111
$Ca \rightarrow CaH_2$	−41.7	−30.4	1.8	4.8	0.085
$LaNi_5 \rightarrow LaNi_5H_6$	−7.6	−26.0	8.3	1.4	0.113
$Ti \rightarrow TiH_{1.97}$	−29.9	−30.0	3.8	4.0	0.147
$V \rightarrow VH_2$	—	—	4.5	2.1	0.171
$H_2(liquid)^b$	—	—	0.071	100	0.070

[a] Adapted from Wiswall.[24]
[b] Data for liquid H_2 is given for reference, showing that all listed hydrides contain more hydrogen per unit volume than liquid hydrogen.

a few significant cases. Hydrogen concentration cells are the simplest power generators that take advantage of a pressure difference between anode and cathode. It is simply a concentration cell that generates power by consuming hydrogen at a high pressure and releasing it at a lower pressure. Such cells are already in use in space vehicles. The use of metal hydrides for providing steady hydrogen pressures for the anode and cathode and thus storing or generating considerable amounts of power in large- and small-size batteries has been proposed.

Production and storage of hydrogen and its reuse in fuel cells may provide a method for electric utility power leveling. A small experimental model was developed by using TiFe for hydrogen storage, and an enlarged research and development program ensued. The results appear to be promising.

Experimental automotive propulsion has been achieved by using $TiFeH_n$. About 7.7 kg of useable H_2, stored in 1016 kg of $TiFeH_n$, has been demonstrated to be capable of providing a range of 121 km, at a sustained speed of about 80 km/hr for a bus weighing about 6.8 metric tons. The hot exhaust gases were used to dissociate the hydride. The spent hydride was recharged to 80% of its capacity in 15 min and full capacity in 1 hr.[24] This performance is far superior to that obtained by acid battery automotive propulsion.

Magnesium hydride, MgH_2, has more than three times the available hydrogen from $TiFeH_{1.9}$ on the basis of hydrogen per unit weight of alloy. It would permit longer range for automobiles with a lighter storage medium. Its dissociation enthalpy and temperature are rather high, demanding special engines with high exhaust temperatures for this purpose.

Ultrapurification of hydrogen, fractionation of H, D, and T, can readily be achieved by using metal diaphragms from hydride-forming metals. Metal hydrides can be used for gas compression for power generation in turbines, for heat storage, and even for air conditioning. Other uses are discussed by Wiswall,[24] Reilly,[25] and Hoffman et al.[26]

Fe–C System

The Fe–C system has been investigated by more than 200 researchers because of the vast industrial importance of iron alloys. We cite a few selected recent papers, summaries, and compilations from which the reader may find all the remaining publications.

The phase diagram as computed by Ohtani et al.[33] from their thermodynamic equations based on selected data for the activity of carbon and phase boundary compositions is shown in Fig. 6.7. This diagram contains small

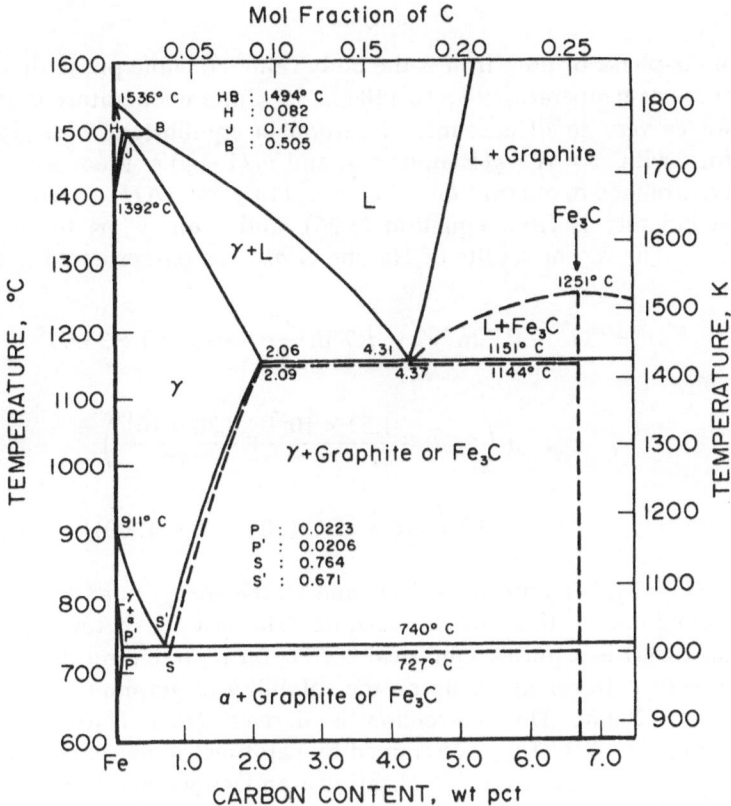

Figure 6.7. Calculated phase diagram for Fe-C(gr) and Fe-Fe$_3$C systems. (Adapted from Ohtani et al.[33] with permission.) Broken lines are for Fe-Fe$_3$C system.

improvements over the previous diagrams,[35-41] but represents the best, and in some respects the most recent results obtained by thermodynamic and metallurgical investigations. The evaluation of the data and the resulting summary in the form of thermodynamic relationships require a great deal of experimental expertise and judgment. The more recent evaluations by Nishizawa and co-workers,[33,34] Ågren,[35] Harvig,[36] and Chipman[37] are required reading for the interpretation, evaluation, and selection of thermodynamic data and the construction of phase boundaries. We assume that the necessary thermodynamic equations are available as given by Ohtani et al.[33] and proceed to compute the phase diagram from these equations. We shall first discuss the Fe-C(gr) [iron-graphite] system in detail and then outline the Fe-Fe$_3$C [iron-cementite] system.

α-Phase

The α-phase of pure iron is the body-centered cubic phase that exists from ambient temperatures up to 1184 K. Its Curie temperature is 1043 K. It dissolves very small amounts of carbon in equilibrium with graphite. Therefore, with $A \equiv A^\alpha$, $\underline{a}$(Raoult) $= 1$, and $y/(1 - y) \approx y$, equation (6.25) can be rearranged to obtain $^{gr}G^\circ - {^\alpha}G^\bullet = RT\ln{^s}y^\alpha = -\Delta H^\bullet + T\Delta S^\bullet$, where the last equality is from equation (6.26), and s on $^s y^\alpha$ is for graphite saturation. The recent results of Hasebe et al.[34], as reinterpreted here, are

$$^{gr}G^\circ - {^\alpha}G^\bullet \approx RT\ln{^s}y^\alpha \approx RT\ln\frac{{^s}x^\alpha}{3} = -99{,}750 + 33.6\,T$$

$$+ \left(8170 - \frac{1.52 \times 10^{10}}{T^2} + \frac{4.80 \times 10^{15}}{T^4}\right);$$

$$[\text{J/g-atom of C } (800 < T < 1200)] \qquad (6.44)$$

where $\Delta H^\bullet = 99{,}750$, and $\Delta S^\bullet = 33.6$, and all the energy units are in joules in the remaining sections of this chapter. The last set of three terms in equation (6.44) is a purely empirical correction representing the effect of paramagnetism-ferromagnetism on the solubility of graphite in the range of 800 to 1200 K. This correction is derived from H_2°(ferromag.) $-$ H_2°(paramag.) $\equiv \Delta H_2^\circ$(f.p.) determined by calorimetric measurements and evaluated and summarized.[42,43] Nishizawa and co-workers[33,34] have proposed that the contribution to equation (6.44) is ΔH_2°(f.p.) $\times (-500/1043)$ where -500 is the rate of change of Curie temperature in kelvins per atomic fraction of carbon, and 1043 is the Curie temperature of pure iron. The solubility of carbon in α-Fe is very small; therefore, -500 cannot be determined accurately; hence, it must be regarded as a reasonable correlation factor that relates ΔH_2°(f.p.)$/1043$ to the Gibbs energy contribution in equation (6.44). Other types of interpretation of magnetic effects on the phase diagrams are discussed by Chuang et al.[44,45] Equation (6.44) is based on the experimental results from 823 K (0.0010 wt% C) to near the eutectoid temperature of 1013 K (0.0206 wt% C) and the extended solubility calculations based on the activity of carbon in the γ-phase as will be seen later. The α-Fe and graphite phase boundary is satisfactorily represented by equation (6.44) from about 800 to 1013 K. At the eutectoid temperature of 1013 K, equation (6.44) yields $^s y^\alpha = 0.000319$ (0.0206 wt% C).

The α-phase ($=$ ferrite) is in equilibrium with the γ-phase ($=$ austenite) from 1013 to 1184 K. The α/γ phase boundary in this temperature range requires the activity of C(gr) in the γ-phase, to be discussed in the next section. The equilibrium relations for dissolved carbon in these phases require $^\alpha\bar{G}$ for carbon in the α-phase in terms of the Henrian standard Gibbs energy $^\alpha G^\bullet$ as follows:

$$^\alpha\bar{G} = {}^\alpha G^\bullet + RT \ln y^\alpha \qquad (6.45)$$

where y^α is so small that the Henrian activity coefficient is unity, and further, the linear term in y^α in equation (6.25) is zero, i.e., $A^\alpha = 0$. The superscript α may be replaced with δ to obtain the corresponding equation for the δ-phase which is also bcc, and stable from 1665 to 1809 K. The superscript α in $^\alpha\bar{G}$ is not essential; it is used here for emphasis.

γ-Phase

The γ-phase is face-centered cubic in structure, constituting the most important area of the iron–carbon diagram. The activity of carbon, $\underline{a}^\gamma$, has been determined[46-49] by equilibration with gaseous mixtures of either carbon monoxide plus carbon dioxide or hydrogen plus methane:

$$CO_2 + [C \text{ in Fe}] = 2CO, \qquad K_p^I = \frac{P_1^2}{\underline{a} \cdot P_2} \qquad (6.46)$$

$$2H_2 + [C \text{ in Fe}] = CH_4, \qquad K_p^{II} = \frac{P_4}{\underline{a} \cdot P^2} \qquad (6.47)$$

Here, P_1, P_2, P_4, and P are the partial pressures of CO, CO_2, CH_4, and H_2, respectively, and $\underline{a}$ is the activity of carbon in the phase under investigation. If the standard state is taken to be pure graphite, then K_p^I and K_p^{II} are known from equilibria only with graphite, as listed in thermodynamic compilations.[50] The results in the range of 1000 to 1800 K are closely represented by

$$\Delta G_I^\circ = -RT \ln K_p^I = 166,774 - 171.695\,T; \text{ (J/mole)} \qquad (6.48)$$

$$\Delta G_{II}^\circ = -RT \ln K_p^{II} = -91,211 + 110.416\,T; \text{ (J/mole)} \qquad (6.49)$$

The values of the activity are obtained from equation (6.46) or (6.47) by using the experimental values of P_1^2/P_2 or P_4/P^2 and K_p^I or K_p^{II} calculated from equation (6.48) or (6.49). The activity data are fitted with equation (6.25) to obtain $(^\gamma G^\bullet - {}^{gr}G^\circ)$ and A^γ, both of which are usually expressed as a linear function of temperature. The results obtained by evaluation of the existing data[33] with equations (6.25) and (6.26) are represented by

$$^\gamma \bar{G} - {}^{gr}G^\circ = RT \ln \underline{a}^\gamma$$

$$= 45,360 - 18.4T + (57,400 + 11.2T)y^\gamma + RT \ln \frac{y^\gamma}{1 - y^\gamma}; \quad \text{(J/mole)}$$

$$(6.50)$$

wherein

$$^\gamma G^\bullet - {}^{gr}G^\circ = 45,360 - 18.4T \tag{6.51}$$

$$A^\gamma = -28,700 - 5.6T \tag{6.52}$$

At the graphite saturation, $^\gamma \bar{G} = {}^{gr}G^\circ$, and the left side of equation (6.50) is zero; then the solution of equation (6.50) for all the assigned temperatures yields the austenite/graphite (γ-phase/graphite) boundary. For example, at 1013 K, the eutectoid temperature, equation (6.50) yields $^s y^\gamma = 0.0314$ from which $^s x^\gamma = {}^s y^\gamma/(1 + {}^s y^\gamma) = 0.03044$ (0.671 wt% C), and at 1424 K, the eutectic temperature, $^s y^\gamma = 0.0977$ (2.06 wt% C).

α/γ Boundary

The partial molar Gibbs energy $^\alpha \bar{G}$ of carbon in α-Fe from equation (6.45) is equal to $^\gamma \bar{G}$ in γ-Fe from equation (6.50) so that $^\alpha G^\bullet - {}^{gr}G^\circ + RT \ln y^\alpha$ is the left side of equation (6.50). The use of equation (6.44) to eliminate ${}^{gr}G^\circ - {}^\alpha G^\bullet$ yields

$$RT \ln y^\alpha = -54,390 + 15.2T + (57,400 + 11.2T)y^\gamma$$

$$+ \left(8170 - \frac{1.52 \times 10^{10}}{T^2} + \frac{4.80 \times 10^{15}}{T^4} \right) + RT \ln \frac{y^\gamma}{1 - y^\gamma}$$

$$(6.53)$$

where y^α is in equilibrium with y^γ at each selected temperature. To check the consistency of y^α and y^γ, we use $y^\gamma = 0.0314$ (0.671 wt% C) at 1013 K to obtain $y^\alpha = 0.000319$ (0.0206 wt% C) from equation (6.53).

Thermodynamic calculations of the α/γ phase boundary require the equality of $^\alpha \bar{G}_2 = {}^\gamma \bar{G}_2$ for iron. For this purpose, we use equation (6.28) to write

$$^\alpha \bar{G}_2 = {}^\alpha G_2^\circ + 3RT \ln(1 - y^\alpha) \tag{6.54}$$

$$^{\gamma}\bar{G}_2 = {}^{\gamma}G_2^{\circ} - (28{,}700 + 5.6T)(y^{\gamma})^2 + RT\ln(1 - y^{\gamma}) \qquad (6.55)$$

where A^{α} is zero in accord with equation (6.45) and A^{γ} is given by equation (6.52). The equality of the left sides in these equations yields

$$^{\gamma}G_2^{\circ} - {}^{\alpha}G_2^{\circ} = 3RT\ln(1 - y^{\alpha}) + (28{,}700 + 5.6T)(y^{\gamma})^2$$
$$- RT\ln(1 - y^{\gamma}) \qquad (6.56)$$

The left side of this equation is the standard Gibbs energy of α to γ transformation for pure iron, which must include the magnetic effects. The equations representing $^{\gamma}G_2^{\circ} - {}^{\alpha}G_2^{\circ}$ are complex; therefore, we present the values in Table 6.4 in close temperature intervals as calculated from the lengthy equations given by Ågren[35] (cf. also Orr and Chipman[43]).

Equation (6.56) with the data in Table 6.4, and equation (6.53) can be solved simultaneously for each temperature to obtain the equilibrium values of y^{α} and y^{γ} and thus calculate the α/γ phase boundary. These equations are transcendental and require a computer or a programmable calculator for iterative computations. Equation (6.44) gives the ferrite/graphite saturation, equation (6.50) with $^{\gamma}\bar{G} = {}^{gr}G^{\circ}$ gives the austenite/graphite saturation, and equation (6.56), based on $^{\alpha}\bar{G}_2 = {}^{\gamma}\bar{G}_2$, gives the equilibrium concentrations of iron in ferrite and austenite. Simultaneous solution of these equations for T, y^{α}, and y^{γ} yields the eutectoid temperature and the compositions of the eutectoid phases. A simple iterative computation for this purpose is to calculate y^{α} and y^{γ} from equations (6.44) and (6.50), respectively, for each temperature in 1 K intervals from 1010 to 1015 K; the set of values for one of the temperatures satisfying equation (6.56) yields the simultaneous solution of equations (6.44), (6.50), and (6.56). Thus, it can be shown that $y^{\alpha} = 0.000319$ (0.0206 wt% C) and $y^{\gamma} = 0.0314$ (0.671 wt% C) at $T = 1013$ K also satisfy equation (6.56) whose left is

Table 6.4. Values of Standard Gibbs Energy of $\alpha \rightarrow \gamma$ Phase Transition, $^{\gamma}G_2^{\circ} - {}^{\alpha}G_2^{\circ} = \Delta G_2^{\circ}$, in J/mole for Pure Iron at Various Temperatures, Calculated from Ågren[35]

T, K	ΔG_2°, J/mole	T, K	ΔG_2°, J/mole	T, K	ΔG_2°, J/mole
800	1367	1000	338	1100	100
840	1116	1020	274	1120	71
880	884	1040	219	1140	45
920	675	1060	173	1160	23
960	491	1080	134	1180	4
980	411				

+296 J/mole as obtained by interpolation of the data in Table 6.4. Excessive numbers of decimal places given here and later are for close checking of the relevant equations; they are not intended to reflect the accuracy of results because the experimental errors are about ±0.001 wt% C for the low ranges of carbon in iron, and up to about 0.01 wt% C for very high ranges of carbon.

Liquid Phase

The γ-liquid and δ-liquid phase boundaries require the activity of carbon in the liquid phase. The liquid is considered to be close-packed in structure for which $c = 1$. The equation for $^l\bar{G}$ recommended by Ohtani et al.,[33] based on selected sets of data on gas-phase and dissolved carbon equilibria,[51-53] in conformity with equation (6.25), is

$$^l\bar{G} = {}^{gr}G° + 24,000 - 16.4T + (52 + 50.6T)y^l + RT \ln \frac{y^l}{1 - y^l} \quad (6.57)$$

where

$$A^l = -26 - 25.3T \quad (6.58)$$

The equality of $^\gamma\bar{G}$ and $^l\bar{G}$ in equations (6.50) and (6.57) correlates y^γ and y^l and establishes the calculated austenite/liquid phase boundary:

$$21,360 - 2.0T + (57,400 + 11.2T)y^\gamma + RT \ln \frac{y^\gamma}{1 - y^\gamma}$$

$$= (52 + 50.6T)y^l + RT \ln \frac{y^l}{1 - y^l} \quad (6.59)$$

The unknowns y^γ and y^l at each selected temperature require an additional equation for their solution. This equation is obtained by setting $^\gamma\bar{G}_2$ of equation (6.55) for γ-Fe equal to $^l\bar{G}_2$ for the liquid phase. The latter is

$$^l\bar{G}_2 = {}^lG_2° - (26 + 25.3T)(y^l)^2 + RT \ln(1 - y^l) \quad (6.60)$$

wherein A^l of equation (6.58) has been used. The result for the austenite/liquid boundary is

$$^lG_2° - {}^\gamma G_2° = (26 + 25.3T)(y^l)^2 - RT \ln(1 - y^l)$$

$$- (28,700 + 5.6T)(y^\gamma)^2 + RT \ln(1 - y^\gamma) \quad (6.61)$$

The left side of this equation is the standard Gibbs energy of melting for pure γ-Fe, which can be obtained from the compiled thermodynamic data[42] by the following steps:

$$^lC_p^\circ = 46.02; \qquad ^\delta C_p^\circ = 41.84; \qquad ^\gamma C_p^\circ = 37.36 \text{ J/mole-K};$$

$$\text{(at 1600 K and assumed constant within } \pm 200 \text{ K)} \qquad (6.62)$$

$$^lH_2^\circ - {}^\delta H_2^\circ = 13{,}807 \text{ J/mole (at 1809 K)}$$

$$^lH_2^\circ - {}^\delta H_2^\circ = 13{,}807 + 4.18(T - 1809) = 6245 + 4.18T$$

$$^lS_2^\circ - {}^\delta S_2^\circ = \frac{13{,}807}{1809} + \int_{1809}^{T} ({}^lC_p^\circ - {}^\delta C_p^\circ)d\ln T = -23.720 + 4.18\ln T$$

$$^lG_2^\circ - {}^\delta G_2^\circ = 6245 + 27.900T - 4.18T\ln T \qquad (6.63)$$

$$^\delta H_2^\circ - {}^\gamma H_2^\circ = 837 \text{ J/mole (at 1665 K)}; \qquad {}^\delta H_2^\circ - {}^\gamma H_2^\circ = -6622 + 4.48T$$

$$^\delta S_2^\circ - {}^\gamma S_2^\circ = -32.728 + 4.48\ln T$$

$$^\delta G_2^\circ - {}^\gamma G_2^\circ = -6622 + 37.208T - 4.48T\ln T \qquad (6.64)$$

Summation of equations (6.63) and (6.64) yields

$$^lG_2^\circ - {}^\gamma G_2^\circ = -377 + 65.108T - 8.66T\ln T \qquad (6.65)$$

The hypothetical melting point of γ-Fe is obtained by setting the left side of this equation to zero and solving for T; the result is 1797 K (1524°C) which will be used later.

Equation (6.65) provides the values for the left side of equation (6.61) which can be solved simultaneously with equation (6.59) to obtain the values of y^γ and y^l in the range of 1424 to 1767 K, and thus calculate the $\gamma/(\gamma + \text{liquid})$ and $(\gamma + \text{liquid})/\text{liquid}$ boundaries. For example, at 1500 K, equation (6.65) yields 2286 J/mole, and it can be shown that $y^\gamma = 0.0810$ (1.71 wt% C) and $y^l = 0.1758$ (3.64 wt% C) satisfy both equations (6.59) and (6.61).

A small and very likely error of ± 50 J/mole in $^lG_2^\circ - {}^\gamma G_2^\circ$ causes large errors in the values of y^γ and y^l calculated by using equation (6.61). Therefore, a greater degree of accuracy is required in equation (6.65) for pure iron to achieve a higher degree of accuracy in the results from equation (6.61). A comparable error in $(21{,}360 + 2.0T)$ of equation (6.59) causes considerably lower errors in the calculated values of y^γ and y^l.

The eutectic temperature and compositions of coexisting phases can now be calculated. There are two alloy phases in equilibrium with graphite and three unknowns, i.e., T, y^γ, and y^l at the eutectic, requiring a simultaneous solution of equations (6.50) and (6.57) with $^\gamma\bar{G} = {}^{gr}G^\circ = {}^l\bar{G}$, and equation (6.61), which was based on $^l\bar{G}_2 = {}^\gamma\bar{G}_2$. A simple iterative method for this purpose is to compute y^γ from equation (6.50) and y^l from equation (6.57) at 1 K intervals from 1420 to 1430 and then substitute each set of results in equation (6.61) to determine the temperature at which equation (6.61) is satisfied. The result is $T = 1424$ K, $y^\gamma = 0.0977$ (2.06 wt% C), and $y^l = 0.2093$ (4.31 wt% C).

Peritectic Equilibrium

The liquid, austenite, and δ phases are in equilibrium at the peritectic temperature. The equalities of partial molar Gibbs energies $^\delta\bar{G}(= {}^\alpha\bar{G}) = {}^\gamma\bar{G} = {}^l\bar{G}$, and $^\delta\bar{G}_2 = {}^\gamma\bar{G}_2 = {}^l\bar{G}_2$ yield four equations to solve for T, y^δ, y^γ, and y^l. The equation for δ-Fe is the same as that for α-Fe without the magnetic correction terms; hence, $^\delta\bar{G} = {}^\gamma\bar{G}$ yields equation (6.53) without the second, third, and fourth terms from the end, i.e.,

$$RT \ln y^\delta = -54{,}390 + 15.2T + (57{,}400 + 11.2T)y^\gamma + RT \ln \frac{y^\gamma}{1 - y^\gamma} \quad (6.66)$$

This is the first equation. Further, $^\gamma\bar{G} = {}^l\bar{G}$ provides equation (6.59), which is the second equation. For iron, $^\delta\bar{G}_2 = {}^\gamma\bar{G}_2$ yields the right side of equation (6.56) with the superscript α replaced with δ, and the left side given by equation (6.64), so that

$$6622 - 37.208T + 4.48T \ln T = 3RT \ln(1 - y^\delta) + (28{,}700 + 5.6T)(y^\gamma)^2$$
$$- RT \ln(1 - y^\gamma) \quad (6.67)$$

which is the third equation. The fourth equation is provided by equation (6.61) whose left side is given by equation (6.65). It can be shown that these equations are satisfied by $T = 1767$ K, $y^\delta = 0.00127$ (0.082 wt% C), $y^\gamma = 0.0079$ (0.170 wt% C), and $y^l = 0.0236$ (0.505 wt% C). If the austenite + liquid field boundaries were extended to zero carbon, they would coincide with the hypothetical melting point of γ-Fe, 1797 K, calculated earlier by using equation (6.65).

The foregoing calculations are very sensitive to small errors in thermodynamic properties of pure iron. Small differences between the calculated phase diagrams of Ohtani et al.[33] and Ågren[35] originate largely in the selection of such data for pure iron (see also Schürmann and Schmid[54]).

The $\delta/\delta +$ austenite/austenite boundaries were obtained by the same procedure as that for the ferrite/(ferrite + austenite)/austenite boundaries, and the $\delta/(\delta +$ liquid)/liquid boundaries, as that for the austenite/ (austenite + liquid)/liquid boundaries by using the relevant preceding equations.

Liquid/Graphite Boundary

The liquidus of graphite saturation, i.e., liquid/liquid + graphite phase boundary in Fig. 6.7, is obtained by setting $^l\bar{G} - {}^{gr}G^\circ$ to zero in equation (6.57) for each selected temperature. Thus, at 1650 K, $y^l = 0.2326$ (4.76 wt% C), and at 1900 K, $y = 0.2526$ (5.15 wt% C). The compositions of phases at the eutectic temperature were calculated earlier.

Cementite

The cementite phase, Fe_3C, denoted as the θ-phase for brevity, is a metastable phase, considered to have no deviation from stoichiometry as the activities and temperature vary. Experimental evidence indicates that deviations from stoichiometry exist, but all the phase diagram calculations are based on perfect stoichiometry. The melting point of Fe_3C is therefore assumed to be congruent. The Curie point is 485 K and the structure is orthorhombic. The solubility data for Fe_3C in austenite and the $CO-CO_2$ equilibrium data[34,55] have been used to obtain the standard Gibbs energy of formation of Fe_3C from pure γ-Fe and graphite. Cementite, as a separate phase, is in equilibrium with austenitic Fe and its carbon; therefore, from the general relationship $G = n_i\bar{G}_i + n_j\bar{G}_j$,

$$G = 3^\gamma\bar{G}_2 + {}^\gamma\bar{G} = {}^\theta G^\circ \tag{6.68}$$

where G is set equal to $^\theta G^\circ$ because the θ-phase is assumed to be a pure stoichiometric compound. Subtraction of $3^\gamma G_2^\circ + {}^{gr}G^\circ$ from both sides yields

$$^\theta G^\circ - 3^\gamma G_2^\circ - {}^{gr}G^\circ \equiv \Delta^\theta G_f^\circ = 3({}^\gamma\bar{G}_2 - {}^\gamma G_2^\circ) + ({}^\gamma\bar{G} - {}^{gr}G^\circ) \tag{6.69}$$

The left side of this equation is the standard Gibbs energy of formation of Fe_3C; the first and second sets of parentheses are given by equations (6.55) and (6.50) respectively, wherein y^γ must refer to the Fe_3C saturation of austenite. The result derived by Ohtani et al.[33] is

$$^\theta G^\circ - 3^\gamma G_2^\circ - {}^{gr}G^\circ = -8900 + 141.1T - 18.99T \ln T \tag{6.70}$$

A similar relation can be obtained for the solubility of cementite in ferrite and the gas condensed-phase equilibria. Ancillary calorimetric data on Fe_3C from near 0 K up to high temperatures and on Fe and C(gr) can also be used[37] to check $\Delta^\theta S°$ in $\Delta^\theta G = \Delta^\theta H° - T\Delta^\theta S°$.

We assume that equation (6.70) has been evaluated by various methods and proceed to outline the method of phase boundary calculation for the Fe-Fe$_3$C system. Substitution of equations (6.50), (6.55), and (6.70) in equation (6.69) and rearrangement of the result yields

$$-54,260 + 159.5T - 18.99T \ln T = (57,400 + 11.2T)[y^\gamma - 1.5(y^\gamma)^2]$$
$$+ RT \ln[y^\gamma(1 - y^\gamma)^2] \qquad (6.71)$$

At the Fe-Fe$_3$C eutectic temperature of 1417 K, this equation is satisfied by $y^\gamma = 0.0991$ (2.09 wt% C) as shown in Fig. 6.7. The austenite-Fe$_3$C boundary can also be calculated from the foregoing equation.

Multiplication of equation (6.65) by 3, and subtraction from equation (6.70) yields

$$^\theta G° - 3^1G_2° - {}^{gr}G° = -7769 - 54.224T + 6.99T \ln T \qquad (6.72)$$

Now, $^\theta G°$ is formed by the liquid phase and the graphite so that equation (6.68) must be rewritten as $^\theta G° = 3^1\bar{G}_2 + {}^1\bar{G}$. Subtraction of $(3^1G_2° + {}^{gr}G°)$ from both sides of this last simple equation gives

$$^\theta G° - 3^1G_2° - {}^{gr}G° = 3(^1\bar{G}_2 - {}^1G_2°) + (^1\bar{G} - {}^{gr}G°) \qquad (6.73)$$

We use the right side of equation (6.72) for the left side of equation (6.73), and then substitute equations (6.60) and (6.57) for their respective terms in parentheses on the right side of equation (6.73), and rearrange the result to obtain

$$-31,769 - 37.824T + 6.99T \ln T = (52 + 50.6T)[y^1 - 1.5(y^1)^2]$$
$$+ RT \ln[y^1(1 - y^1)^2] \qquad (6.74)$$

This equation represents the liquidus for the cementite saturation of liquid iron, as shown by the broken curve above 1417 K in Fig. 6.7. At 1475 K, $y^1 = 0.2516$ (5.13 wt% C) is obtained from the preceding equation. At $y^1 = 1/3$ ($x^1 = 0.25$), which is the composition of Fe$_3$C, this equation is satisfied at 1524.1 K, and this is the melting point of Fe$_3$C. The eutectic temperature, y^γ, and y^1 are calculated from equations (6.71), (6.74), and (6.61); the results are $T = 1417$ K, $y^\gamma = 0.0991$ (2.09 wt%), and $y^1 = 0.2124$ (4.37 wt%).

The cementite/ferrite equilibrium relationships can be obtained by a similar procedure. For this purpose, it is necessary to add $3(^\gamma G_2^\circ - {}^\alpha G_2^\circ)$ to the left side of equation (6.70), and the numerical value of $3(^\gamma G_2^\circ - {}^\alpha G_2^\circ)$ from Table 6.4 to the right side of equation (6.70). The left side then becomes $^\theta G^\circ - 3^\alpha G_2^\circ - {}^{gr}G^\circ$ which is the standard Gibbs energy of formation of cementite from α-Fe and graphite. The θ-phase is now formed by (Fe + C) in the α-phase, i.e., $^\theta G^\circ = 3^\alpha \bar{G}_2 + {}^\alpha \bar{G}$; subtraction of $3^\alpha G_2^\circ + {}^{gr}G^\circ$ from both sides of this last simple equation gives

$$^\theta G^\circ - 3^\alpha G_2^\circ - {}^{gr}G^\circ = 3(^\alpha \bar{G}_2 - {}^\alpha G_2^\circ) + (^\alpha \bar{G} - {}^{gr}G^\circ) \tag{6.75}$$

The terms in the first set of parentheses are given by equation (6.54). The remaining terms require writing equation (6.45) as $^\alpha \bar{G} - {}^{gr}G^\circ = {}^\alpha G^\bullet - {}^{gr}G^\circ + RT \ln y^\alpha$, wherein $^\alpha G^\bullet - {}^{gr}G^\circ$ is given by equation (6.44). Then the calculation of the ferrite/cementite boundary follows the same procedure as that for the austenite/cementite boundary. The ferrite/austenite boundaries remain unaltered since they do not involve either graphite or the cementite above the eutectoid temperature. The calculation of the eutectoid temperature and compositions with Fe_3C is similar to that with graphite.

Despite enormous numbers of investigations on the Fe-C and Fe-Fe_3C systems, new and more precise thermodynamic and metallurgical data are needed in various regions of these systems. Improved modern techniques invite reinvestigation of discordant results recently discussed by various investigators.[33-38]

Wagner Model for Ternary Interstitial Alloys

The Wagner model[56] deals with dilute solutions of an interstitially dissolved component C in ideal binary solutions of A and B. We use C as an interstitial, including carbon and $C_2(g)$ as the diatomic gas. The interstitial dissolution requires not only a favorable site-size but also the ability of the solvent atoms to stretch out to accommodate the solute. The liquid metals and alloys probably have a close-packed structure easily capable of stretching apart to dissolve the interstitial solutes. It is therefore not surprising that the solubilities of interstitials are generally higher in the liquid phase than in a solid phase of a metal or alloy. The interstitials in liquids are assumed to occupy the octahedral site for which $Z = 6$. Additional qualitative considerations are also possible. For example, according to Pauling,[57] the radius of oxygen atom is 0.066 nm, whereas that of copper is 0.128 nm, and Ag, 0.144 nm, so that, with a slight stretching of the lattice, oxygen may occupy the interstitial sites. When the interstitials are dissolved, only a partial transfer of electrons occurs because the metal or the alloy contains a fairly high density of conduction electrons. For a given oxygen pressure below the formation of an oxide phase, more oxygen is dissolved in liquid copper than in liquid silver.[58-61] Wagner[56] interprets that this phenomenon as due to the greater transfer of electrons to oxygen in copper than in silver. As copper is added in silver in increasing amounts, the extent of electron transfer to oxygen also increases, thus increasing the solubility of oxygen.

The dissolution of a gaseous diatomic element, C_2, such as H_2, N_2, or O_2 may be written as the transfer of their monatomic species $C(g)$, i.e.,

$$C(g) + V^i = C^i; \qquad (\Delta\varepsilon = \varepsilon_C^i - \varepsilon_C^g) \qquad (6.76)$$

where $C(g)$ is the monatomic gas, V^i is the vacant quasi-lattice site with i atoms of B and $Z - i$ atoms of A, and C^i is the interstitial atom occupying this vacancy. The monatomic gaseous C is in equilibrium with its pre-dominating diatomic species according to equation (6.16) for $C_2(g)$:

$$0.5C_2(g) = C(g) \qquad (6.77)$$

The change in energy for reaction (6.76) is simply $\varepsilon_C^i - \varepsilon_C^{gas}$, where ε_C^{gas} is taken as zero; hence, the energy with respect to gaseous atomic C is simply ε_C^i. If C^i at site i moves to site $i + 1$, with $i + 1$ atoms of B, Wagner writes

$$C^i + V^{i+1} = V^i + C^{i+1} \qquad (6.78)$$

The energy change for this reaction is simply

$$\varepsilon_C^{i+1} - \varepsilon_C^i = \Delta\varepsilon(i \to i + 1) \qquad (6.79)$$

Likewise, from $i + 1$ to $i + 2$, we have $\Delta\varepsilon(i + 1 \to i + 2)$. Wagner assumes that

$$\Delta\varepsilon(i + 1 \to i + 2) - \Delta\varepsilon(i \to i + 1) = h \qquad (6.80)$$

where h is a constant. Equation (6.80) signifies that as the interstitial atom moves from one site to another site with one more B atom, it gains energy as prescribed by h. Prior to this section in this book, it was assumed that the energy per bond for each type of bond is equal, e.g., every ε_{CA} *per bond* is equal to other ε_{CA} whether a C atom has one CA bond or as many CA bonds as possible. Therefore, energy ε_C^i of a C atom surrounded by i atoms of B and $Z - i$ atoms of A is

$$\varepsilon_C^i = i\varepsilon_{CB} + (Z - i)\varepsilon_{CA} \qquad (6.81)$$

where all ε_{CB} are equal to each other, and all ε_{CA} are also equal to each other. Recall that A and B form an ideal solution; hence, ε_{AA}, ε_{AB}, and ε_{BB} are equal or they may be taken as zero without affecting our argument. Substitution of equation (6.81) in equations (6.79) and (6.80) gives

$$[(i+2)\varepsilon_{CB} + (Z-i-2)\varepsilon_{CA} - (i+1)\varepsilon_{CB} - (Z-i-1)\varepsilon_{CA}]$$

$$- [(i+1)\varepsilon_{CB} + (Z-i-1)\varepsilon_{CA} - i\varepsilon_{CB} - (Z-i)\varepsilon_{CA}] = h = 0 \qquad (6.82)$$

Unfortunately, $h = 0$ does not realistically represent the activities of interstitials in binary metal solvents. This means that the bond energy for each type of bond has to vary according to the number of each type of bond surrounding the interstitial solute, i.e., the equality of ε_{CA} bond energy in one configuration to ε_{CA} in another configuration must be abandoned.

The Wagner equation for the energy of reaction $\Delta\varepsilon = \varepsilon_C^i$ for reaction (6.76) must therefore be written in terms of h, after using equations (6.79) and (6.80); thus,

$$(\varepsilon_C^{i+2} - \varepsilon_C^{i+1}) - (\varepsilon_C^{i+1} - \varepsilon_C^i) = \varepsilon_C^{i+2} - 2\varepsilon_C^{i+1} + \varepsilon_C^i = h \qquad (6.83)$$

For pure A and pure B as solvents, two boundary conditions exist for this equation, i.e., when $i = 0$ we have pure A and when $i = Z$ we have pure B, and therefore,

$$\varepsilon_C^{i=0} \equiv \varepsilon_C^\circ; \quad \text{(C in pure A)}; \qquad \varepsilon_C^Z; \quad \text{(C in pure B)} \qquad (6.84)$$

Chiang and Chang[62] define a new variable y_i by

$$y_i = \varepsilon_C^{i+1} - \varepsilon_C^i \qquad (i = 0, 1, \ldots, Z-1) \qquad (6.85)$$

where y_i is a very convenient parameter for the derivation of the Wagner equation. Here $i = Z$ would give ε_C^{Z+1}, which is physically impossible; hence, the upper limit of i is $Z - 1$ for i insofar as y_i is concerned. Equations (6.83) and (6.85) yield

$$y_{i+1} - y_i = h \qquad (6.86)$$

Each successive value of this equation from $i = 0$ to $i - 1$ gives $y_1 - y_0 = h$, $y_2 - y_1 = h$, $y_3 - y_2 = h$, ...; hence, the summation from $i = 0$ to $i - 1$ yields

$$y_i - y_0 = \sum_{i=0}^{i-1} (y_{i+1} - y_i) = ih \qquad (6.87)$$

Substitution of equation (6.85) here and rearrangement of the result yields

$$\varepsilon_C^{i+1} - \varepsilon_C^i = ih + y_0 \tag{6.88}$$

Resummation of both sides for $i = 0$ to $i - 1$ leads to

$$\sum_{i=0}^{i-1} (\varepsilon_C^{i+1} - \varepsilon_C^i) \equiv \varepsilon_C^i - \varepsilon_C^\circ = \left(\sum_{i=0}^{i-1} ih \right) + iy_0 \tag{6.89}$$

The sum of ih is the sum of a simple arithmetic progression which is equal to $hi(i - 1)/2$; therefore,

$$\varepsilon_C^i = \frac{i(i - 1)h}{2} + iy_0 + \varepsilon_C^\circ \tag{6.90}$$

We note that for $i = 0$, this equation is obviously satisfied, and for $i = Z$,

$$\varepsilon_C^Z = \frac{Z(Z - 1)h}{2} + Zy_0 + \varepsilon_C^\circ \tag{6.91}$$

Solving for y_0 and substituting in equation (6.90), we obtain

$$\varepsilon_C^i = \frac{Z - i}{Z} \varepsilon_C^\circ + \frac{i}{Z} \varepsilon_C^Z - \frac{(Z - i)}{2} ih \tag{6.92}$$

[For the mathematically minded reader, it is to be noted that the result is quadratic in i because the left-hand side of equation (6.83) is the differential of two differences, and it is evident from calculus that a double differential which is equal to a constant, leads to a quadratic function upon double summation or double integration.] Equation (6.92) is derived by a different procedure by Wagner, but the detailed derivation in this section is based on the easier procedure by Chiang and Chang.[62]

Equation (6.92) is equivalent to that postulated by Mathieu et al.,[63] that the bond energies vary by the numbers of different bonds surrounding an atom. This is, at best, a postulate, but without such a postulate it is difficult to account for the variation of activity coefficients in such alloys.

We follow a somewhat different procedure but use the essentials of Wagner's arguments to obtain the equation for the activity coefficient of the interstitial C as a function of the mole fraction x_A of component A in alloys of A–B. The partition function q_C for C may be obtained by regarding the C atoms distributed in N interstitial sites as in boxes with various energies [see equation (3.23)]. The probability of finding i atoms of A and $Z - i$ atoms of B at an interstitial site is $D_Z^i x_A^i x^{Z-i}$, and the total number of such sites (or degeneracy) is N times this product where $N = N_A + N_B$. The energy of each such site is ε_C^i, and the partition function is

$$q_C = N \sum_{i=0}^{z} D_Z^i x_A^{Z-i} x_B^i \cdot e^{-\varepsilon_C^i / kT}; \qquad \left[D_Z^i = \frac{Z!}{(Z-i)! \, i!} \right] \quad (6.93)$$

The molar Gibbs energy of component C is given by equation (3.53) by setting N_C equal to Avogadro's number, i.e.,

$$G_C = -kT \ln \left(\frac{q_C}{N_C} \right)^{N_C} = -RT \ln \frac{q_C}{N_C}; \qquad (N_C kT = RT) \quad (6.94)$$

When ε_C^i in equation (6.93) is zero, C atoms also distribute themselves randomly; the summation then becomes unity and then $q_C = N$ so that

$$G_C(\text{random}) = +RT \ln \frac{N_C}{N} = RT \ln x_C; \qquad (\varepsilon_C^i = 0) \quad (6.95)$$

It is now necessary to express the mole fraction of C in pure A, x_C^A for the general case when ε_C^i is not zero. Equation (6.93) for pure A is obtained by setting $i = 0$:

$$q_C^A = N e^{-\varepsilon_C^o / kT} \quad (6.96)$$

Equation (6.94) for the molar Gibbs energy of C in pure A, G_C^A, is now

$$G_C^A = -kT \ln \left(\frac{q_C^A}{N_C} \right)^{N_C} = RT \ln x_C^A + N_C \varepsilon_C^o \quad (6.97)$$

This G_C^A is equal to $G_C(\text{gas})$ of the monatomic gas in reaction (6.76) as well as $0.5 G_{C_2}$ for the diatomic gas in reaction (6.77) at equilibrium. Since

$G_{C_2}(gas) = G_{C_2}^\circ + RT \ln P_{C_2}$, from $0.5 G_{C_2}(gas) = G_C^A$ we obtain

$$RT \ln \frac{x_C^A}{P_{C_2}^{0.5}} = 0.5 G_{C_2}^\circ - N_C \varepsilon_C^\circ \qquad (6.98)$$

If we take any arbitrary pressure of $C_2(g)$ in equilibrium with x_C^A in pure A as unity, i.e., $P_{C_2} = 1$, which can be any pressure that introduces a very limited number of C atoms in the alloy, and noting that $0.5 G_{C_2}^\circ$ is also a constant for a given temperature, we obtain

$$x_C^A = K' e^{-N_C \varepsilon_C^\circ / RT} = K' e^{-\varepsilon_C^\circ / kT}; \qquad [K' = \exp(0.5 G_{C_2}^\circ / RT)] \qquad (6.99)$$

where K' is a constant equal to $\exp(0.5 G_{C_2}^\circ / RT)$. For C dissolved in pure B at the same pressure of C_2, an identical procedure gives

$$x_C^B = K' e^{-\varepsilon_C^Z / kT} \qquad (6.100)$$

For C dissolved in the binary alloy AB, we equate G_C in equation (6.94) to $0.5 G_{C_2} = 0.5 G_{C_2}^\circ = RT \ln K'$, since $P_{C_2} = 1$, and then write

$$x_C = K' q_C / N = K' \sum_{i=0}^{z} D_z^i x_A^{Z-i} x_B^i \cdot e^{-\varepsilon_C^i / kT} \qquad (6.101)$$

The reference state for a dilute solution of C in a binary alloy AB is usually taken to be the dilute solution of C in one of the components; therefore, we take dilute C in A as the reference state so that the activity of C, $\overset{\bullet}{a}_C^A$, is equal to x_C^A, i.e., $f_C^A = 1$ in $a_C^A = f_C^A x_C^A$. Combination of reaction (6.77) with $C(g) = [C$ in alloy$]$ gives

$$0.5 C_2(g) = [C \text{ in alloy}]; \qquad K_p = x_C^A = f_C x_C; \qquad (P_{C_2} = 1) \qquad (6.102)$$

where x_C and f_C without superscripts refer to the alloy. This equation shows that the activity of C over the entire range of binary composition must also be the same for a fixed pressure of gaseous C_2; therefore, the activity coefficient f_C in the alloy is defined by

$$\overset{\bullet}{a}_C^A = x_C^A = f_C x_C, \qquad \text{or} \quad f_C = \frac{x_C^A}{x_C}, \qquad f_C^B = \frac{x_C^A}{x_C^B} \qquad (6.103)$$

Equations (6.99), (6.100), and (6.103) give

$$f_C^B = e^{-(\varepsilon_C^\circ - \varepsilon_C^Z)/kT} \qquad (6.104)$$

We substitute equation (6.92) in (6.101), and regroup the terms to obtain

$$x_C = K' \sum_{i=0}^{z} D_Z^i x_A^{Z-i} x_B^i [e^{i(\varepsilon_C^\circ - \varepsilon_C^Z)/ZkT}](e^{-\varepsilon_C^\circ/kT})[e^{(Z-i)ih/2kT}] \quad (6.105)$$

The term inside the first set of brackets is $(f_C^B)^{-i/Z}$ from equation (6.104), and that in the first set of parentheses is x_C^A/K' from equation (6.99). Substitution of these terms in equation (6.105) with $1/f_C = x_C/x_C^A$, yields the following Wagner equation:

$$\frac{1}{f_C} = \sum_{i=0}^{z} D_Z^i x_A^{Z-i} x_B^i (f_C^B)^{-i/Z} [e^{(Z-i)ih/2kT}] \quad (6.106)$$

For $Z = 4$ this equation is

$$\frac{1}{f_C} = x_A^4 + 4x_A^3 x_B (f_C^B)^{-1/4} \cdot e^{3h/2kT} + 6x_A^2 x_B^2 (f_C^B)^{-1/2} \cdot e^{2h/kT}$$
$$+ 4x_A x_B^3 (f_C^B)^{-3/4} \cdot e^{3h/2kT} + x_B^4 (f_C^B)^{-1} \quad (6.107)$$

The parameter h is dependent on x_A or x_B in a complex way. It is empirically determined by using a suitable experimental value of f_C at such a composition that f_C can be represented throughout the compositional range x_A for A–B.

If in equation (6.103) we write $\mathring{a}_C^A = f_C^A x_C^A$ for pure A, assuming that f_C^A is not unity, then f_C^A would enter in equation (6.103) as $f_C' = f_C^A x_C^A/x_C$, where f_C' is designated with a prime to distinguish it from f_C in the preceding equations. The resulting equation for f_C' is then

$$\frac{1}{f_C'} = \sum_{i=0}^{z} D_Z^i \left[\frac{x_A}{(f_C^A)^{1/Z}}\right]^{Z-i} \left[\frac{x_B}{(f_C^B)^{1/Z}}\right]^i [e^{(Z-i)ih/2kT}] \quad (6.108)$$

This is the original Wagner equation for which we shall have no specific application in this book, because it is much simpler and preferable to use $f_C^A = 1$ as the reference state, taken to be an infinitely dilute solution of C in pure A. An unusual and interesting derivation of equation (6.108) is also given by Blander and Saboungi.[64]

Application of Equation (6.106)

The activity coefficient of an interstitial is conveniently determined by using diatomic gases such as H_2, N_2, and investigating[65] the equilibria in reaction (6.102). For carbon, reactions (6.46) and (6.47) are investigated as

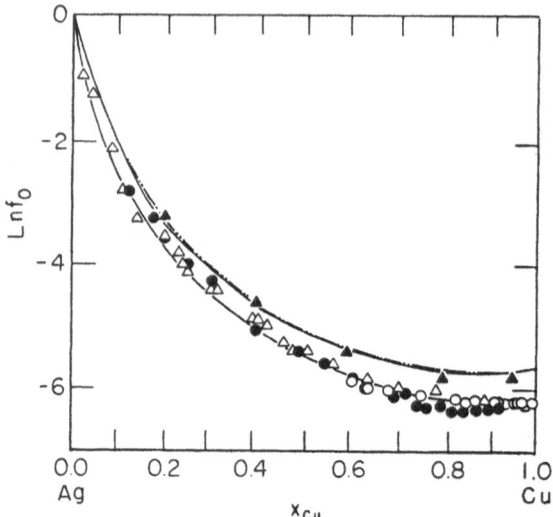

Figure 6.8. Variation of activity coefficient of sparingly dissolved oxygen, f_o, in liquid Ag–Cu alloys at 1373 and 1473 K. Reference state is oxygen in pure Ag for which $f_o^{Ag} = 1$. -▲, Tankins and Gokcen[60]; -···-, Block and Stüwe[58]; solid curve is for $h/k = 683.4$ in equation (6.106); all at 1473 K. -●, Tankins and Gokcen[60]; △, Fruehan and Richardson[59]: ○, Jacob and Jeffes[61]; solid curve is for $h/k = 653.7$; all at 1373 K. (Adapted from Chiang and Chang[62] with permission.)

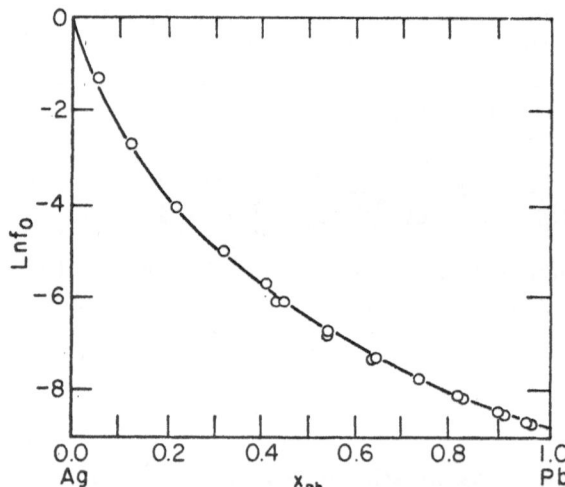

Figure 6.9. Variation of activity coefficients of sparingly dissolved oxygen, f_o, in liquid Ag–Pb alloys at 1273 K. Reference state is oxygen in pure Ag for which $f_o^{Ag} = 1$. Data of Jacob and Jeffes[61]; solid curve is for $h/k = 278.8$ in equation (6.106). (Adapted from Chiang and Chang[62] with permission.)

discussed earlier, and for oxygen and sulfur, the following reactions are investigated:

$$H_2O(g) \rightarrow [O^i] + H_2(g); \qquad K_p = \frac{P_{H_2} f_o^i x_o^i}{P_{H_2O}} \qquad (6.109)$$

$$H_2S(g) \rightarrow [S^i] + H_2(g); \qquad K_p = \frac{P_{H_2} f_s^i x_s^i}{P_{H_2S}} \qquad (6.110)$$

The details of experimental techniques and an analysis of requirements for attainment of equilibrium are presented elsewhere.[66,67]

Extensive applications of the Wagner equation to oxygen in all the binary alloys for which data are reasonably good have been carried out by Chiang and Chang.[62] Selected examples are shown in Figs. 6.8 and 6.9. The results, obtained by linear regression analysis on a computer and by using $Z = 6$, show that the Wagner equation is reasonably successful. As an example, consider Fig. 6.8 for the Ag–Cu–O system studied by equilibration of reaction (6.109) at 1473 K. The value of f_o can be obtained by writing

$$K_p(\text{for Ag}) \equiv \frac{P_{H_2} x_o^{Ag}}{P_{H_2O}} = K_p(\text{alloy}) \equiv \frac{P_{H_2} f_o x_o^i}{P_{H_2O}} \qquad (6.111)$$

The value of K_p for Cu is the same; therefore, for the same value of H_2/H_2O, f_o^{Cu} is given by

$$f_o^{Cu} = x_o^{Ag}/x_o^{Cu} = 0.0036; \qquad (1473 \text{ K})$$

Equation (6.106) can now be used with $Z = 6$ to obtain f_o at $x_{Ag} = x_{Cu} = 0.5$. The result is $f_o(x_{Cu} = 0.5) = 0.00652$, which is located on the upper curve in Fig. 6.8. Similar calculations for the Ag–Pb–O system are shown in Fig. 6.9.

Limitations of Wagner Model

We present Wagner's discussions on the limitations of his equation (6.108):

1. Equation (6.80) for the difference of differences in energy changes is a heuristic approximation leading to ε_C^i as a parabolic function of i as in equation (6.92).

2. The binary solvent alloy A–B is assumed to be random; therefore, large deviations from ideality would make $D_Z^i x_A^{Z-i} x_B^i$ in equation (6.93) a poor approximation. However, other unpredictable factors may affect the results favorably or unfavorably.

3. The effect of the second solvation shell surrounding the atoms around the first solvation shell is ignored; this is also the case in nearly all the statistical treatments.

4. Electron transfer from A–B in the interstitial site to the interstitial atom C may change the interaction of A–B atoms surrounding C in the first and second solvation shells (see Jacob and Alcock[68]).

5. If the molar volumes of components A and B are significantly different, then the dependence of ε_C^i on i, the number of B atoms around C, becomes more complicated. This effect might be largely due to the change in the distance between C and the first solvation atoms, and in addition, it might affect the value of Z. For example, the molar volume ratio of Cu/Sn is 7.1/16.25 for the solid elements at ambient temperatures, and the volume effect is thus expected to be considerable in Cu–Sn alloys.

6. The effects of changes in the conduction electrons per gram atom of an alloy, due to the changes in composition, have not been specifically considered, but these may be at least partially attributed to h in equation (6.92). Such effects are dominant in the solubility of hydrogen in Ag–Pd and Cu–Ni alloys wherein the additional electrons contributed by Pd and Ni increase the solubility of hydrogen.[69-71]

7. The effects due to the changes in the vibrational changes of C–A, C–B, and A–B bonds with the changes in the composition of A–B have not been considered. We add, in summary, that the most important effect due to the deviations from randomness is not considered. The Wagner model is nevertheless an important step in statistical thermodynamics of interstitials in solvent alloys deviating to small extents from ideality. In fact, deviations from the ideality of the binary solvent are partly accounted for by selecting the experimental value of h appropriate for the system.

The data for nitrogen in Co–Fe, Co–Ni, Fe–Ni, and a number of other systems are in agreement with the Wagner interpretation.[62]

Sulfur atoms are larger than carbon, nitrogen, and oxygen atoms and slow diffusional data[72] for sulfur in solid Ag, Cu, Fe, and Ni indicate that sulfur occupies the substitutional sites whereas rapid diffusion of H, C, N, and O shows that they occupy the interstitial sites. Therefore, the value of Z for sulfur in the liquid metals and their alloys should be 8 or 12 instead of 6.

The foregoing treatment is not applicable to hydrogen according to Wagner because the changes in the electronic constitution of the alloy greatly affect the solubility of hydrogen[69,70,73,74] in Ag–Pd, Al–Cu, Cu–Ni, and Cu–Zn, and in Laves phases such as $MgCu_2$, $MgZn_2$, $MgNi_2$, and $MgZn_2$. This effect is due to the decrease in ε_C^i with increasing screening of protons (H^+) by electrons in the alloys:

$$H(g) = H^+ + e^-(\text{conduction electrons}) \qquad (6.112)$$

Thus, when hydrogen enters in a metal having a greater density of electrons, the proton H^+ is more easily screened and a greater number of H^+ are accommodated. There are other factors involved in dissolution of hydrogen that make the solubility in various metals and alloys difficult to explain by simple attractive forces of the nearest neighbors.

References

1. G. Alefeld and J. Voelkl, editors, *Hydrogen in Metals I, and II*, Springer-Verlag, Berlin (1978).
2. G. A. Lewis, *The Palladium Hydrogen System*, Academic Press, New York (1967).
3. W. M. Mueller, J. P. Blackledge, and G. G. Libowitz, *Metal Hydrides*, Academic Press, New York (1968).
4. H. Frieske and E. Wicke, *Ber. Bunsenges. Phys. Chem.* **77**, 50 (1973).
5. E. Wicke and H. Brodowsky, with H. Zuchner, in *Hydrogen in Metals II*, edited by G. Alefeld and J. Voelkl, Springer-Verlag, Berlin (1978).
6. T. B. Flanagan, S. Kishimoto, and G. E. Biehl, in *Chemical Metallurgy—A Tribute to Carl Wagner*, edited by N. A. Gokcen, Metall. Soc. AIME, p. 471 (1981).
7. E. Wicke and G. H. Nernst, *Ber. Bunsenges. Phys. Chem.* **68**, 224 (1964).
8. J. F. Lynch and T. B. Flanagan, *J. Phys. Chem.* **77**, 2628 (1973).
9. D. M. Nace and J. G. Aston, *J. Am. Chem. Soc.* **79**, 3619, 3623, 3627 (1957); J. G. Aston, *Engelhard Ind. Tech. Bull.* **7**, 14 (1966).
10. J. D. Clewley, T. Curran, T. B. Flanagan, and W. A. Oates, *J. Chem. Soc. Faraday Trans. 1* **69**, 449 (1973).
11. S. Schmidt, in *Hydrogen in Metals II*, edited by G. Alefeld and J. Völkl, Springer-Verlag, Berlin (1978).
12. G. Sicking, *Ber. Bunsenges. Phys. Chem.* **76**, 790 (1972).
13. J. R. Lacher, *Proc. R. Soc. London Ser. A* **161**, 525 (1937).
14. M. Shamsuddin and O. J. Kleppa, *J. Chem. Phys.* **71**, 5154 (1979); W. A. Oates and R. Ramanathan, in *Proceedings, 2nd International Congress on Hydrogen in Metals, Paris, 1977*, Paper 2A11, Pergamon Press, Elmsford, New York (1978); G. Bourreau, O. J. Kleppa, and K. C. Hong, *J. Chem. Phys.* **67**, 3437 (1977).
15. M. Hillert and L.-I. Staffansson, *Acta Chem. Scand.* **24**, 3618 (1970); see also M. Hillert and M. Jarl, *Metall. Trans. AIME* **6A**, 553 (1975).
16. C. Wagner, *Z. Phys. Chem. Abt. A* **193**, 386, 407 (1944).
17. See, e.g., B. Baranowski, Part II in *Hydrogen in Metals*, edited by G. Alefeld and J. Völkl, Springer-Verlag, Berlin (1978), p. 157.
18. M. J. B. Evans and D. H. Everett, *J. Less-Common Met.* **49**, 123 (1976).
19. T. Takeshita, W. E. Wallace, and R. S. Craig, *Inorg. Chem.* **13**, 2283 (1974).
20. U. Köbler and J. M. Welter, *J. Less-Common Met.* **84**, 225 (1984).

21. U. Köbler and T. Schöber, *J. Less-Common Met.* **60**, 101 (1978); see also T. Schöber and H. Wenzl, in *Hydrogen in Metals II*, edited by G. Alefeld and J. Völkl, Springer-Verlag, Berlin, p. 12 (1978).

22. T. N. Veziroglu and J. B. Taylor, editors, *Hydrogen Energy Progress V: Proceedings of the 5th World Hydrogen Energy Conference, Toronto, Canada, 15-20 July 1984*, Pergamon Press, Elmsford, New York (1984).

23. J. O. Bockris, *Energy: The Solar Hydrogen Alternative*, Wiley, New York (1977).

24. R. Wiswall, in *Hydrogen in Metals II*, edited by G. Alefeld and J. Voelkl, Springer-Verlag, Berlin, p. 201 (1978).

25. J. J. Reilly, *Z. Phys. Chem.* **117**, 155 (1979).

26. K. C. Hoffman, J. J. Reilly, C. H. Waide, R. H. Wiswall, and W. E. Winsche, *Int. J. Hydrogen Energy* **1**, 133 (1976).

27. G. G. Libowitz, H. F. Hayes, and T. R. P. Gibb, *J. Phys. Chem.* **62**, 76 (1958).

28. J. H. N. van Vucht, F. A. Kuijpers, and H. C. Bruning, *Philips Res. Rep.* **25**, 33 (1970); H. H. van Mal, *Philips Res. Rep. Suppl. 1* (1976).

29. H. H. van Mal, K. H. J. Buschow, and F. A. Kuijpers, *J. Less-Common Met.* **32**, 289 (1973).

30. J. L. Anderson, T. C. Wallace, A. L. Bowman, C. L. Radosevich, and M. L. Courtney, Hydrogen Absorption by AB_5 Compounds, Los Alamos Sci. Lab., Rep. LA-5320-MS (1973).

31. F. A. Kuijpers and H. H. van Mal, *J. Less-Common Met.* **23**, 395 (1971).

32. G. Bambakidis, editor, *Metal Hydrides*, Plenum Press, New York (1981).

33. H. Ohtani, M. Hasebe, and T. Nishizawa, *Trans. Iron Steel Inst. Jpn.* **24**, 857 (1984).

34. M. Hasebe, H. Ohtani, and T. Nishizawa, *Met. Trans.* **16A**, 913 (1985).

35. J. Ågren, *Metall. Trans. AIME* **10A**, 1847 (1979).

36. H. Harvig, *Jernkontorets Ann.* **155**, 157 (1971).

37. J. Chipman, *Metall. Trans. AIME* **3**, 55 (1972).

38. O. Kubaschewski, *Iron Binary Phase Diagrams*, Springer-Verlag, Berlin (1982).

39. R. Hultgren, P. D. Desai, D. T. Hawkins, M. Gleiser, and K. K. Kelley, *Selected Values of the Thermodynamic Properties of Binary Alloys*, ASM, Metals Park, Ohio (1973).

40. M. Benz and J. F. Elliott, *Trans. Metall. Soc. AIME* **221**, 323 (1961).

41. M. Hansen and K. Anderko, *Constitution of Binary Alloys*, McGraw-Hill, New York (1958); First Supplement by R. P. Elliott (1965); Second Supplement by F. A. Shunk (1969).

42. R. Hultgren, P. D. Desai, D. T. Hawkins, M. Gleiser, K. K. Kelley, and D. D. Wagman, *Selected Values of the Thermodynamic Properties of the Elements*, ASM, Metals Park, Ohio (1973).

43. R. L. Orr and J. Chipman, *Trans. Metall. Soc. AIME* **239**, 630 (1967).

44. Y.-Y. Chuang, Y. A. Chang, and R. Schmid, *Acta Metall.* in press.

45. Y.-Y. Chuang, R. Schmid, and Y. A. Chang, *Acta Metall.* in press.

46. R. P. Smith, *J. Am. Chem. Soc.* **68**, 1163 (1946).

47. H. Schenk, M. G. Frohberg, and E. Jaspert, *Archiv. Eisenhütenw.* **36**, 683 (1965).

48. K. Bungardt, H. Preisendanz, and G. Lehnert, *Arch. Eisenhuettenwes.* **35**, 999 (1964).

49. S. Ban-ya, J. F. Elliott, and J. Chipman, *Trans. Metall. Soc. AIME* **245**, 1199 (1969); *Met. Trans.* **1A**, 1313 (1970).

50. L. B. Pankratz, J. M. Stuve, and N. A. Gokcen, *Thermodynamic Data for Mineral Technology*, Bureau of Mines Bulletin 677 (1984).

51. F. D. Richardson and W. E. Dennis, *Trans. Faraday Soc.* **49**, 171 (1953).

52. S. Ban-ya and Y. Matoba, *Physical Chemistry of Process Metallurgy*, Interscience, New York, pp. 373-402 (1961).

53. T. Mori, K. Fujimura, H. Okajima, and A. Yamanchi, *Tetsu To Hagane* **54**, 321 (1968).

54. E. Schürmann and R. Schmid, *Arch. Eisenhuettenwes.* **50**, 101 (1971).

55. E. Scheil, T. Schmidt, and J. Wünning, *Arch. Eisenhuettenwes.* **32**, 251 (1961).

56. C. Wagner, *Acta Metall.* **21**, 1297 (1973).

57. L. Pauling, *The Nature of the Chemical Bond*, Cornell University Press, Ithaca, New York (1960).

58. U. Block and H. P. Stüwe, *Z. Metallkd.* **60**, 709 (1969).

59. R. J. Fruehan and F. D. Richardson, *Trans. Metall. Soc. AIME* **245**, 1721 (1969).

60. E. S. Tankins and N. A. Gokcen, *High Temp. Sci.* **4**, 393 (1972); E. S. Tankins, *Metall. Trans. AIME* **1**, 2637 (1970).

61. K. T. Jacob and J. H. E. Jeffes, *Trans. Inst. Min. Metall.* **C80**, 32 (1971); see also *J. Chem. Thermodyn.* **3**, 433 (1971), **5**, 365 (1973).

62. T. Chiang and Y. A. Chang, *Metall. Trans. AIME* **7B**, 453 (1976).

63. J.-C. Mathieu, F. Durand, and E. Bonnier, *J. Chim. Phys.* **62**, 1289, 1297 (1965); B. Brion, J.-C. Mathieu, P. Hicter, and P. Desré, *J. Chim. Phys.* **66**, 1238, 1745 (1970).

64. M. Blander and M.-L. Saboungi, in *Chemical Metallurgy—A Tribute to Carl Wagner*, edited by N. A. Gakcen, *Metall. Soc. AIME*, p. 223 (1981).

65. N. A. Gokcen, *Thermodynamics*, Techscience, Hawthorne, California (1975).

66. N. A. Gokcen, *Trans. Metall. Soc. AIME* **206**, 1558 (1956); **197**, 191 (1953).

67. M. R. Baren and N. A. Gokcen, in *Advances in Sulfide Smelting, V. 1 Basic Principles*, edited by Y. H. Sohn, D. B. George, and A. D. Zunkel, AIME, Warrendale, Pennsylvania (1983) p. 41; see also G. Urbain, W. Burgmann, and M. G. Frohberg, *C. R. Acad. Sci. Ser. C* **263**(8), 595 (1966).

68. K. T. Jacob and C. B. Alcock, *Acta Metall.* **20**, 221 (1972).

69. H. Brodowsky and E. Poeschel, *Z. Phys. Chem.* **44**, 143 (1965).

70. H. Brodowsky and H. Husemann, *Ber. Bunsenges. Phys. Chem.* **70**, 626 (1966).

71. F. G. Jones and R. D. Pehlke, *Metall. Trans. AIME* **2**, 2655 (1971).

72. S. J. Wang and H. J. Grabke, *Z. Metalld.* **61**, 597 (1970).

73. W. Siegelin, K. H. Lieser, and H. Witte, *Z. Elektrochem.* **61**, 359 (1957).

74. H. Schnabl, *Ber. Bunsenges. Phys. Chem.* **68**, 549 (1964).

7

Semiconductors

Introduction

The discovery of semiconductors and their practical applications have played the greatest scientific and industrial role in this century. It is therefore fitting to devote this chapter to the physical and thermodynamic properties of semiconductors, and the simplest devices manufactured from them, with a limited emphasis on solar cells.

The classical definition of semiconductors is that their resistivities are about 10^{-2} to 10^9 ohm-cm, whereas good metallic conductors have resistivities on the order of 10^{-6} ohm-cm. Insulators have much greater resistivities than semiconductors, i.e., roughly in excess of 10^{14} ohm-cm. Resistivities of semiconductors and insulators decrease, and those of metals increase, with increasing temperature. These properties are much more satisfactorily explained in terms of the band theory of solids.

The properties and usefulness of semiconductors were fully understood when germanium and silicon were ultrapurified to obtain these elements with impurity levels on the order of a few parts per billion by weight. The background required for the physical properties of semiconductors is presented in standard texts[1-3] and monographs[4-9]; it is therefore sufficient to present a brief outline of their physical metallurgy, and then delve into their statistical thermodynamics, and a few interesting uses.

Crystal Structure

The unit cell of a crystalline substance is the smallest volume of a solid that contains all the crystallographic information of a macroscopic crystal as shown in Fig. 7.1(a). The unit cell in this case is a cube, whose edge-length is called the lattice constant. Consecutive repetition of unit cells of a solid in three dimensions generates the macroscopic crystal. The planes most frequently encountered in semiconductor technology and indicated by Miller indices in parentheses are as follows: the (100)-planes are perpendicular to the x axis as in Fig. 7.1(b);.the (111)-planes intersect the x, y, and z axes at equal distances from the origin as in Fig. 7.1(c); the (110)-planes are perpendicular to the x–y plane and intersect the x and y axes at equal distances from the origin as in Fig. 7.1(d). The straight lines perpendicular to these planes are called the crystal directions, and indicated by the same numbers in brackets; e.g., the [100]-direction is perpendicular to (100)-plane

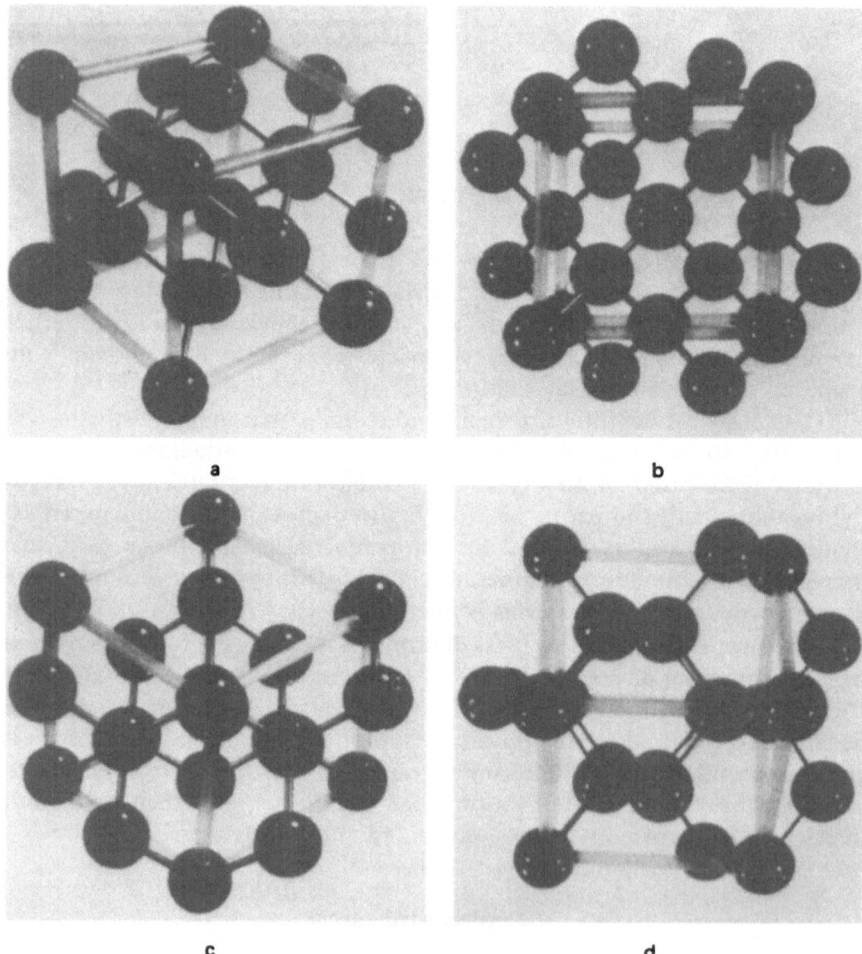

a   b

c   d

Figure 7.1. (a) Diamond lattice structure with outline of unit cell; (b) view of unit cell in [100] direction; (c) view in [111] direction; (d) view in [110] direction. (From Green[7] with permission.)

shown in Fig. 7.1(b). Two other useful directions, [111] and [110], are shown by (c) and (d), respectively. The crystal structure for Ge and Si is called the diamond structure as indicated in Figs. 7.1(a) and 7.2(A). For most of the useful Group III and Group V compounds, such as GaAs, the crystal structure is quite similar, but designated as the zinc blende structure. Figure 7.1(b) indicates that these semiconductors are tetrahedral, i.e., each atom has four closest neighbors and therefore the structure can be represented as shown in Fig. 7.2. Other interesting semiconductors such as CdS and ZnS are in the würtzite structure, and PbS and PbTe are in the rock-salt structure, but we shall not be concerned with the last two compounds in this chapter since we wish to limit ourselves to the structures shown in Figs. 7.1 and 7.2.

Band Structure

Silicon and germanium are the semiconductors to be discussed in greater detail in this chapter. Each atom of these elements has four outer electrons, two in the s-level and two in the p-level in the gaseous state. There can be two states (occupancy) in the s-level and six states in the p-level in each free atom. A possible band structure of the solid, formed

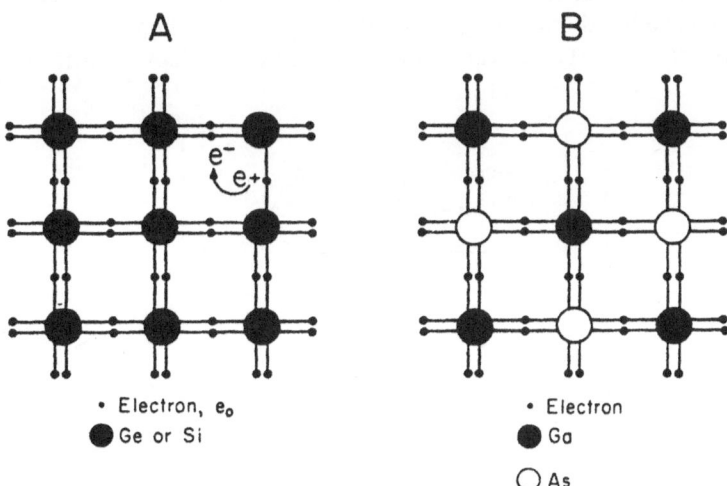

A   B

• Electron, e_0
● Ge or Si

• Electron
● Ga
○ As

Figure 7.2. (A) Structure of germanium and silicon; (B) structure of GaAs; CdS and ZnS are identical with GaAs. Each atom shares four electrons with its neighbors. When an electron, e_0, jumps to the conduction band, it forms a free electron e^- and leaves a hole e^+ in the valence band as shown in (A). Double lines from two neighboring atoms signify one bond of previous chapters.

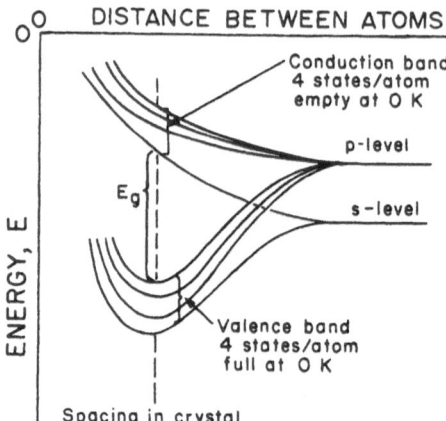

Figure 7.3. Schematic energy levels of Ge and Si as a function of distance between atoms. Band-gap energy, E_g, separates valence and conduction bands.

by condensed atoms, is shown in Fig. 7.3. Half of the *s*- and *p*-levels merge into a lower band filled with four electronic states, completely full at 0 K; the resulting lower band is called the valence band. The remaining four electronic states formed by the remaining four split levels merge to form the conduction band. As the temperature increases beyond 0 K, or photons of proper energy levels are injected into the semiconductor, electrons are excited into the conduction band. The energy E_g required to excite an electron into the conduction band is called the forbidden band-gap energy, or briefly the band-gap energy, and often band-gap when a specific number follows this term. The covalent bond energy of semiconductors such as Si

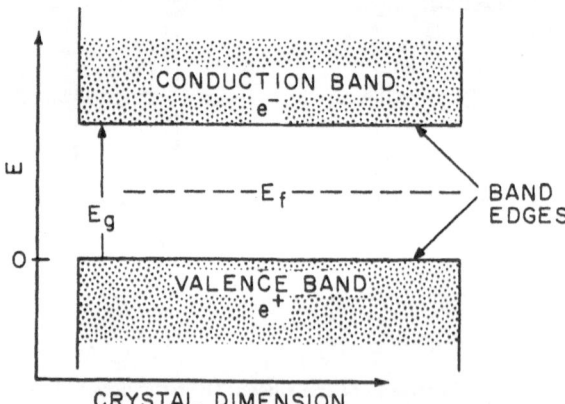

Figure 7.4. Simple band diagram for a semiconductor. Energy is measured from valence band edge; hence, E_g, E_f, and energy of e^- are positive, and energy of e^+ is negative.

and Ge is high; hence, their entropies of fusion are much larger than those of metals since each atom in Fig. 7.2 contributes four electrons and shares four electrons with four neighbors, and thus completes the octet. The structure of most interesting semiconductors is tetrahedral, or the diamond lattice type. Covalent compounds of Group II and Group VI (II-VI), e.g., CdS and ZnS, or III-V, such as GaAs and InSb, also have four electrons per atom as in Si and Ge. For simplicity, the band structure is drawn in block diagrams as shown in Fig. 7.4. The upper band is the conduction band for occupancy by the electrons that carry electric current. The lower band is for the valence electrons, and as will be seen later, the lower band is also for the holes, which also carry electric current. Here the horizontal axis is not important, but it is sometimes designated as the crystal dimension, but the important point is the vertical axis which represents the energy required to excite an electron to the conduction band.

Distribution of Electrons

The energy level of electrons at 0 K is less than the Fermi energy E_f, i.e., $E_- < E_f$ (see Chapter 3). Therefore, all the electrons occupy the valence band at 0 K as shown in Fig. 7.5. Some electrons gain energy in excess of E_f as the temperature increases and occupy the state corresponding to E_-. The probability $\phi_- = \phi_-(E_-)$ of occupation of an energy level E_- is given by

$$\frac{N}{g} \equiv \phi_-(E_-) = \frac{1}{1 + \exp[(E_- - E_f)/kT]} \approx \exp[-(E_- - E_f)/kT] \quad (7.1)$$

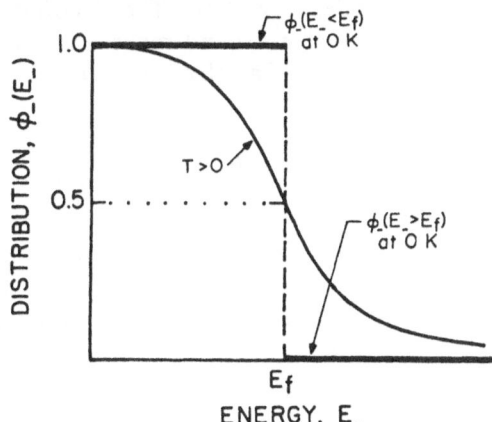

Figure 7.5. Distribution of fermions as a function of energy at 0 K and at $T > 0$ K. At $E_- = E_f$ and $T > 0$, $\phi_-(E_f) = 0.5$. At $T = 0$ K, $\phi_-(E_-) = 1$ for $E_- < E_f$, and $\phi_-(E_-) = 0$ for $E_- > E_f$.

where the equality holds for the general case, and the approximate equality holds for the values of $E_- - E_f$ several times larger than kT so that "1" in the denominator can be neglected. The electrons have a finite probability of occupying the conduction band when their energy E is equal to or greater than the band-gap energy E_g. The states for $0 < E_- < E_g$ are forbidden; therefore, electrons with $E_- < E_g$ have no allowed states and they must return to the valence band. For pure and mildly alloyed (i.e., doped) semiconductors, $E_- - E_f \gg kT$ is always satisfied, and the approximate equality in equation (7.1) is valid; such semiconductors are called nondegenerate semiconductors. Conduction of electricity is due to the electrons in the conduction band and the holes in the valence band. As the temperature increases, the number of electrons and holes, hence the conductivity of semiconductors increases, whereas the conductivity of metals decreases with increasing temperature. When an electron leaves the valence band of a semiconductor to occupy the conduction band, it leaves a positive charge in the valence band that acts in every respect as a positively charged electron, e^+. Attempts to explain the properties of holes by any other scheme than the positively charged particles have not been successful. The electrons and holes are collectively called the carriers since both carry electricity. The band-gap of a semiconductor such as Si is 1.12 eV, whereas the band-gap of an insulator such as diamond is 7 eV. In a metal, such as Na or K, E_f is within one of the allowed bands for the valence electrons, and nearly all the valence electrons move freely as an electron gas and conduct electricity.

Motion of Electrons and Holes

The treatment of the motion of electrons and holes is similar to that of free particles in vacuum. The electrons in the conduction band and the holes in the valence band move according to Newton's law; hence, the kinetic energy E_- of an electron is a quadratic function of its momentum p_-:

$$E_- - E_c = \frac{(p_- - p_c)^2}{2m_-} \equiv E_-(\text{kin}) \qquad (7.2)$$

where E_- is the energy of an electron corresponding to p_-, E_c is the energy at the minimum point p_c in the conduction band, and $E_-(\text{kin})$ is the net kinetic energy. The minimum momentum in the conduction band is p_c, and m_- is called the effective mass of electrons. Equation (7.2) and the corresponding band-gap are illustrated in Fig. 7.6. The energy of holes is measured

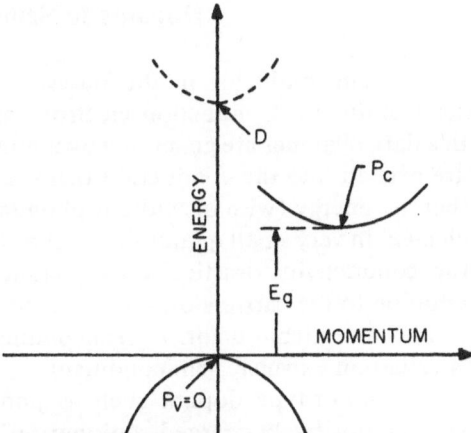

Figure 7.6. Energy–momentum diagram for indirect band-gap semiconductors. For semiconductors such as silicon, a direct band-gap may also exist as between $P_v = 0$ and D.

in the opposite direction with a relationship similar to equation (7.2):

$$E_v - E_+ = \frac{p_+^2}{2m_+} \equiv E_+(\text{kin}) \tag{7.3}$$

where E_v is the energy of the valence band edge, E_+ is the energy of the hole, and m_+, the effective mass of the hole. Here p is measured from the top of the valence band which is taken to be zero. Note that the band-gap energy E_g is given by

$$E_g = E_c - E_v \tag{7.4}$$

Solid curves in Fig. 7.6 are for an indirect band-gap semiconductor, e.g., silicon and germanium, for which p_c is not zero. In a direct band-gap semiconductor such as GaAs, $p_c = 0$, i.e., the extrema of the two curves occur at the same value of p, but again with $E_g = E_c - E_v$. The vertical band-gap or direct band-gap energy for Si or Ge is about twice as high as its indirect band-gap energy E_g, but for direct-gap type semiconductors such as GaAs, vertical band-gap is identical with the usual band-gap energy, E_g. We shall have no particular application for the vertical band-gap energy for indirect gap semiconductors; hence, the band-gap will always refer to the lowest value of E_g which may be direct (vertical) or indirect (nonvertical). Actual momentum–energy diagrams for the possible values of p in different crystal directions are complex and need not be discussed here for our purposes. The electron and hole masses are the average masses that fit the properties under consideration as will be seen later.

Dopants in Semiconductors

A semiconductor in the purest possible state is called an intrinsic semiconductor. Conduction electrons and holes in such a semiconductor in a dark chamber are generated by thermal excitation, i.e., valence electrons are excited into the conduction band, and holes, into the valence band by thermal energy (with or without phonon energy as necessary). An alloying element in very small quantities is called a dopant. Ionizing dopants increase the conductivity drastically by yielding positive or negative carriers in addition to the carriers originating from the excitation of valence electrons from the semiconductor. A semiconductor containing an ionizing dopant is called an extrinsic semiconductor.

A donor-type dopant such as phosphorus in the silicon lattice site yields a positively charged stationary p^+ ion and an electron e^-, as shown in Fig. 7.7. Phosphorus has five $(s + p)$ electrons, and four of these are used in bonding with the neighboring atoms, leaving one electron to move freely in the conduction band and thus increasing the conductivity of silicon. For a given temperature, and for light doping, the product $n_- n_+$, to be discussed later in detail, is nearly the same as that for pure or intrinsic Si, but the total number of carriers $(n_- + n_+)$ increases dramatically by doping with P. For example, $n_- n_+ \approx 10^{20}$, or $n_- = n_+ \approx 10^{10}$ particles/cm^3 for pure Si at room temperature, but Si doped with 10^{16} of P atoms/cm^3 has $n_- \approx 10^{16}$ and $n_+ \approx 10^4$, so that $n_- + n_+$ is increased by a factor of about 500,000 upon doping with P. The silicon doped with Group V elements is called n-doped or n-type silicon and P is called an n-dopant. In an n-doped semiconductor,

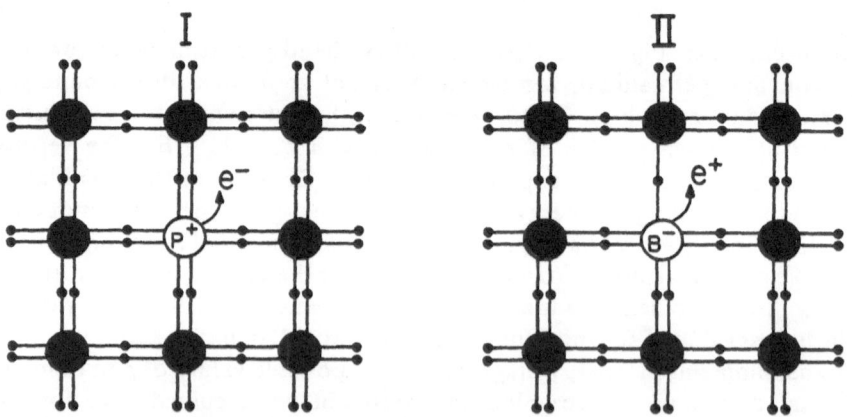

Figure 7.7. (I) Phosphorus as ionized n-dopant; free electron e^- given up by P is in the conduction band. (II) Boron as ionized dopant; free hole e^+ in the valence band is formed upon acquisition of an electron from top of the valence band by B, which then forms B$^-$ ion.

the electrons, e^-, are the majority carriers, and the holes, e^+, are the minority carriers.

Addition of a dopant such as boron from Group III into a substitutional lattice site has an opposite effect on the free electrons. Boron has only three $(s + p)$ electrons, and in bonding with the neighboring silicon atoms, a boron atom attempts to borrow one electron from its neighbors and generate a positive particle free to move in silicon. Thus, the borrowed electron on B leaves a deficiency of one electron in the neighboring Si atoms and this deficiency is the same as generation of a hole in the silicon valence band. A p-dopant is also called an acceptor because it accepts electrons from the semiconductor and thus creates holes in the valence band. Again, for a given temperature and for light doping with B, the product n_-n_+ is nearly the same as that for pure Si but the total number of carriers $(n_- + n_+)$ increases dramatically by doping with B. The silicon doped with Group III elements is called p-doped or p-type silicon and B is called a p-dopant. Some elements may act as either p-, or n-dopant; e.g., Si as dopant in GaAs may act amphoterically, i.e., if Si occupies the Ga lattice site in GaAs, it is an n-dopant, whereas when it occupies the As lattice site, it is a p-dopant. This can be paraphrased by $Si \rightarrow Si^+ + e^-$ when Si is an n-dopant in GaAs and by $Si \rightarrow Si^- + e^+$ when Si is a p-dopant. In both cases, the dopant ionizes to yield an ion and a charge carrier. A Group IV element such as carbon as a dopant may act as a neutral atom in Si. The ionization of a dopant is a function of temperature, but beyond some range of temperature, ionization is very nearly complete as will be discussed later. At 0 K, each dopant captures its carrier and becomes neutral. For example, P^+ captures an electron and B^- captures a hole to become neutral substitutional atoms. The ionized dopant atoms above 0 K cannot move rapidly in a semiconductor to contribute significantly to electrical conductivity.

Energy of Electrons and Holes

The net kinetic energy of an electron was given by equation (7.2). The electrons in the conduction band are free particles, and according to quantum mechanics of free particles, they are subject to the Pauli exclusion principle; this energy is given by

$$E_-(\text{kin}) = \frac{h^2}{8\pi^2 m_-} (3\pi^2 N_-)^{2/3} \qquad (7.5)$$

where h is Planck's constant and N_- is the density of quantum states for electrons, or briefly, the density of states available for occupancy by electrons in numbers per cubic centimeter. This equation is obtained by replacing the terms in the parentheses of equation (3.55) with $(3N/\pi)^{2/3}$, which is very close to $N_-^{2/3}$. A similar equation for the kinetic energy of a hole in the valence band is

$$E_+(\text{kin}) = -\frac{h^2}{8\pi^2 m_+}(3\pi^2 N_+)^{2/3} \tag{7.6}$$

where N_+ is the density of states for holes. The negative sign arises from the fact that the energy of holes is measured downward from the top of the valence band; hence, it is always negative.

Concentration of Electrons

The energy E_- of electrons in the conduction band has an additional term which is the band-gap energy E_g; i.e., for the electrons in the conduction band, E_g must be added to its net kinetic energy because the energy is measured from the valence band, E_v; hence,

$$E_- = E_g + \frac{h^2}{8\pi^2 m_-}(3\pi^2 N_-)^{2/3} \tag{7.7}$$

Differentiation of this equation after writing the left side as $(E_- - E_g)^{1.5}$, and rearrangement of the result yields

$$\frac{dN_-}{dE_-} \equiv \rho(E_-) = \frac{4\pi}{h^3}(2m_-)^{1.5}(E_- - E_g)^{0.5} \tag{7.8}$$

where $\rho(E_-)$ is the density of states per unit energy at E_- so that $dN_- = \rho(E_-)\,dE_-$ is the number of states within a narrow energy dE_-. The concentration of electrons n_- in the conduction band is then the integral of $\phi_- dN_-$, i.e.,

$$n_- = \int \phi_-\, dN_- = \int_{E_g}^{\infty} \rho(E_-)\phi_-\, dE_- = \int_{E_g}^{\infty} \frac{4\pi}{h^3}(2m_-)^{1.5}$$
$$\times (E_- - E_g)^{0.5}\exp[-(E_- - E_f)/kT]\,dE \tag{7.9}$$

Note that at energies $E_- < E_g$, the electrons do not have enough energy to hop in the conduction band. The integration is carried out by substituting

$$-(E_- - E_f) \equiv -(E_- - E_g) + (E_f - E_g)$$

and observing that $(E_f - E_g)$ is constant for a given temperature, and further, $dE_- = d(E_- - E_g)$. Substitution of these relationships in equation (7.9) yields

$$n_- = \frac{4\pi}{h^3}(2m_-)^{1.5} \exp\left(\frac{E_f - E_g}{kT}\right) \int_{E_- = E_g}^{\infty} \exp\left(\frac{E_g - E_-}{kT}\right)$$
$$\times (E_- - E_g)^{0.5} d(E_- - E_g) \qquad (7.10)$$

The integrand is of the standard form $[\exp(-\alpha x)]x^{0.5}dx$, whose integral in mathematical tables is given as $(\pi/\alpha)^{0.5}/2\alpha$; consequently, equation (7.9) becomes

$$n_- = \frac{2}{h^3}(2\pi m_- kT)^{1.5} \exp[(E_f - E_g)/kT] \qquad (7.11)$$

where the density of state N_- is the coefficient of the exponential term, i.e.,

$$N_- = \frac{2}{h^3}(2\pi m_- kT)^{1.5} \qquad (7.12)$$

and N_- is equal to q/V of equation (3.58).

Concentration of Holes

The Fermi distribution function for holes ϕ_+ is symmetric to that for electrons and symmetric with respect to E_f in energy so that $\phi_+ = \phi_+(E_+)$ is given by

$$\phi_+ = \frac{1}{1 + \exp[(E_f - E_+)/kT]} \approx \exp(E_+ - E_f)/kT \qquad (7.13)$$

Here E_+ is measured from the top of the valence band; hence, the numerical value of E_+ is negative, and the approximate equality is valid when $E_f - E_+ \gg kT$. The energy of a hole is the same as its kinetic energy because holes need not overcome a barrier such as the band-gap energy to acquire additional energy. The density of states ρ_+ within a narrow energy difference $d(E_+)$ is obtained by differentiating equation (7.6); the result is

$$\rho_+ = \frac{dN_+}{dE_+} = \frac{4\pi}{h^3}(2m_+)^{1.5}(-E_+)^{0.5} \qquad (7.14)$$

The density of holes, n_+, is obtained by integration as in equation (7.9), i.e.,

$$n_+ = \int_{-\infty}^{0} \rho_+ \phi_+ dE_+ = \frac{4\pi}{h^3}(2m_+)^{1.5} \int_{-\infty}^{0} (-E_+)^{0.5} \exp[(E_+ - E_f)/kT]\, dE_+$$

$$= \frac{2}{h^3}(2\pi m_+ kT)^{1.5} \exp(-E_f/kT) \tag{7.15}$$

where the integral is from $-\infty$ to 0 because the permissible energies of holes are measured from the top of the valence band in the negative direction. The product term before the exponential term is called the density of states for holes, i.e.,

$$N_+ = \frac{2}{h^3}(2\pi m_+ kT)^{1.5} \tag{7.16}$$

This expression is analogous to equation (7.12) for N_-.

In an intrinsic semiconductor n_+ is equal to n_-; hence, the equality of equations (7.11) and (7.15) leads to

$$1.5kT \ln \frac{m_+}{m_-} + E_g = 2E_f$$

At moderate temperatures, e.g., $T = 500$ K, with $k = 8.617 \times 10^{-5}$ eV/particle-K, and $m_+/m_- \approx 1.2$ for GaAs so that the first term on the left is about 0.012 eV, and since E_g for GaAs is about 1.35 eV, it is evident that

$$E_f \approx E_g/2 \tag{7.17}$$

For other semiconductors, the correction from the logarithmic term might be an order of magnitude higher, but the preceding approximate equality is still valid.

Equilibrium Constant for Charge Carriers

The charge carriers from an intrinsic semiconductor obey an equilibrium relationship similar to the ionization of water. The reaction for the formation of the charge carriers is

$$e_0 \text{ (in valence band)} = e^- + e^+ \tag{7.18}$$

where e_0 is the electron bound in the valence band that generates e^- (i.e., electron) in the conduction band and e^+ (i.e., hole) in the valence band,

and e_0 acts in the same way as a 1-1 electrolyte in water. The electron e_0 will always be referred to as the bound electron to avoid confusion with e^-.

The band-gap of a semiconductor can be measured by the injection of photons whose energy is gradually increased toward the band-gap of the semiconductor. No change in the conductivity of semiconductors is detectable when the photon energy is less than the band-gap energy, but when the photon energy is equal to or greater than the band-gap energy, the conductivity of semiconductors sharply increases due to the generation of charge carriers. The number of quanta of photons may be kept to a minimum during the measurements so that the concentrations of e^- and e^+ at thermal equilibrium are not significantly disturbed, and further, in principle, reaction (7.18) can generate a photon or corresponding energy upon the reversal of reaction (7.18) so that equilibrium can be assumed. The energy thus involved is therefore directly related to the formation of the reaction products from one reactant e_0; hence, $\Delta G = E_g$. Another convincing argument that E_g is a chemical potential or a Gibbs energy per particle is presented by Thurmond,[10] and paraphrased by the following equation:

$$E_g = \left(\frac{\partial h\nu N_0^\circ}{\partial n_-} \right)_{S,V,x} = h\nu = \Delta \bar{G} \qquad (7.19)*$$

where N_0° is the number of photon quanta and n_- is the number of electrons generated at constant entropy (S), volume (V), and composition of e^-, denoted by x; $\Delta \bar{G}$ is therefore a chemical potential per particle.[11] Since each photon among N_0° generates one electron, e^-, it is evident that $\partial N_0^\circ / \partial n_-$ is unity. Equation (7.19) is correct because in the limit, the photon energy is absorbed adiabatically without changing the volume and the carrier concentration, and hence E_g must represent a Gibbs energy per particle. The equality of E_g to the standard Gibbs energy change, ΔG°, however, can be proven rigorously by considering the equilibrium in reaction (7.18) as will be seen later.

The usual derivation of the equilibrium constant is to equate the sum of the molar Gibbs energies of one electron and one hole, $G_- + G_+$, to the molar Gibbs energy of one electron in the valence band, $G(e_0)$. We shall return to this procedure later, but first show that the required relationship can be obtained very simply by eliminating E_f from equations (7.11) and (7.15); the result is

$$\ln \frac{n_-}{N_-} \cdot \frac{n_+}{N_+} = -\frac{E_g}{kT} \equiv -\frac{\Delta G^\circ}{kT} \qquad (7.20)$$

where ΔG° is the standard Gibbs energy for reaction (7.18).

*Precise measurement of band-gap energies may involve phonons, which will not be discussed here.

A rigorous treatment of the equilibrium for reaction leads to new interesting ramifications.[12] This procedure requires rewriting the density of states in terms of the partition functions as follows [see equation (3.60)]:

$$N_- = 2(2\pi m_- kT/h^2)^{1.5} \equiv q_-/V \qquad (7.21)$$

$$N_+ = 2(2\pi m_+ kT/h^2)^{1.5} \equiv q_+/V \qquad (7.22)$$

where q_- and q_+ are the partition functions, and V is the molar volume of the semiconductor in cubic centimeters. The concentrations of electrons and holes may be expressed by n_-/N_- and n_+/N_+ which are dimensionless properties similar to mole fractions.[10] The particle Gibbs energies G_- and G_+ as functions of these concentrations are

$$G_- = G_-^\circ + kT \ln \frac{n_-}{N_-} \qquad (7.23)$$

$$G_+ = G_+^\circ + kT \ln \frac{n_+}{N_+} \qquad (7.24)$$

where G_-° and G_+° are the standard Gibbs energies on the Henrian scale for dilute solutions.[11] It should be noted that n/N for electrons and holes are considerably smaller than unity for intrinsic and lightly doped semiconductors so that Henry's law is obeyed, or the activity coefficients are unity. This treatment is not concerned with heavily doped semiconductors for which the activity coefficients differ from unity, and modified statistical thermodynamic treatments become necessary. The standard state is the hypothetically existing state for which n/N is unity and $G = G^\circ$ for $-$ and $+$ particles. It is important to stress here that in these hypothetical standard states, both charge carriers are free particles, or boltzons, obeying classical statistics because e^- is above the Fermi level and e^+ is below the valence band edge.

It is necessary to assume here that the bound electrons participating in reaction (7.18) are those right on the valence band edge, and their Gibbs energy is $G_0 = G_0^* + kT \ln x_0$ where G_0^* is a standard Gibbs energy and x_0 is a very small fraction of all valence electrons. If x_0 is of the order of 10^{-4}, then a small fraction, such as $10^{-5}x_0$, ending up in the conduction band in an intrinsic semiconductor cannot change the value of x_0 significantly. Therefore, the term $kT \ln x_0$ can be added to G_0^* to write $G_0 = G_0^\circ$ for the electrons participating in reaction (7.18). The equilibrium in reaction (7.18) requires that

$$G_0^\circ = G_- + G_+ \qquad (7.25)$$

Substitution of equations (7.23) and (7.24) in equation (7.25) yields

$$G_-^\circ + G_+^\circ - G_0^\circ \equiv \Delta G^\circ = -kT \ln \frac{n_- n_+}{N_- N_+} \tag{7.26}$$

Equations (7.11) and (7.15) can be used to obtain the logarithmic term in equation (7.26), eliminate E_f, and thus show that

$$\Delta G^\circ = E_g \tag{7.27}$$

This equation confirms that $\Delta \bar{G}$ of equation (7.19) is the same as ΔG° of equation (7.26) for reaction (7.18). Consequently, equilibrium in reaction (7.18) is necessary to prove equation (7.27).

The terms G_-° and G_+° in equations (7.23) and (7.24), or briefly $G_\pm^\circ$, refer to $G_\pm = G_\pm^\circ$ when $n_\pm / N_\pm = 1$; hence, their partition functions for the standard state are given by equations (7.21) and (7.22), so that $G_\pm^\circ = -kT \ln(q_\pm / N_0)$ per particle according to equation (3.53), and

$$G_\pm^\circ - G_\pm^\circ(\text{at } 0 \text{ K}) = -kT \ln \frac{q_\pm}{N_0} \tag{7.28}$$

where $\pm$ is for "either $+$ or $-$" but not both, N_0 is Avogadro's number, and $G_\pm^\circ(\text{at } 0 \text{ K})$ is zero as T approaches zero in $-T \ln(q_\pm / N_0)$. Substitution of appropriate values in $q_\pm$ yields

$$G_-^\circ + G_+^\circ - G_-^\circ(0 \text{ K}) - G_+^\circ(0 \text{ K})$$
$$= -2kT(74.753 + 1.5 \ln\langle m \rangle + 1.5 \ln T + \ln V) \tag{7.29}$$

where $\langle m \rangle = (m_- m_+)^{0.5}$ is the geometric mean of the carrier masses, which usually increases very slowly with temperature. The carrier masses used here are called the density of state carrier masses. It will be assumed here that the temperature dependence of $\langle m \rangle$ is negligible, in reasonable agreement with the summary presented by Thurmond.[10]

The electrons in the valence band move in a highly restricted volume bound by the atomic core on one side and the electron cloud of the neighboring atoms on the other side. The partition function of this restricted cloud of particles is difficult to formulate but it is proposed here that

$$q_v = 2V_v(2\pi m_0 kT / h^2)^{1.5}; \qquad (V_v = \lambda V) \tag{7.30}$$

where V_v is the restricted volume available for bound electrons very close to the valence band edge, which is assumed to be a very small fraction λ of the semiconductor volume V. Further, V_v may contain other corrections such as electron spin effects, and those due to the splitting of bands in the semiconductor, and interactions of electrons within each band. It was shown in Chapter 3 that the contributions of rotation of a bound electron about an atomic core and its vibration with the core are quite negligible because the moment of inertia of an electron and the distance between the electron and the shielded proton are both very small. Thus, λV consists of very thin peripheral ribbonlike volumes joined to each other in such a way that the motion of bound electrons is very much limited and may involve occasional exchange of positions with the neighboring peripheral volumes. The corresponding Gibbs energy for a valence electron from equation (7.30) is

$$G_0^\circ - G_0^\circ(\text{at } 0 \text{ K}) = -kT(74.753 + 1.5 \ln m_0 + 1.5 \ln T + \ln \lambda V) \quad (7.31)$$

Subtraction of this equation from equation (7.29) as required by the left side of equation (7.26) and then substitution of the numerical value of m_0 as the electron rest mass yields

$$\Delta G^\circ - \Delta G^\circ(0 \text{ K}) \equiv E_g - E_g(0 \text{ K})$$

$$= kT\left(18.642 - 3 \ln \frac{\langle m \rangle}{m_0} + \ln \frac{\lambda}{V} - 1.5 \ln T\right) \quad (7.32)$$

It is well to remember that ΔG° refers to reaction (7.18) at constant pressure and constant volume, which is generally the condition for the reactions that take place in solids at ambient pressures.

The fraction of volume occupied by the electrons near the top of the valence band is difficult to estimate. For silicon, it is tentatively estimated here that $\lambda = 1.5 \times 10^{-4}$, which is not unreasonable since e_0 is highly constrained. Substitution of this value as well as $\langle m \rangle / m_0 = 1.2$ from the summary of data for silicon by Thurmond[10] and $V = 12.05 \text{ cm}^3/\text{mole}$ yields

$$E_g - E_g(0 \text{ K}) \equiv \Delta E_g = kT(6.801 - 1.5 \ln T)$$

$$= 5.860 \times 10^{-4} T - 1.293 \times 10^{-4} T \ln T \quad (7.33)$$

where E_g is in electron volts and $\Delta C^\circ = 1.5k$—the coefficient of $-T \ln T$—is the change of heat capacities for reaction (7.18). The values of ΔE_g calculated from this equation are within 6% of the values from a smooth curve representing the experimental values ranging from 200 to 420 K for silicon. Precise values are not yet available above 420 K.

Equation (7.33) is based on only one parameter, λ, and it is quite likely that the errors involved in $\langle m \rangle$, m_0, and their temperature dependence as well as V, V_v, and their temperature dependence, could introduce corrections for the terms in equations (7.32) and (7.33). Therefore, it is assumed that the coefficients in equation (7.33) are adjustable within reason if the accuracy of band-gap measurements are indeed as accurate as correctly stated by Thurmond[10]; consequently,

$$\Delta G^\circ - \Delta G^\circ(0 \text{ K}) = \Delta E_g = AT - BT \ln T \tag{7.34}$$

where A and B are adjustable parameters within the magnitudes prescribed in equation (7.33). The success of this correlation with the best-fit curve for the experimental data for each semiconductor will now be considered.

Correlation of Band-Gap Data

The data obtained by various investigators on the band-gap energies of Si, Ge, GeAs, and GaP, and summarized by Thurmond, will be used to illustrate the remarkable utility of equation (7.34). Two or more sets of independent investigations for the same semiconductor agree so well that a curve passing through or between the points on a plot of temperature versus ΔE_g represents the data remarkably well for each semiconductor. The set of data for GaP has slightly larger deviations above 800 K. Other sets of data for diamond, SiC, InAs, InP, and so on are not sufficiently accurate to test equation (7.34). The data for Si and GaP are plotted in Fig. 7.8.

The experimental data for silicon and germanium were obtained from near 0 K to approximately 420 K and the analytical correlation presented in this chapter is expected to be valid up to possibly 500 K. For Si, the resulting analytical equation, based on equation (7.34), is

$$\Delta E_g(\text{Si}) = E_g - 1.1700 \text{ eV} = 4.546 \times 10^{-4} T - 1.060 \times 10^{-4} T \ln T;$$

$$(\Delta C^\circ = 1.230k) \tag{7.35}$$

This equation is represented in Fig. 7.8. Likewise for Ge, the corresponding equation is

$$\Delta E_g(\text{Ge}) = E_g - 0.7437 \text{ eV} = 4.218 \times 10^{-4} T - 1.205 \times 10^{-4} T \ln T;$$

$$(\Delta C^\circ = 1.398k) \tag{7.36}$$

The coefficients of $-T \ln T$ as $\Delta C°$ are presented in parentheses in equations (7.35) and (7.36). Both equations have been obtained by using the smooth curve based on experimental data and selecting the values of ΔE_g at 200 and 400 K.

The equations for ΔE_g of GaAs and of GaP are as follows:

$$\Delta E_g(\text{GaAs}) = E_g - 1.519 \text{ eV} = 3.086 \times 10^{-4}T - 1.099 \times 10^{-4}T \ln T;$$

$$(\Delta C° = 1.275k) \tag{7.37}$$

$$\Delta E_g(\text{GaP}) = E_g - 2.338 \text{ eV} = 6.318 \times 10^{-4}T - 1.537 \times 10^{-4}T \ln T;$$

$$(\Delta C° = 1.784k) \tag{7.38}$$

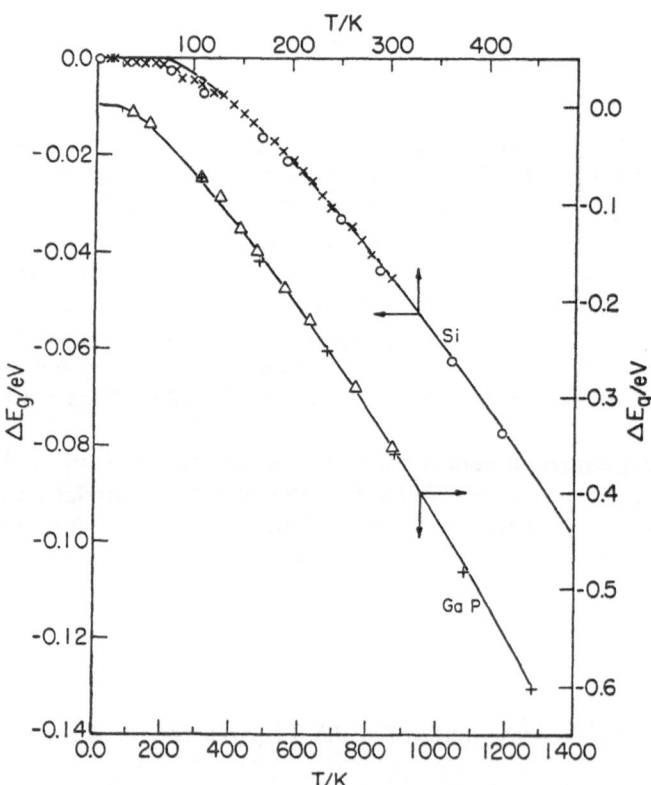

Figure 7.8. Variation of band-gap energy with temperature. Points are experimental data from various sources, evaluated and presented by Thurmond[10]; curves are from equation (7.35) for Si, and equation (7.38) for GaP.

Equation (7.38) is plotted in Fig. 7.8. Equations (7.37) and (7.38) have been obtained by substituting the experimental best-fit data at 300 and 800 K from a smooth curve. The experimental results for GaAs were obtained from 0 to 1000 K, and for GaP, from 0 to 1273 K. Both sets of data are perfectly represented by equations (7.37) and (7.38). Accurate representations of data are important in deriving the thermodynamic properties obtained by differentiation of the equations for ΔE_g as functions of temperature. However, the form of the selected equation is also very important.[12]

Further justification for equation (7.34) will be dealt with in the remaining sections. It is now significant to point out that below approximately 73 K, ΔE_g becomes zero for Si according to equation (7.35), in agreement with experimental data, and similarly, ΔE_g is zero for Ge, GaAs, and GaP below 33, 17, and 61 K, respectively. The data for Si particularly support this conclusion as shown in Fig. 7.8. We call these points "characteristic temperatures" for temperature independence of band-gap energies. This is reminiscent of the disappearance of temperature dependence of a number of other physical properties in solids. The lower temperature limit of physical validity of equation (7.34) is therefore when the experimental values of ΔE_g become independent of temperature. Nevertheless, equation (7.34) contributes very little to ΔE_g as 0 K is approached because T appears as a coefficient in both terms.

Derived Thermodynamic Properties

Differentiation of equation (7.34) with respect to temperature leads to

$$-\frac{\partial \Delta G°}{\partial T} = \Delta S° = B - A + B \ln T \qquad (7.39)$$

where $\Delta S°$ is the standard entropy change for reaction (7.18). The results for the foregoing semiconductors are

$$\Delta S°(\text{Si}) = -3.486 \times 1.060 \times 10^{-4} + 1.060 \times 10^{-4} \ln T \qquad (7.40)$$

$$\Delta S°(\text{Ge}) = -3.013 \times 10^{-4} + 1.205 \times 10^{-4} \ln T \qquad (7.41)$$

$$\Delta S°(\text{GaAs}) = -1.987 \times 10^{-4} + 1.099 \times 10^{-4} \ln T \qquad (7.42)$$

$$\Delta S°(\text{GaP}) = -4.781 \times 10^{-4} + 1.537 \times 10^{-4} \ln T \qquad (7.43)$$

It is evident that as the temperature decreases below $\exp[(A - B)/B]$, e.g., below 27 K for Si and below 12 K for Ge, the numerical value of ΔS° changes from positive to negative. Further, ΔS° approaches $-\infty$ as T approaches zero in accord with thermodynamic computation for free electrons[13] and the fact that the standard states must be chosen as dictated by the partition functions for boltzons in equations (7.21) and (7.22).

The values of ΔC° for the standard heat capacities are given in equations (7.35)-(7.38), and they indicate that ΔC° is not greatly different from $1.5k$ as required for the free particles or boltzons at constant volume.

The equation for the standard enthalpy change, ΔH°, can be readily obtained from the definitional equation $\Delta G^\circ = \Delta H^\circ - T\Delta S^\circ$. The result from equations (7.34) and (7.39) is

$$\Delta H^\circ - \Delta H^\circ(0\ \text{K}) \equiv \Delta\Delta H^\circ = BT \equiv \Delta C^\circ T \qquad (7.44)$$

where $\Delta H^\circ(0\ \text{K})$ is the same as $\Delta G^\circ(0\ \text{K})$ or $E_g(0\ \text{K})$ since $T\Delta S^\circ$ is zero at 0 K, and the identity sign $\equiv$ defines $\Delta\Delta H$.

Equation (7.34) represents the statistical thermodynamic interpretation of the variation of E_g with temperature. This representation can be reconciled with other interpretations of temperature dependence of E_g, based on various models and mechanistic interactions,[10,14,15] but not the derived properties such as $\Delta\Delta H^\circ$, ΔS°, and ΔC°. The main thesis of equation (7.34) is its basis on sound statistical thermodynamic arguments, and its remarkable success in representing the experimental data. Additional justifications for equation (7.34) are given in Ref. 12.

It is feasible to estimate closely the unknown values of the band-gap energy E_g of a semiconductor at various temperatures from a single value of E_g at any temperature by using equation (7.33). Further, one more reliable value of E_g at another temperature would permit the calculation of A in equation (7.34) while retaining $B = 1.5k$. The adjustment required for the refinement of B necessitates at least one more well-spaced value of E_g.

Ionization Equilibria

It was stated earlier that a dopant atom ionizes to yield a free or moving carrier and a stationary charged ion. The energy levels of both the donors and acceptors lie within the band-gap of semiconductors. The energy required to obtain an electron is measured from the donor level to the conduction level as a positive quantity, and this energy is a small fraction of the band-gap, usually about 0.1 eV or less, as shown in Fig. 7.9. The energy required to obtain a hole is measured from the valence band to the

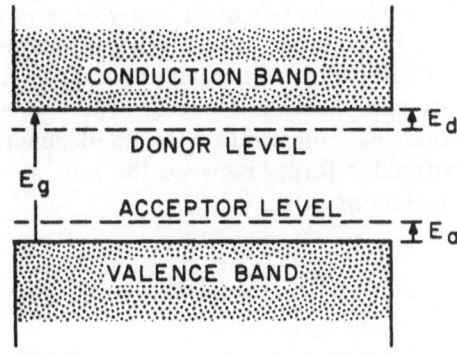

Figure 7.9. Band-gap of a semiconductor with donor or acceptor levels. Each level may have closely spaced levels but for simplicity they are generally shown as single levels. Donors ionize to yield electrons in the conduction band and acceptors take up an electron from the top of the valence band to leave a hole in the valence band. These levels are shallow, i.e., much closer to band edges than indicated here.

acceptor level as shown in Fig. 7.9 and it is usually about 0.1 eV or less. Actually, each of these levels has closely spaced levels, but for simplicity it is usually shown as a single level.

The ionization equilibria, as typified by the following reactions for phosphorus and boron in silicon, are as follows:

$$P^\circ \to P^+ + e^- \tag{7.45}$$

$$B^\circ \to B^- + e^+ \tag{7.46}$$

where the neutral phosphorus atom P° is partially ionized to yield P^+ and an electron in the conduction band whereas the neutral boron atom B° takes on an electron from the valence band of silicon and thus leaves a hole (e^+) also in the valence band. The corresponding equilibrium constants are related to the donor level E_d for P and for the acceptor level E_a for B, in the same way as equation (7.26), i.e.,

$$\Delta G_d^\circ = E_d = -kT \ln\left(\frac{n_-}{N_-}\right)\left(\frac{D_P^+}{D_P^\circ}\right)\gamma_- \tag{7.47}$$

$$\Delta G_a^\circ = E_a = -kT \ln\left(\frac{n_+}{N_+}\right)\left(\frac{D_B^-}{D_B^\circ}\right)\gamma_+ \tag{7.48}$$

where D_P^+/D_P° is the ratio of the concentrations of ionized and nonionized P, both in the same concentration units, and γ_- is the product of the activity coefficients of n_-, D_P^+, and D_P°. Equation (7.48) and its terms are similar to equation (7.47). As the dopant concentration decreases, γ_- or γ_+ approaches unity as expected. The dopant concentration is often known from chemical or other analyses, and the carrier concentration (e^+ or e^-) can be determined by electrical methods, e.g., conductivity measurements. At sufficiently high temperatures, all dopants are ionized and at 0 K, they are completely nonionized. The values of $\gamma_\pm$ are usually less than one at ambient temperatures for nondegenerate semiconductors, indicating that attractive forces between the ions and the charge carriers cause negative deviations in $\gamma_\pm$.[16]

Fermi Energy in Doped Semiconductors

It was shown by equation (7.17) that the Fermi level (Fermi energy) in an intrinsic semiconductor lies in the close neighborhood of $E_g/2$. In n-doped semiconductors, the Fermi level varies with both doping level and temperature according to equation (7.11) with (7.12) so that

$$E_f - E_g = kT \ln \frac{n_-}{N_-} \qquad (7.49)$$

Increasing the n-doping level, hence increasing n_-, increases n_-/N_- toward unity and E_f increases toward E_g. Further, as temperature decreases, the right side of equation (7.49) decreases because n_-/N_- is less than unity for reasonable degrees of doping and $\ln(n_-/N_-)$ varies slower than T; therefore, decreasing temperature also increases E_f toward E_g.

An increase in the p-doping level has the reverse effect as shown by using equation (7.15), i.e.,

$$E_f = -kT \ln \frac{n_+}{N_+} \qquad (7.50)$$

As n_+ increases, E_f decreases, and as temperature decreases, E_f tends to zero, i.e., E_f tends to coincide with the valence band.

$p-n$ Junctions

A junction of p- and n-type semiconductors is called a $p-n$ junction. A silicon wafer doped with boron during the crystal growth is a p-type semiconductor, and if a layer of phosphorus is diffused on one of its surfaces, a homogeneous $p-n$ junction is formed. Cadmium sulfide is an n-type semiconductor, and if Cu_2S, a p-type semiconductor, is deposited on CdS, it forms a heterogeneous junction.

Two separate blocks of p- and n-type semiconductors are shown in Fig. 7.10(A). The difference between the Fermi levels of these semiconductors is called the contact potential or Helmholtz energy, which is usually 1 volt or less. The Fermi level of an n-doped semiconductor is higher than that of a p-type semiconductor as shown by comparison of equations (7.49) and (7.50), such as n-type Si and p-type Si; therefore, the electrons would flow from the n-type to the p-type when these two blocks are joined together. The electrons will continue to flow until the Fermi levels become the same as shown in Fig. 7.10(B). As a result, the band edge of the n-side is depressed with respect to the p-side as indicated in Fig. 7.10(B). A perfect junction is made when the interface region is very thin, usually on the order of 1 μm.

The migration of electrons from the n-side to the p-side leaves an excess of positively charged ions on the n-side and creates an excess of negatively charged ions on the p-side as shown in Fig. 7.10(B), and this

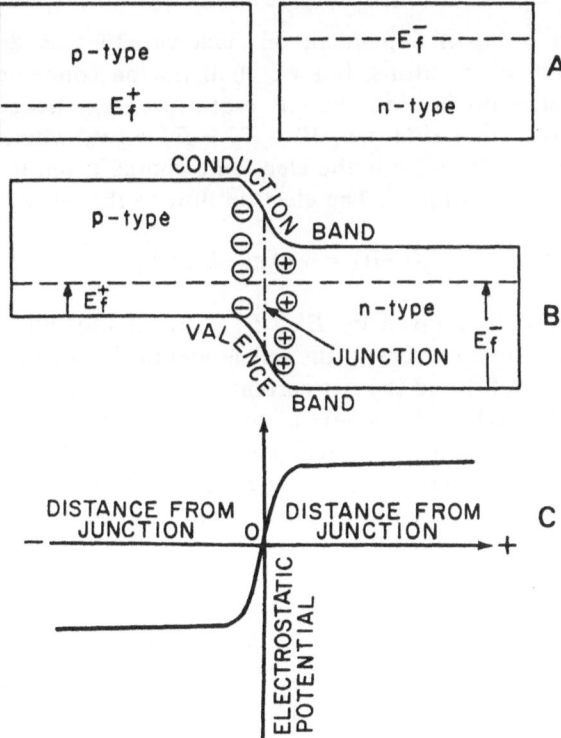

Figure 7.10. (A) Fermi levels of n- and p-type semiconductors. (B) Fermi levels coincide after junction formation; conduction and valence band are bent as shown. Near the junction, negative ions $\ominus$ and positive ions $\oplus$ prevent excessive migration of carriers by establishing the electrostatic potential shown in (C).

condition creates a built-in electrostatic potential that prevents further migration of electrons to the left side. The electric potential of a $p-n$ junction is shown in Fig. 7.10(C).

Effects of Applied Current on $p-n$ Junctions

The net current flowing through an unbiased (bias = applied voltage) $p-n$ junction is evidently zero because the Fermi energy is equal on both sides of the junction. This can be analyzed as follows: The electron on the p-side needs to jump from its Fermi level to the valence band by gaining $E = E_g^+ - E_f^+$ as shown in Fig. 7.11, and roll to the right so that the electron flow to the right, $J(\text{right})$, is

$$J(\text{right}) = \alpha \exp[-(E_g^+ - E_f^+)/kT]; \qquad (E = E_g^+ - E_f^+) \qquad (7.51)$$

where α is a constant dependent on such variables as diffusivity, and diffusion length of electrons, but we shall not be concerned with these details. The electron flow to the left requires energy acquisition by the electrons on the other side, i.e., $E' = E_g^- - E_f^- + eV_{bi}$ where V_{bi} is called the built-in potential and e is the electronic charge taken to be a positive quantity to avoid writing $|e|$. The electron flow to the left is

$$J(\text{left}) = \alpha \exp(-E'/kT) \qquad (7.52)$$

The energy $E' = E$ is given by $E' = E_g^+ - E_f^+$ as indicated in Fig. 7.11; hence, the electronic current to the left is identical with that to the right, i.e., $J(\text{left}) = J(\text{right})$ and the net current is zero as expected.

The Fermi level of the n-side is depressed with respect to the p-side when an external current is applied as shown in Fig. 7.12(A). In this case, the $p-n$ junction is reverse-biased and E_f^- is depressed with respect to E_f^+ by eV and where e is taken to be positive and the numerical value of V is negative.

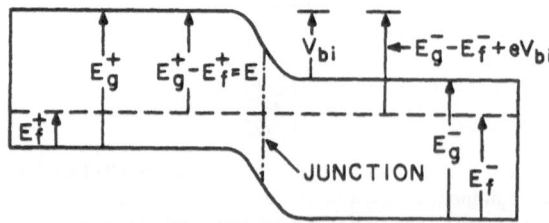

Figure 7.11. Energy levels of electrons and built-in potential V_{bi} across an unbiased junction.

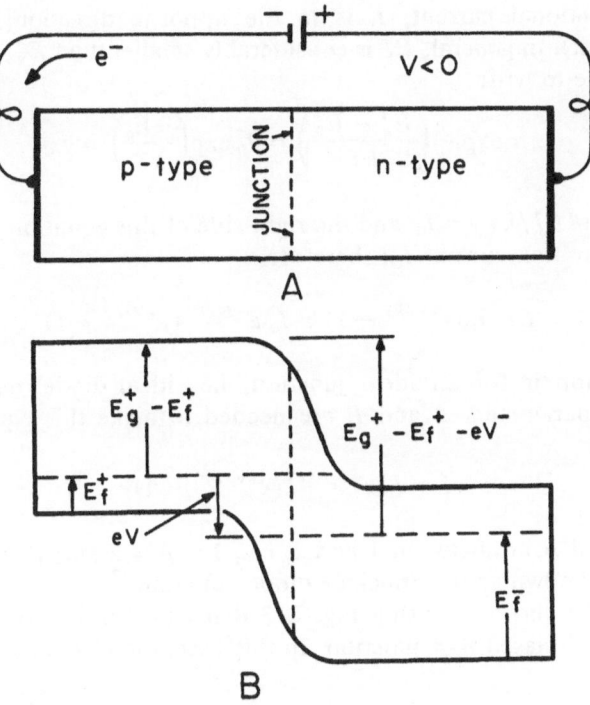

Figure 7.12. Energies in a reverse-biased p–n junction. Applied voltage, V, is considered to be negative in this configuration; eV is the applied energy.

The electronic current flowing to the right, $J(\text{right})$, is again given by equation (7.51):

$$J(\text{right}) = \alpha \, \exp[-(E_g^+ - E_f^+)/kT] \qquad (7.53)$$

where the electrons gain enough energy over E_f^+ to move to the right. The electrons on the n-side need to gain $(E_g^+ - E_f^+ - eV)$ above their Fermi level E_f^- to move to the left as indicated in Fig. 7.12(B). The electronic current flowing to the left is therefore

$$J(\text{left}) = \alpha \, \exp[-(E_g^+ - E_f^+ - eV)/kT] \qquad (7.54)$$

The net electronic current to the right side of the p–n junction is obtained by subtracting equation (7.54) from equation (7.53); thus,

$$J(\text{net, right}) = -\alpha \, \exp[-(E_g^+ - E_f^+)/kT][\exp(eV/kT) - 1] \quad (7.55)$$

The conventional current, I, is in the opposite direction; hence, $I = -J(\text{net, right})$. In general, E_f^+ is considerably smaller than E_g^+, and further, it is possible to write

$$\alpha \exp -\left(\frac{E_g^+ - E_f^+}{kT}\right) = I_0 \exp\left(\frac{-E_g}{kT}\right) \equiv I_{rs} \qquad (7.56)$$

where $\alpha \exp(E_f^+/kT) = I_0$, and the right side of this equation is called the reverse saturation current I_{rs}; therefore,

$$I = I_{rs}(e^{eV/kT} - 1) = I_0\, e^{-E_g/kT}(e^{eV/kT} - 1) \qquad (7.57)$$

This equation is for an ideal junction, i.e., ideal diode; however, two adjustable parameters A and B are needed to make this equation more general, i.e.,

$$I = I_0\, e^{-E_g/AkT}(e^{eV/BkT} - 1) \qquad (7.58)$$

where A and B are between 1 and 2, i.e., $1 \leq A \leq 2$ and $1 \leq B \leq 2$. This equation is known as the Shockley diode equation.[17]

It can be shown by using Fig. 7.13 that equation (7.55) is also valid for a forward-biased p–n junction. In this case, the electrons on the right

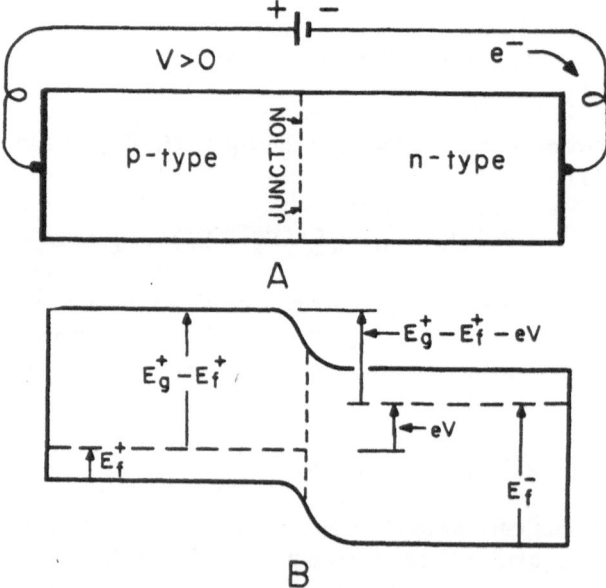

Figure 7.13. Forward-biased p–n junction.

side must overcome the energy from the Fermi level E_f^- on the right to the conduction band level of the left side, and this energy is equal to $E_g^+ - E_f^+ - eV$ because E_f^- for the right side is now raised by eV.

The coefficient I_0 in equation (7.58) is largely a property of electrons; hence, it may be assigned[18] a value of 6.03×10^9 milliamp/cm^2, and for a simple ideal diode, i.e., $A = B = 1$, equation (7.58) becomes

$$I \text{ (in mA/cm}^2) = 6.03 \times 10^9 \, e^{-E_g/kT}(e^{eV/kT} - 1);$$

$$(I_{rs} = 6.03 \times 10^9 \, e^{-E_g/kT}) \qquad (7.59)$$

(The assumption that $A = B = 1$ is used up to page 235.) For example, I_{rs} at 300 K for a semiconductor having $E_g = 1.5$ eV is

$$I_{rs} = 6.03 \times 10^9 \exp(-1.5 \times 1.60219 \times 10^{-19}/1.38062 \times 10^{-23} \times 300)$$

$$= 3.808 \times 10^{-16} \text{ mA/cm}^2 \qquad (7.60)$$

where the conversion factor of $1 \text{ eV} = 1.60219 \times 10^{-19}$ J, and $k = 1.38062 \times 10^{-23}$ J/K both have been used on a per particle basis. The term

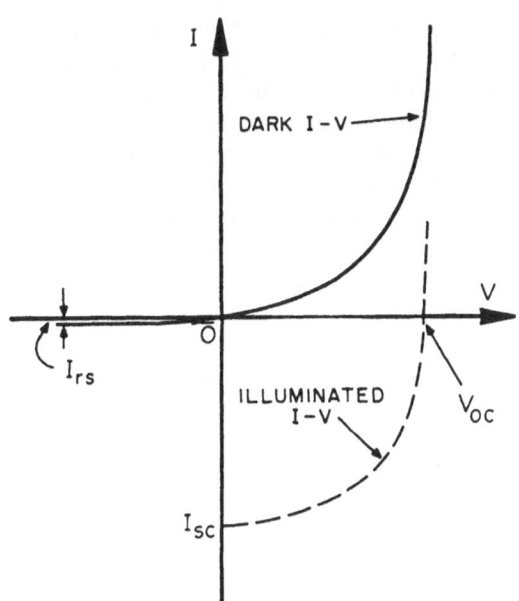

Figure 7.14. Current-voltage characteristics of a p-n junction. Solid curve is for I-V of a dark (nonilluminated) junction; dashed curve is for I-V of an illuminated junction.

$\exp(eV/kT) - 1$ quickly approaches -1 for reverse bias of approximately -0.2 to voltages often in the neighborhood of -50 volts. Thus, at $V = -0.2$ volt,

$$\exp(-0.2 \times 1.60219 \times 10^{-19}/1.38062 \times 10^{-23} \times 300) - 1 = -0.9996$$

hence, the negative value of reverse saturation current of 3.808×10^{-16} mA/cm^2 is quickly attained. For voltages in the range of -0.2 to -50 volts, the reverse current is the same. For a forward bias of $+1$ volt, the term $\exp(eV/kT) - 1 = 6.315 \times 10^{16}$; and $I = 24.03$ mA/cm^2, which is dramatically larger than $I = -3.808 \times 10^{-16}$ mA/cm^2 for -1 volt of reverse bias. As the value of E_g decreases, I_{rs} increases and this has important ramifications for solar cells to be discussed later. Thus, for $E_g = 0.6$ eV, $I_{rs} = 0.502$ mA/cm^2 which is considerably larger than that for $E_g = 1.5$ eV. The $I\text{-}V$ characteristic plot of equation (7.57) is shown by the solid curve in Fig. 7.14. Since the reverse current is very small for $E_g \geq 0.6$, a $p\text{-}n$ junction is ideally suited for current rectification, switching, and other remarkably useful purposes.

Solar Cells

The solar cells are $p\text{-}n$ junction devices that convert solar energy directly into electricity by the photovoltaic effect (PVE). This occurs when the energy of photons is equal to or higher than the band-gap of the semiconductor. Such photons are said to be ionizing radiation. Pure single-crystal semiconductors are transparent to photons with energies smaller than the band-gap energies. The ionizing radiation generates holes and electrons in excess of those corresponding to the thermal equilibrium. The excess electrons on the p-side in Fig. 7.10 are attracted by the positive ions on the n-side of the junction and the excess holes generated on the n-side are attracted by the negative ions on the p-side of the function and thus the charges are separated at the $p\text{-}n$ junction where a dipole layer (or space-charge region) exists.

The charge carriers thus separated are the minority carriers, and their separation increases the overall concentration of holes on the p-side and electrons on the n-side. Consequently, the p-type semiconductor becomes a source of positive electricity, and the n-type semiconductor, a source of electrons. A shorted solar cell yields a current that is called the short-circuit current, I_{sc}, but when a cell has no external load, it generates a voltage called the open-circuit voltage V_{oc}. The maximum power obtainable from a cell is the product of maximum power current I_{mp} and maximum power voltage V_{mp}, which are situated at the knee of the I-V curve shown in Fig. 7.14. Note that this curve is in the fourth quadrant because the external current flows in the same way as in the reverse-biased junction in Fig. 7.12 while the measured potential is opposite in direction. The potential difference between the two ends of the space-charge region sets an upper limit to the magnitude of voltage produced by photons. This difference is determined by the difference in the Fermi levels of p- and n-semiconductors and, hence, by the doping levels. The potential difference is high when the doping levels are high and the junction profile is sharp in concentration gradient. The lifetimes of the photogenerated carriers should be sufficiently long so that they can be collected by the conductors attached to the p- and n-sides.

A diagram of a typical silicon solar cell and a photograph of a commercial cell are shown in Figs. 7.15 and 7.16. The top layer, exposed to the solar radiation, has collector bars attached to the surface by various techniques for collecting the electrons. The surface area covered by the bars is usually of the order of 3% of the top surface area. The bottom layer can be made to have collector bars for tandem or stacked-up cells, but for single cells, it is usually covered or soldered with a metallic layer. A space-qualified silicon solar cell of 2×2 cm can generate about 55 mW at approximately 0.55 volt, and 100 mA at air mass zero (AM0) radiation outside the atmospheric region of the earth. Various aspects of solar cells, their manufacture and uses are discussed well in a number of monographs,[5-9] and in detail in the *IEEE Photovoltaic Specialists Conference Records*.[19] The remaining sections of this chapter are largely based on a paper by Gokcen and Loferski.[18]

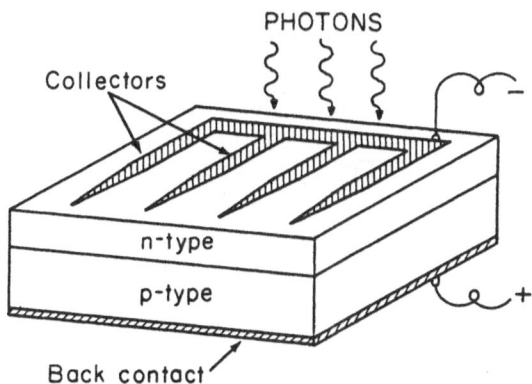

Figure 7.15. Diagram of a solar cell.

Figure 7.16. Silicon solar cell. (Courtesy of Solarex Corporation.)

Simple Equivalent Circuit

The I-V characteristics of a solar cell in the fourth quadrant are generally represented in the first quadrant for the convenience of photovoltaic specialists as shown in Fig. 7.17. The line amperage I_1 of a cell with a resistive load R is given by

$$I_1 = I_{sc} - I_{rs}(e^{\Lambda V} - 1); \qquad (\Lambda = e/kT = 1{,}1604.5/T) \qquad (7.61)$$

where Λ will be used hereafter for e/kT for brevity. The derivation of this equation is based on the fact that the I-V curve of a nonilluminated junction in Fig. 7.14 is displaced downward by I_{sc}, but otherwise the curve maintains its shape. When $I_1 = 0$, $V = V_{oc}$, and I_{sc} is then correlated with V_{oc} from

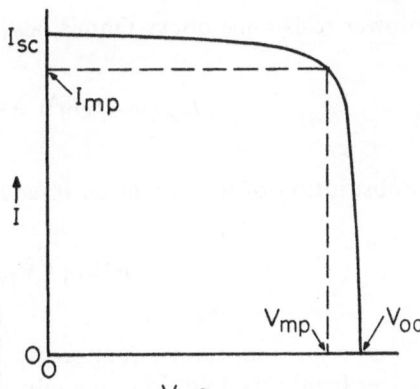

Figure 7.17. Current-voltage characteristics of a solar cell. Maximum power is $I_{mp}V_{mp}$, which is located at the knee of the curve. $I_{mp}V_{mp}/I_{sc}V_{oc}$ is called the fill factor. I_{sc} here is $-I_{sc}$ of Fig. 7.14.

equation (7.61) by

$$e^{\Lambda V_{oc}} = \frac{I_{sc}}{I_{rs}} + 1 \tag{7.62}$$

The maximum power Π_{mp} is delivered to the external load when R_{mp} matches what is known as the dynamic impedance R_D of the cell, given by $\partial V/\partial I$, i.e.,

$$R_D = -\frac{\partial V}{\partial I} = \frac{1}{I_{rs}\Lambda \, e^{\Lambda V}} \tag{7.63}$$

which is obtained from equation (7.61) by differentiation. The value of R_{mp} is adjusted to be the same as R_D for which V in equation (7.63) becomes the maximum power voltage V_{mp}, i.e.,

$$R_{mp} = \frac{1}{I_{rs}\Lambda \, e^{\Lambda V_{mp}}} \tag{7.64}$$

The equivalent circuit of such a cell is shown in Fig. 7.18. The maximum

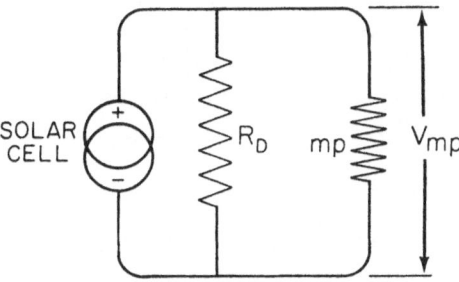

Figure 7.18. Simple equivalent circuit of a solar cell. Dynamic impedance of cell is R_D. For maximum power voltage V_{mp}, resistance at mp should match R_D.

power resistance obeys Ohm's law, i.e.,

$$I_{mp} \, (\mathrm{mA/cm^2}) = \frac{V_{mp}}{R_{mp}} = V_{mp} I_{rs} \Lambda \, e^{\Lambda V_{mp}} \qquad (7.65)$$

Substitution of this equation in equation (7.61) with $I_l = I_{mp}$ yields

$$e^{\Lambda V_{mp}} (\Lambda V_{mp} + 1) = \frac{I_{sc}}{I_{rs}} + 1 \qquad (7.66)$$

This is an important equation for our subsequent purposes. Solution of $e^{\Lambda V_{mp}}$ from this equation and substitution in equation (7.65) yields

$$I_{mp} = \frac{\Lambda V_{mp}}{\Lambda V_{mp} + 1} (I_{sc} + I_{rs}) \qquad (7.67)$$

The maximum power Π_{mp} is I_{mp} from this equation multiplied by V_{mp} in volts; the result is

$$\Pi_{mp} \, (\mathrm{mW/cm^2}) = I_{mp} V_{mp} = \frac{\Lambda V_{mp}^2}{\Lambda V_{mp} + 1} (I_{sc} + I_{rs}) \qquad (7.68)$$

The ratio $\Pi_{mp}/(I_{sc} V_{oc})$ is called the fill factor. It is now appropriate to proceed to tandem or stacked solar cells.

Tandem Solar Cells

The efficiency of photovoltaic solar energy conversion can be increased substantially by using a tandem solar cell system[18,20-22] which is defined as an arrangement of solar cells based on semiconductors having different band-gaps with the largest band-gap cell exposed directly to the sun and acting as a transparent window for the photons of smaller energy than its band-gap; the next cell of lower band-gap in the sequence acts as a transparent window for the succeeding cell of still lower band-gap, and so on. The band-gaps, E_g (cell i), are therefore arranged as follows:

$$\mathrm{Sun} \rightarrow E_g(\mathrm{cell}\ 1) > E_g(\mathrm{cell}\ 2) > E_g(\mathrm{cell}\ 3) > \cdots \qquad (7.69)$$

In order to simplify the notation in this chapter, the band-gap of the cell being considered is denoted by E_g, and that of the cell acting as a filter, by E. It is evident that for a selected cell, E_g is fixed and E may assume various values in excess of E_g.

The reason for the efficiency increase provided by tandem solar cell systems lies in the fact that they bring about a closer match between the maximum energy photons $h\nu_{max}$ absorbed in a given cell and the energy gap E_g of the absorbing semiconductor. The excess energy $h\nu - E_g$ of an absorbed photon degenerates into heat. However, as the number of cells increases, the photon flux available for absorption in any one semiconductor in the stack decreases which leads to a lowering of photovoltaic conversion efficiency of each cell. The total photon flux, incident on the tandem cell stack, can be increased by recourse to concentrating mirrors and lenses, thus circumventing this last-named efficiency reduction. It becomes evident, therefore, that the tandem cell concept is particularly compatible with solar concentrator systems. This is potentially of great importance in view of studies of large-scale solar energy conversion systems which show that the combination of high levels of concentration with high-efficiency cells is one possible way to achieve economically competitive photovoltaic solar cell systems.[20]

The next section explores the theoretical limits on solar energy conversion efficiencies at various temperatures, various concentration ratios, and various numbers of cells in the tandem cell stack. Four temperatures ($T =$ 200, 300, 400, and 500 K), four concentration ratios ($C = 1$, 100, 500, and 1000 suns), and cell numbers up to 24 are considered. The energy gaps of the photovoltaically active semiconductors which result in optimum efficiencies are also presented.

Calculation Procedure

Solar Spectrum and Short-Circuit Current

The calculation for photovoltaic power generation is carried out for the air mass zero (AM0) solar spectrum as described in Ref. 22, and summarized here. The solar radiation is very close to that from the black body radiation for 5750 K. Table 7.1 shows the integrated AM0 photon flux $N_{ph}(E)$ per cm^2 per second for energies between $h\nu = \infty$ and $h\nu = E$, obtained from

$$N_{ph}(E) = \int_{h\nu = E}^{\infty} n_{ph}(h\nu)\,d(h\nu) \tag{7.70}$$

where $n_{ph}(h\nu)$ is the photon flux (per cm^2) per $h\nu$. Table 7.1 also includes the short-circuit current which would flow in a cell having an absorption cutoff at $h\nu = E$ if the collection efficiency were unity, i.e., if every photon

Table 7.1. Limiting Short-Circuit Current as a Function of Energy Gap E_g for Air Mass Zero (AM0) Solar Radiation[a]

$E = h\nu$	λ (μm)	$10^{-14}N_{ph}(E)$	$I_{sc}(E)$ (mA/cm²)	Solar energy (mW/cm²)
7.000	0.1772	0	0	0
5.000	0.2480	2.9	0.046	0.25
4.000	0.3100	31.8	0.510	2.24
3.900	0.3179	41.1	0.569	2.84
3.800	0.3263	53.4	0.856	3.59
3.700	0.3351	68.9	1.104	4.51
3.600	0.3444	85.9	1.377	5.51
3.500	0.3542	104.7	1.678	6.57
3.400	0.3647	125.2	2.006	7.71
3.300	0.3757	149.3	2.392	8.99
3.200	0.3875	174.8	2.800	10.30
3.100	0.3999	203.9	3.267	11.79
3.000	0.4133	249.2	3.993	14.00
2.900	0.4275	301.7	4.833	16.46
2.800	0.4428	358.5	5.743	19.09
2.700	0.4592	433.5	6.945	21.37
2.600	0.4769	518.6	8.309	25.99
2.500	0.4959	611.7	9.800	29.78
2.400	0.5166	711.8	11.404	33.71
2.300	0.5391	818.9	13.120	37.84
2.200	0.5636	935.5	14.987	42.08
2.100	0.5904	1068.3	17.116	46.66
2.000	0.6199	1223.8	19.606	51.53
1.900	0.6526	1382.7	22.152	56.61
1.800	0.6888	1568.6	25.134	61.87
1.700	0.7293	1757.3	28.154	67.32
1.600	0.7757	1966.9	31.512	72.94
1.500	0.8273	2198.8	35.227	78.65
1.400	0.8862	2445.0	39.171	84.43
1.300	0.9539	2715.9	43.512	90.34
1.200	1.0353	3020.5	48.391	96.35
1.100	1.1289	3335.6	53.439	102.23
1.000	1.2418	3674.9	58.875	107.89
0.900	1.3789	4011.4	64.367	113.34
0.800	1.5514	4413.5	70.709	118.60
0.700	1.7724	4817.7	77.184	123.42
0.600	2.0675	5175.4	82.915	127.14
0.500	2.4803	5513.2	88.327	130.17
0.400	3.0996	5854.2	93.790	132.64
0.300	4.1333	6122.0	98.080	134.14
0.200	6.2249	6325.6	101.343	134.95
0.100	12.4177	6426.9	102.966	135.25
0.000	0.0000	6476.6	103.762	135.30

[a] The photon flux $N_{ph}(E)$, defined as the number of photons per cm² per second having energy greater than E, is taken to be zero at $E = 7$ eV. Short-circuit current $I_{sc}(E) = eN_{ph}(E)$. The column labeled "Solar energy" is the solar energy density delivered by photons having energies in excess of $E = h\nu$.

having an energy $h\nu > E$ contributed one pair of minority carriers to the short-circuit current. Thus,

$$I_{sc}(E) = eN_{ph}(E) \qquad (7.71)$$

where e is the charge on the electron in coulombs. For example, Table 7.1 shows that for a cell in which the AM0 sunlight is incident on a photovoltaically active semiconductor with an energy gap $E_g = 3.00$ eV, the limiting value of I_{sc} is 3.993 mA/cm^2.

If a cell based on a photovoltaically active semiconductor having an energy gap E_g has interposed between itself and the solar source a filter which cuts off all photons with energy greater than E, where $E > E_g$, then the limiting short-circuit current which this cell can produce is obtained from the relation

$$I_{sc}(E_g, E) = I_{sc}(E_g) - I_{sc}(E); \qquad \text{(for cell with } E_g\text{)} \qquad (7.72)$$

For example, again referring to Table 7.1, if $E = 3.40$ eV, and $E_g = 3.00$ eV, then $I_{sc}(3.00 \text{ eV}, 3.40 \text{ eV}) = (3.993 - 2.006) \text{ mA/cm}^2 = 1.987 \text{ mA/cm}^2$. For the required calculations, $I_{sc}(E_g, E)$ will be used in equations (7.66) and (7.68) instead of I_{sc}, and E_g will be used in the relationship for I_{sc} in equation (7.59) to calculate I_{rs}. The solar energy conversion efficiency η of a single cell is

$$\eta = \Pi_{mp}/135.3C \qquad (7.73)$$

where 135.3 mW/cm^2 is the AM0 solar spectrum intensity, which is the last entry in Table 7.1, and C is the concentration ratio expressed by the number of suns.

Calculation for an example with $E_g = 1.5$ at 300 K for a single cell and $C = 1$ requires first I_{rs} from equation (7.59), which was found earlier to be $I_{rs} = 3.808 \times 10^{-16} \text{ mA/cm}^2$. The value of $I_{sc} = 35.227 \text{ mA/cm}^2$ is listed in Table 7.1 for $E = E_g = 1.5$ eV; therefore, $I_{sc}/I_{rs} = 35.227/3.808 \times 10^{-16} = 9.2508 \times 10^{16}$; also, $\Lambda = e/kT = 11,604.9/300 = 38.6829$, and substitution of these values in the logarithm of equation (7.66) yields

$$38.6829 V_{mp} + \ln(38.6829 V_{mp} + 1) = (38.0661 + 1) \qquad (7.74)$$

This equation can be solved by successive approximation, but to make the required steps shorter, we set $V_{mp} = 1.00$ for the second term on the left in order that $\ln(\Lambda V + 1) = 3.6809$, and solve for V_{mp} of the first term to obtain $V_{mp} = 0.9147$. This procedure can be repeated to obtain $V_{mp} = 0.9170$, and one more repetition yields $V_{mp} = 0.9169$ which is close enough for computation with equation (7.68). The result from this equation yields $\Pi_{mp} = 31.414 \text{ mW/cm}^2$, and then efficiency η

Table 7.2. Efficiency (%) of Tandem Solar Cells[a]

E_g (eV)

E (eV)	0.60	0.70	0.80	0.90	1.00	1.10	1.20	1.30	1.40	1.50	1.60	1.70	1.80	1.90	2.00	2.10	2.20	2.30	2.40	2.50	2.60	2.70	2.80	2.90	3.00	3.20	3.40	3.60	3.80	4.00	6.90
0.60	0.00																														
0.70	0.10	0.00																													
0.80	0.33	0.48	0.00																												
0.90	0.60	1.10	0.89	0.00																											
1.00	0.86	1.66	1.75	1.09	0.00																										
1.10	1.14	2.25	2.66	2.31	1.47	0.00																									
1.20	1.41	2.82	3.54	3.50	2.96	1.71	0.00																								
1.30	1.68	3.38	4.41	4.67	4.44	3.48	1.99	0.00																							
1.40	1.93	3.89	5.19	5.73	5.78	5.10	3.87	2.07	0.00																						
1.50	2.17	4.36	5.91	6.59	7.01	6.59	5.62	4.05	2.16	0.00																					
1.60	2.39	4.81	6.60	7.64	8.19	8.01	7.28	5.95	4.28	2.31	0.00																				
1.70	2.60	5.22	7.23	8.48	9.26	9.30	8.79	7.69	6.23	4.45	2.29	0.00																			
1.80	2.79	5.60	7.80	9.25	10.22	10.47	10.16	9.26	8.01	6.42	4.46	2.31	0.00																		
1.90	2.97	5.97	8.36	10.01	11.18	11.63	11.53	10.82	9.77	8.38	6.61	4.62	2.47	0.00																	
2.00	3.14	6.29	8.84	10.66	12.01	12.63	12.69	12.16	11.28	10.06	8.46	6.64	4.65	2.29	0.00																
2.10	3.30	6.60	9.32	11.30	12.81	13.60	13.84	13.48	12.76	11.71	10.28	8.63	6.80	4.59	2.42	0.00															
2.20	3.44	6.87	9.73	11.85	13.51	14.44	14.82	14.60	14.04	13.13	11.85	10.33	8.65	6.58	4.54	2.21	0.00														
2.30	3.56	7.11	10.08	12.33	14.12	15.17	15.68	15.59	15.16	14.38	13.22	11.84	10.28	8.33	6.41	4.20	2.07	0.00													
2.40	3.67	7.33	10.41	12.78	14.68	15.85	16.48	16.51	16.21	15.53	14.49	13.22	11.78	9.95	8.15	6.05	4.02	2.03	0.00												
2.50	3.78	7.54	10.72	13.19	15.20	16.49	17.22	17.36	17.15	16.61	15.68	14.52	13.19	11.47	9.77	7.78	5.86	3.96	2.01	0.00											
2.60	3.88	7.73	11.01	13.58	15.69	17.08	17.92	18.16	18.05	17.61	16.78	15.73	14.50	12.88	11.29	9.40	7.58	5.78	3.92	1.98	0.00										
2.70	3.97	7.90	11.28	13.94	16.14	17.62	18.55	18.89	18.87	18.53	17.80	16.83	15.70	14.18	12.68	10.88	9.15	7.44	5.67	3.82	1.90	0.00									
2.80	4.05	8.06	11.51	14.25	16.53	18.10	19.11	19.53	19.60	19.34	18.69	17.81	16.76	15.32	13.90	12.19	10.54	8.92	7.23	5.44	3.61	1.76	0.00								
2.90	4.11	8.18	11.69	14.49	16.83	18.46	19.54	20.02	20.15	19.95	19.37	18.55	17.57	16.19	14.83	13.18	11.60	10.04	8.41	6.69	4.91	3.12	1.40	0.00							
3.00	4.17	8.29	11.85	14.71	17.11	18.79	19.93	20.47	20.66	20.52	19.99	19.24	18.31	16.99	15.69	14.10	12.58	11.07	9.50	7.84	6.12	4.38	2.71	1.35	0.00						
3.20	4.25	8.44	12.09	15.02	17.50	19.27	20.49	21.11	21.38	21.32	20.88	20.21	19.36	18.13	16.92	15.41	13.96	12.54	11.06	9.48	7.83	6.18	4.58	3.30	2.01	0.00					
3.40	4.30	8.54	12.24	15.23	17.77	19.59	20.86	21.54	21.87	21.86	21.47	20.86	20.07	18.89	17.73	16.28	14.89	13.52	12.09	10.57	8.98	7.38	5.83	4.60	3.37	1.45	0.00				
3.60	4.35	8.63	12.36	15.40	17.97	19.84	21.15	21.87	22.25	22.29	21.94	21.37	20.63	19.49	18.38	16.97	15.62	14.30	12.91	11.43	9.89	8.33	6.83	5.64	4.45	2.61	1.24	0.00			
3.80	4.38	8.69	12.47	15.53	18.15	20.05	21.40	22.15	22.56	22.64	22.33	21.80	21.09	19.99	18.91	17.54	16.23	14.95	13.59	12.15	10.64	9.12	7.65	6.50	5.34	3.58	2.27	1.10	0.00		
4.00	4.40	8.74	12.53	15.62	18.26	20.18	21.56	22.34	22.77	22.87	22.59	22.08	21.39	20.32	19.27	17.92	16.63	15.37	14.05	12.63	11.14	9.64	8.20	7.08	5.94	4.22	2.97	1.84	0.78	0.00	
6.90	4.44	8.81	12.63	15.76	18.43	20.39	21.80	22.61	23.08	23.22	22.97	22.50	21.85	20.81	19.79	18.48	17.23	16.01	14.71	13.33	11.88	10.42	9.01	7.92	6.82	5.17	3.99	2.93	1.94	1.23	0.00

[a] $T = 300$ K; $A = 1$, $B = 1$; solar concentration $= C = 1$.

$$\eta = (31.414/135.3)100\% = 23.22\%$$

This value is listed in Table 7.2 at the bottom of the column for $E_g = 1.5$. The same value is also listed in Table 7.3 for a single cell. Table 7.2 has been obtained by the foregoing procedure with I_{sc} of equation (7.72) calculated from Table 7.1. Each column after the first column gives the efficiency of each cell for the impinging energy E on the first column. For example, the efficiency of a cell with $E_g = 1.5$, having a filter of $E \geqslant 2$, is 10.06 as this filter would absorb radiation greater than 2.0 eV. A trial-and-error method for two tandem cells on a computer would yield $E = 2.0$, and $E_g = 1.2$ for the highest efficiency, E being the band-gap of the semiconductor which acts as a filter for the cell with $E_g = 1.2$, the latter receiving radiation equal to or smaller than 2.0 eV. The efficiency for the filter cell is 19.79 from Table 7.2; moving horizontally to the left from the point $E_g = 2.0$ at the top of the column (efficiency 0.000), and reaching $E_g = 1.2$, we obtain 12.69; therefore, the total efficiency of two tandem cells is 19.79 + 12.69 = 32.48, as listed in Table 7.3. Any other combination of cells would yield a lower efficiency.

Table 7.3 shows how sensitive the overall efficiency is to variations in the selected band-gaps and the number of cells. The contribution to the efficiency is dramatic up to six cells, after which it levels off as shown in Table 7.3 and represented in Fig. 7.19.

Increasing the operating temperature of the cells decreases their efficiencies as shown in Fig. 7.19. The contribution to the efficiency from the smaller E_g semiconductors decreases more rapidly with increasing temperature than for larger E_g semiconductors, because the reverse saturation current, I_{rs},

Table 7.3. Variation of Optimum Efficiency with Numbers of Cells[a]

No. of cells	1	2	3	3	4	5
E_g(eV)	1.5	2.0, 1.2	2.3, 1.6, 1.0	2.4, 1.6, 1.0	2.5, 1.8, 1.3, 0.9	2.6, 2.0, 1.6, 1.2, 0.8
Efficiency	23.22	32.48	37.42	37.39	40.45	42.45

No. of cells	6	6	6	7
E_g(eV)	2.7, 2.3, 1.9, 1.5, 1.1, 0.8	2.8, 2.4, 2.0, 1.6, 1.2, 0.8	2.6, 2.2, 1.8, 1.4, 1.0, 0.8	2.9, 2.5, 2.1, 1.8, 1.5, 1.1, 0.8
Efficiency	43.82	43.67	43.64	44.86

No. of cells	7	7	13	26
E_g(eV)	3.0, 2.6, 2.2, 1.8, 1.4, 1.0, 0.8	2.8, 2.4, 2.1, 1.8, 1.5, 1.2, 0.8	0.8 to 3.0 in intervals of 0.2, 3.4	0.7 to 3.0 in intervals of 0.1, 3.2, 3.4
Efficiency	44.70	44.67	48.13	50.09

[a]Band-gap range: 0.7 to 3.4; $A = B = 1$; 300 K; 1 sun.

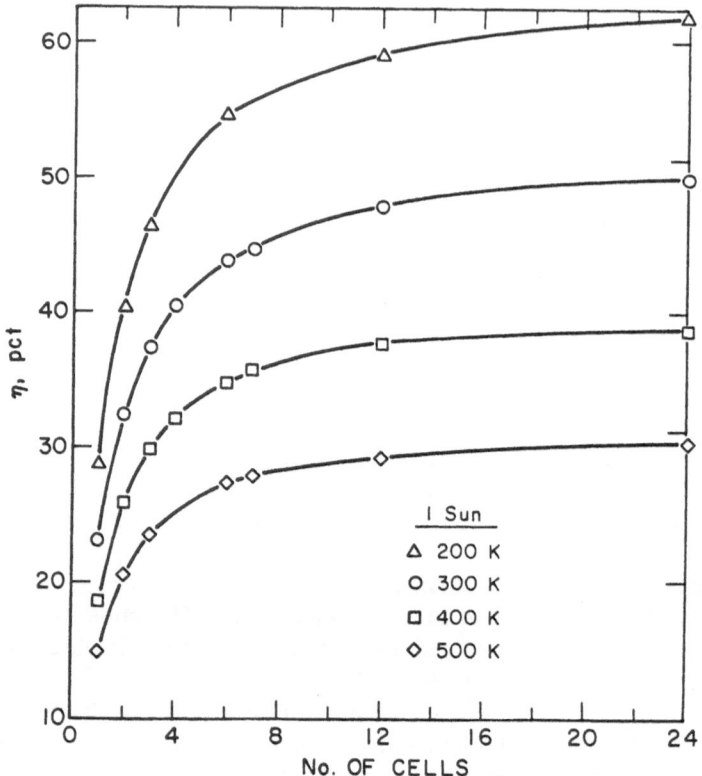

Figure 7.19. Variation of efficiency, η (%), with number of cells at various temperatures. Solar intensity is one sun.

increases more rapidly for smaller values of E_g. Therefore, the range of band-gaps useful for operating at the higher temperatures is shifted to higher E_g values. Thus, for 1 sun, the useful band-gap range is 0.7 to 3.4 eV at 300 K, 0.9 to 3.4 eV at 400 K, and 1.1 to 3.6 eV at 500 K. Figures 7.20 and 7.21 show the effects of 500-fold ($C = 500$ suns) and 1000-fold solar concentration ($C = 1000$), respectively, on the efficiency. Comparison with Fig. 7.19 is striking because not only the efficiency is higher, but also a correspondingly smaller cell area is required for the same amount of power generation, and further, the solar concentrators are very much cheaper than solar cells.

Additional conclusions may be drawn from Fig. 7.22 where the effects of temperature and numbers of cells (as 6 and 24 cells) on the efficiency are illustrated. For large concentrations ($C \geq 1000$) and large numbers of cells, e.g., 24 cells, the efficiency approaches 70% at 200 K, a temperature

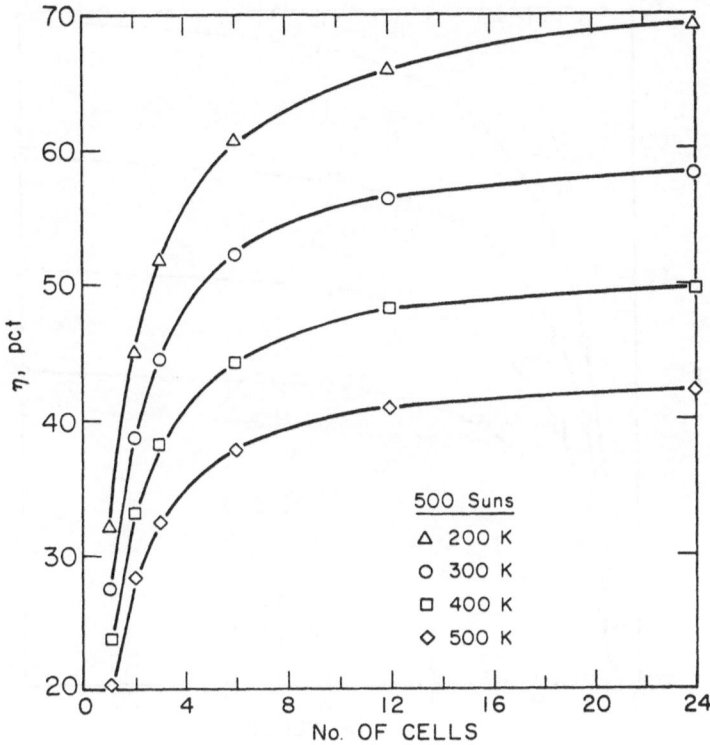

Figure 7.20. Variation of efficiency with number of cells at various temperatures. Solar intensity is 500 suns.

achievable for solar cells on satellites, space laboratories, and space stations. The efficiency of a Carnot engine operating between 5750 K and 200 K is $[(5750 - 200)/5750]100\% = 96.5\%$, indicating that tandem solar cells are not far from ideally achievable efficiencies. At solar concentrations greater than 1000 suns and fewer than a dozen tandem cells, the minority carrier concentration becomes comparable to the majority carrier concentration and equation (7.58) and the derived relations are no longer valid.

The foregoing results are for AM0 solar radiation. If the AM1 or AM2 spectrum had been used (AM1 being the radiation when the sun is vertically above the horizontal cells at sea level), the results would have been similar in the sense that high solar concentration and low cell temperature would lead to efficiencies approaching 70%. However, the range of useful band-gaps on the high band-gap side would have been cut off at smaller values of E_g and the combination of E_g values required for optimized efficiency would be different from that appropriate to the AM0 spectrum, but the

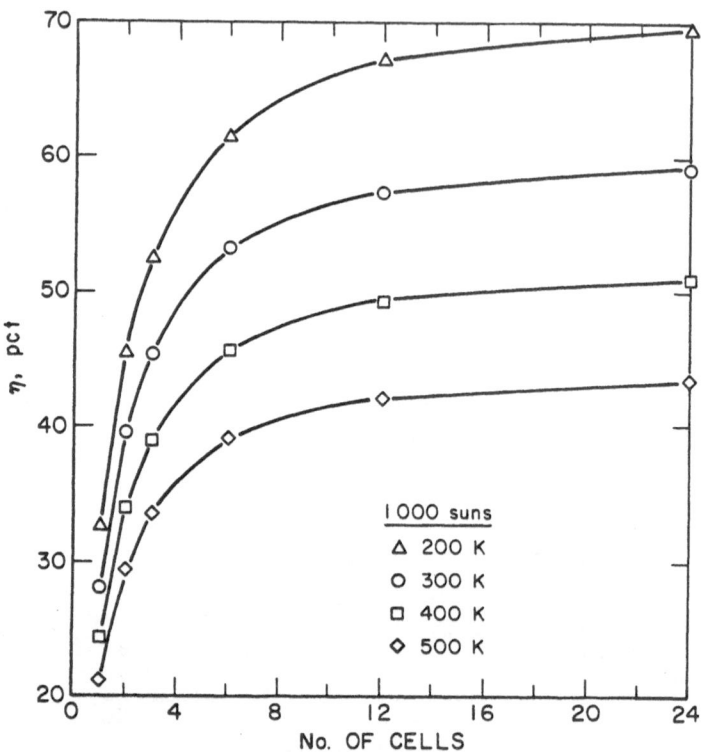

Figure 7.21. Variation of efficiency with number of cells at various temperatures. Solar intensity is 1000 suns.

Table 7.4. Variation of Efficiency with Number of Cells, Temperature, and Solar Concentration, $A = B = 2$

	No. of cells				
	1	2	3	6	12
1 sun					
200 K	25.3	35.4	40.6	47.6	51.4
300 K	19.2	26.7	30.6	35.3	37.9
400 K	14.4	19.7	22.7	25.7	27.3
100 suns					
200 K	30.1	42.0	48.3	56.2	60.0
300 K	25.2	35.1	40.1	46.7	49.9
400 K	21.1	29.2	33.3	38.3	40.6
500 suns					
200 K	32.1	42.5	51.1	59.5	63.2
300 K	27.7	38.5	44.0	51.0	54.2
400 K	24.0	33.2	37.8	43.5	46.2

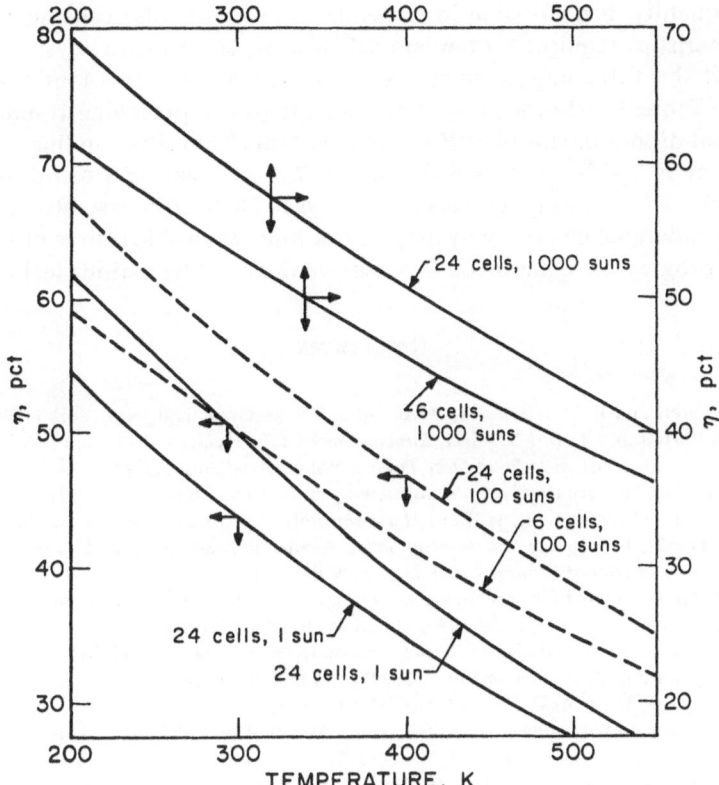

Figure 7.22. Variation of efficiency with temperature. For clarity, results for 500 suns are not shown.

achievable efficiency becomes less sensitive to the E_g values as the numbers of cells increase. This means that the variation of the solar spectrum which occurs in the course of a day can be accommodated reasonably well by any of a number of carefully selected energy gap combinations.

The efficiencies calculated here are limit efficiencies based on the assumptions that there are no reflection losses, no cell internal series resistance losses, no collection efficiency losses, and so on. Solar cells can be designed to minimize such losses.

Fabrication of tandem cells of the type described here requires the availability of semiconductors having continuously varying energy gaps as those which can be achieved in alloys. Antypas and Moon[23] have described such systems composed of $A^{III}B^V$ semiconductors while Loferski et al.[24] have shown how this principle can be extended to chalcopyrite semiconductors of the type $A^{I}B^{III}C^{VI}$. In both of these systems, energy gap, lattice constant, and electron affinities can be varied independently, within limits.

Consequently, it is possible in principle to produce solar cells having the characteristics required by tandem cell systems from such alloys.

All the foregoing calculations are based on $A = B = 1$ in equation (7.58). Table 7.4 shows calculations similar to the preceding results with nonideal diodes having $A = B = 2$ in equation (7.58). Even in this case the efficiency is 63.2% for $C = 500$ and $T = 200$ K. Solar cells could provide as much as 30% of energy needs by the year 2020, with very little upkeep and considerable efficiency by using a free and renewable source of energy. Rapid progress is being made[19] to achieve this goal by various techniques.

References

1. C. A. Wert and R. M. Thomson, *Physics of Solids*, McGraw-Hill, New York (1970).
2. C. S. Barrett and T. B. Massalski, *Structure of Metals*, McGraw-Hill, New York (1980).
3. C. Kittel, *Introduction to Solid State Physics*, Wiley, New York (1976).
4. J. S. Blakemore, *Semiconductor Statistics*, Pergamon Press, New York (1962).
5. A. L. Fahrenbruch and R. H. Bube, *Fundamentals of Solar Cells*, Academic Press, New York (1983); R. H. Bube, *Electrons in Solids*, Academic Press, New York (1981).
6. S. M. Sze, *Physics of Semiconductor Devices*, Wiley-Interscience, New York (1981).
7. M. A. Green, *Solar Cells*, Prentice-Hall, Englewood Cliffs, New Jersey (1982).
8. C. Hu and R. M. White, *Solar Cells*, McGraw-Hill, New York (1983).
9. S. J. Fonash, *Solar Cell Device Physics*, Academic Press, New York (1981).
10. C. D. Thurmond, *J. Electrochem. Soc.* **122,** 1133 (1975); see also M. B. Panish and H. C. Casey, Jr., *J. Appl. Phys.* **40,** 163 (1969).
11. N. A. Gokcen, *Thermodynamics*, Techscience, Hawthorne, California, Chapter 6 (1975).
12. N. A. Gokcen, *J. Chem. Phys.*, **83,** 1240 (1985).
13. M. W. Chase, Jr., J. L. Curnutt, J. R. Downey, R. A. McDonald, A. N. Syverud, and E. A. Valenzuela, *J. Phys. Chem. Ref. Data* **11,** 795 (1982).
14. M. L. Cohen and D. J. Chadi, in *Handbook of Semiconductors*, Volume 2, edited by M. Balkanski, North-Holland, Amsterdam, p. 155 (1980).
15. P. B. Allen and M. Cardona, *Phys. Rev. B* **27,** 4760 (1983).
16. S. I. Chikichev, *Russ. J. Phys. Chem.* (English translation of *Z. Fiz. Khim.*) **57,** 989 (1983).
17. A. K. Jonscher, *Principles of Semiconductor Device Operation*, Wiley, New York (1960).
18. N. A. Gokcen and J. J. Loferski, *Sol. Energy Mater.* **1,** 271 (1979).
19. *IEEE Photovoltaic Specialists Conference Records* (1964-1984), 20 volumes.
20. J. J. Loferski, in *12th IEEE Photovoltaic Specialists Conference, 1976,* p. 957 (1976).
21. N. S. Alvi, C. E. Backus, and G. W. Masden, in *12th IEEE Photovoltaic Specialists Conference, 1976,* p. 948 (1976).
22. *Solar Array Design Handbook*, JPL-SP43-38, Volume 1, NASA, Washington D.C. (1976).
23. G. Antypas and R. L. Moon, *J. Electrochem. Soc.* **120,** 1579 (1973).
24. J. J. Loferski, J. Shewchun, B. Roessler, R. Beaulieu, J. Piekoszewski, M. Gorska, and G. Chapman, in *13th IEEE Photovoltaic Specialists Conference, 1978,* p. 190 (1978).

A

Engel–Brewer Theories

The electronic theories of metals explain numerous characteristics of metals and alloys, and the existence of some of the phases at various compositions. The Engel–Brewer (E–B) theories, or more appropriately the E–B rules, attempt to correlate a very large number of phases existing in metals and alloys with the electronic configurations, and to predict the existence of phases for which experimental data are not yet available. Earlier work of Engel[1] during 1939–1949 was subsequently clarified, amended, and greatly expanded by Brewer.[2-11]

The band theories of metals assume that the valence energy levels of the free atoms in their ground states split into nearly continuous bands upon condensation and crystallization. The electrons in a band of energy levels are mobile and not localized on individual atoms. While the E–B theory, based on the valence bond theory, does not directly contradict the band theory, it assumes that the free atoms of an element may be first excited to place their electrons to certain higher levels by absorption of energy (promotion energy) and then the resulting atoms are condensed to form a metal having an appropriate number of electrons in these excited states and thus dictate the crystal structure. The electrons in their excited and unpaired states are considered to play a major role in binding the atoms in their condensed states.

The electronic structures of atomic gaseous elements in their ground states are shown in Appendix G. (See also the first entry for each of the selected elements in Table A.1.) The number of electrons exceeding one is denoted by superscript numbers in this section (e.g., s, s^2, p, p^2, p^3). The s-orbitals are filled when there are two electrons whose spins are paired. The p-orbitals are filled with six electrons, and the d-orbitals, with ten electrons. The electrons in the p-orbitals begin to be paired when the number of *p-electrons exceeds three electrons*, and those in the d-orbitals, when the

Table A.1. Electronic Configurations, Promotion Energies, and Crystal Structures

Element	Configuration	Promotion energy, kcal/mole	Crystal structure[b]	Element	Configuration	Promotion energy, kcal/mole	Crystal structure
Na	s	0^a	bcc	Nb	d^4s	0	bcc
Mg	s^2	0	—		d^3sp	48	—
	sp	63	cph	Mo	d^5s	0	bcc
Al	s^2p	0	—	Tc	d^5s^2	0	—
	sp^2	83	fcc		d^6s	7	—
K	s	0	bcc		d^5sp	47	cph
Ca	s^2	0	—	Ru	d^7s	0	—
	sp	43	—		d^6sp	72	cph
	ds	51	bcc	Rh	d^8s	0	—
Sc	ds^2	0	—		d^6sp^2	128	fcc
	dsp	45	cph	Pd	d^{10}	0	—
	d^2s	33	bcc		d^7sp^2	140	fcc
Ti	d^2s^2	0	—	Ag	$d^{10}s^1$	0	—
	d^2sp	45	cph		d^8sp^2	133	fcc
	d^3s	19	bcc	Cd	$d^{10}s^2$	0	—
V	d^3s^2	0	—		$d^{10}sp$	87	cph
	d^3sp	47	—	Ba	s^2	0	—
	d^4s	6	bcc		ds	26	bcc
Cr	d^5s	0	bcc		sp	35	—
	d^4sp	71	—	La	ds^2	0	—
	d^4sp^2	165	—		dsp	38	cph
Mn	d^5s^2	0	—		d^2s	8	bcc
	d^5sp	53	—		sp^2	44	fcc
	d^6s	49	bcc	Hf	d^2s^2	0	—
Fe	d^6s^2	0	—		d^3s	40	bcc
	d^6sp	55	—		d^2sp	51	cph
	d^7s	20	bcc	Ta	d^3s^2	0	—
	d^5sp^2	92	fcc		d^4s	28	bcc
Co	d^7s^2	0	—	W	d^4s^2	0	—
	d^6sp^2	119	fcc		d^5s	8	bcc
	d^7sp	67	cph	Re	d^5s^2	0	—
	d^8s	10	—		d^6s	34	—
Ni	d^8s^2	0	—		d^5sp	54	cph
	d^9s	1	—	Os	d^6s^2	0	—
	d^8sp	74	—		d^7s	15	—
	d^7sp^2	120	fcc		d^6sp	67	cph
Cu	$d^{10}s$	0	—	Ir	d^7s^2	0	—
	d^8sp^2	120	fcc		d^7sp	75	—
Zn	$d^{10}s^2$	0	—		d^6sp^2	162	fcc
	$d^{10}sp$	93	cph	Pt	d^9s	0	—
Sr	s^2	0	—		d^7sp^2	156	fcc
	ds	52	bcc	Au	$d^{10}s$	0	—
	sp	41	—		d^8sp^2	159	fcc
Y	ds^2	0	—	Hg	$d^{10}s^2$	0	—
	d^2s	31	bcc		$d^{10}sp$	106	—
	dsp	43	cph				
Zr	d^2s^2	0	—				
	d^3s	14	bcc				
	d^2sp	42	cph				

[a] Ground state has zero promotion energy; other values are the lowest energies for each configuration as given by Brewer.
[b] Stable crystal structure at ambient pressure; see page 318.

number of *d-electrons exceeds five electrons*. For example, $3d^6$ structure for iron signifies that 3 is the quantum shell and that there are six electrons in the *d*-orbital, and further, one electron in excess of five is paired with one of the five electrons, leaving four unpaired electrons for bonding upon condensation. The paired electrons in pure elements are assumed to play no role in bonding in solid state according to the valence bond theory. For the first 20 elements in the periodic table (Appendix G), each orbital is completed in sequence as an electron is added, but from elements 21 to 29 the *s*-orbital accepts electrons before the inner *d*-orbital is filled. Elements 21 through 28, for which the *d*-orbital accepts electrons in this manner, are called the first series of the transition elements; they lie in the fourth series in the periodic table and occupy Sc through Ni (IIIA–VIIIA). The second and third series of transition elements lie below the first series and their *s*-orbitals also accept electrons before their *d*-orbitals are filled. In the lanthanide and actinide series, the inner *d*- and *f*-orbitals are filled in the same irregular way as the *d*-orbitals in the remaining transition elements.

The ground states of the monatomic gaseous elements considered in this section are listed in Table A.1. The energy required to go from the ground state to an excited state is called the promotion energy, and the lowest of several values of the promotion energies corresponding to each electronic structure are listed in Table A.1. For example, the ground state of Ti is d^2s^2 and the promotion energy to d^2sp is 45 kcal/mole, and to d^3s, 19 kcal/mole.

Basic Principles

The basic principle of the E–B theory lies in correlating the unpaired *s*- and *p*-electrons with the crystal structures. An element crystallizes in body-centered cubic (bcc) structure when it has one outer unpaired electron. The alkali elements and Cr and Mo have one outermost *s*-electron each, and they crystallize in bcc structure. In Cr and Mo, the ground state is d^5s; the next lowest level with an unpaired *s*-electron and one *p*-electron, d^4sp, requires 71 and 80 kcal/mole for promoting one *d*-electron to the *p*-orbital for Cr and Mo, respectively. These are considered to be high promotion energies, particularly because of no change in the total number of unpaired *d*- plus *s*-electrons in d^5s and in d^4sp configurations; hence, Cr and Mo cannot crystallize in any other structure than bcc, in agreement with observations.

For magnesium, the picture is different, because its ground state outside the closed shell is $3s^2$, or briefly s^2 (Table A.1); this state is not suitable for bonding because s^2 is a paired group of two electrons, unavailable for bonding. The sp configuration obtained by promoting one s-electron to the p-orbital with 63 kcal of promotion energy, E_{prom}, yields two electrons for bonding. The excited atoms are then condensed to form close-packed hexagonal (cph) crystal structure for Mg. The energy of sublimation of crystalline Mg to monatomic gaseous atoms, ΔE_{atomiz}, is called the atomization energy by Brewer, despite the fact that it is the *bond energy* in solid-state physics, and the sublimation from the solid state into the sp gaseous atomic state, E_{sp} (or E_{dsp} when d-electrons are involved), is called the *bonding energy* by Brewer,[2-4] not to be confused with the *bond energy*. These processes are summarized as follows:

$$\text{solid} \rightarrow \text{gas } (s^2); \qquad \Delta E_{atomiz}$$
$$\text{gas } (s^2) \rightarrow \text{gas } (sp); \qquad E_{prom}$$

$$\text{solid} \rightarrow \text{gas } (sp); \qquad E_{sp} = \Delta E_{atomiz} + E_{prom} = \text{bonding energy} \quad \text{(A.1)}$$

The solidification process in sequence is therefore $s^2 \rightarrow sp \rightarrow$ solid. This process is favored by a low value of E_{prom} and a high value of E_{sp} with as many available bonding electrons as possible. In fact, the promotion to a reasonably high energy is justified if the promotion increases the number of unpaired electrons available for metallic bonding in the solid; however, the promotion of electrons from the completed noble gas shell is not permitted because the required energy is prohibitively high.

The ground state for Al is s^2p, and promotion of one of the two paired s-electrons to the p-orbital yields three electrons for bonding with a promotion energy of 83 kcal/mole for obtaining sp^2 configuration. The resulting *three electrons are then necessary for obtaining face-centered cubic (fcc) crystal structure for aluminum.*

The ground state for Si is $(3s^2)(3p^2)$, or briefly s^2p^2, a state considered to be unsuitable for bonding because s^2 is a paired group of electrons. The promotion of one s-electron to p yields sp^3 configuration with *four unpaired electrons available for binding, and this configuration gives the tetrahedral or diamond crystal structure.*

Engel postulated that all the unpaired d-electrons in transition elements also take part in metallic bonding but they do not determine the crystal structure; only the outer unpaired s- and p-electrons control the type of crystal. This postulate has been criticized[12-14] on the basis that the d-electrons in transition elements have strong bonding and directional properties; hence, they should also play a dominant role in determining the crystal structure. However the d-electrons control the crystal structure indirectly by contributing to the s- and p-shells according to the E–B postulates.[3] (See also Altmann et al.[14a])

The foregoing considerations lead to the following E–B rules. (1) The crystal structure depends on the total number of unpaired $(s + p)$ electrons; the crystal structures and the numbers of unpaired $(s + p)$ electrons are as follows:

- bcc: 1 electron
- cph: 2 electrons
- fcc: 3 electrons
- diamond structure: 4 electrons

We shall see later that these integral numbers of electrons per atom will be modified to cover a range of fractional numbers for each crystal structure encountered in some metals and in all alloys. For brevity, Brewer used I, II, and III for bcc, cph, and fcc, respectively. (2) The bonding energy depends on the number of *unpaired d-, s-,* and *p*-electrons available for bonding. (3) The electronic structure in solid phases is close to that in the gaseous atomic element promoted as necessary to yield the appropriate numbers of $(s + p)$ electrons. The electrons are promoted with as small amounts of energy as possible for a given crystal structure with as many bonding *d-, s-,* and *p*-electrons as feasible. The promotion is mainly justified (i) on the basis of appropriate numbers of $(s + p)$ electrons and (ii) on the basis of maximum number of unpaired electrons obtained by reasonable promotion energies.

The basic assumption that the excited atoms in gaseous state also exist in condensed state cannot be proven by existing experimental techniques. The cph crystal structure in Li and Na at low temperatures cannot be explained by this theory because no reasonable promotion energy can provide two $(s + p)$ electrons for cph structure; to do so would require promoting one electron from the closed noble gas shell to the *p*-orbital beyond the outer *s*-orbital. Likewise, for Ca and Sr, fcc structure requiring three $(s + p)$ electrons cannot be reconciled with E–B rules. Despite these and other shortcomings, the E–B concepts have achieved remarkable success in correlating the behavior of metals and alloys, and in estimating the compositional ranges of alloy phases as will be seen later.

Nonintegral Electronic Configurations

In its refined form, the E–B theory considers that the structures of all the elements, and particularly the transition elements and their alloys, will tolerate some deviation from the integral numbers of 1, 2, and 3, $(s + p)$ electrons for bcc, cph, and fcc structures, respectively. The postulated values of $(s + p)$ electrons per atom, based on observation of phase boundaries in binary systems, are as follows:

- bcc(I): 1.0 to 1.5
- cph(II): 1.7 to 2.1
- fcc(III): 2.5 to 3.0

These ranges are in general agreement with the Hume-Rothery rules; however, they are not rigid, and may be varied somewhat as justified by the available data for binary and multicomponent systems as will be discussed later. Nonintegral ranges of numbers for electrons are an amendment to the first E–B rule presented earlier, and essential for the alloy phases. For the bcc structure of the elements, the fractional electronic structure is $d^{n-0.5}sp^{0.5}$ for all the elements listed in Table A.1 with $d^{n-1}s$ structure, except for Mo and W for which the usual structure is d^5s, and for Nb, d^4s (here n is the same as the number of electrons outside the noble gas shell). For the cph structure of the elements, only Ru is assigned $d^{6.3}sp^{0.7}$ configuration, and for the fcc structure, Mn is assigned $d^{4.5}sp^{1.5}$, and Fe, $d^{5.5}sp^{1.5}$. The bonding energies for the fractional configurations are computed from Fig. A.1 as will be seen later.

Transition Elements

The unpaired d-electrons of the transition elements are generally less bonding than the s- and p-electrons on a per electron basis. However, the empty d-orbitals of one element may provide a sink for the excess electrons from another alloying element and thus contribute strongly to bonding. The d-electrons of the first transition series are less bonding than the remaining two series. In fact, the contribution of d-electrons to bonding is so poor in Fe, Co, and Ni that they are left largely undisturbed in the solid and thus assumed to contribute to ferromagnetism. The promotion energies for the first two elements of each series for s, and sp configurations are close enough to justify qualitatively the existence of cph and bcc forms of Sc, Ti, Y, Zr, and Hf. For V, Nb, and Ta, the promotion energies for the d^4s configuration are 6, 0, and 28 kcal/mole, respectively, and these values are low enough to permit bcc structure (see Table A.1). The bcc crystal structure with d^5s configuration for Cr and Mo does not require promotion energies, and W requires only 8 kcal/mole. The promotion energy of 53 kcal/mole for d^5sp for Mn suggests that cph structure should exist, but it does not; however, for d^6s, the promotion energy is 49 kcal, and indeed bcc structure does exist both at low and high temperatures. The observed existence of fcc Mn requires d^4sp^2 configuration, which is difficult to reconcile[15] with the E–B theory since the promotion energy is very high. For Tc and Re, the promotion energies in Table A.1 for d^5sp are not too large, considering the fact that there are seven bonding or unpaired electrons for d^5sp instead of five for d^6s; hence, cph structure is favored, in agreement with observations. For Fe, d^7s and d^6sp configurations have acceptable promotion energies but cph structure corresponding to d^6sp does not exist at ambient pressures. For Ru, d^7s configuration should yield bcc as in Fe, but this form of Ru does not exist. However, the promotion energy for the d^6sp configuration for Ru is rather high, in fact higher than that for Fe, but the corresponding cph structure is the observed form of Ru at all temperatures, whereas cph structure is not observed for Fe at ordinary pressures. The bonding of extra d- and p-electrons offset the high promotion energy of d^5sp structure because $4d$ electrons have much greater bonding capability than $3d$ electrons as shown in Fig. A.1.

The configuration for the cph form of Co is evidently d^7sp since the ground state d^7s^2 is not suitable for bonding. The promotion energy for d^6sp^2 for fcc is high and at high temperatures, solid Co is in fcc structure. For Rh and Ir, the promotion energy for the d^6sp^2 configuration is high but promotion to d^6sp^2 is necessary for the fcc structure that is observed at all temperatures. The configuration proposed by Brewer for Cu, Ag, and Au is d^8sp^2, i.e., three $(s + p)$ electrons, for which the promotion energy is fairly high. These metals exist only in fcc structure.

Correlation of Bonding Energies

The bonding energies of s- and p-electrons (in kilocalories per mole of electrons) versus the elements, referring to the vertical and upper horizontal axes, as devised by Brewer, are represented by the upper curves in Fig. A.1. The ends of the upper curve in Fig. A.1(A) refer to Ca and Zn, which have no bonding d-electrons; therefore, the bonding energy from equation (A.1) divided by two $(s + p)$ electrons yields these end points. Thus, for Ca, E_{sp} is the sum of $E_{prom} = 43$ and $E_{atomiz} = 42$, i.e., 85, and this value divided by two $(s + p)$ electrons is equal to 42.5 kcal/mole electrons $= E_{sp}$. The corresponding value for Zn is 62 kcal. The initial sharp curvature on the left can be explained on the basis that the bonding energies of s- and p-electrons vary approximately with the inverse ratios of covalent atomic radii in the periodic table. Thus, for Ca and Sc, the covalent radii are 1.74 and 1.44 Å, respectively, and $E_{sp} = 42.5$ (for Ca) $\times 1.74/1.44 = 51$ kcal/mole electrons for Sc. The remaining portion of the curve up to Zn is a smoothly joined and very nearly straight line ending at Zn. The influence of the nuclear charge on $(s + p)$ electrons is complicated by the screening effect of the inner electrons; however, any error in the upper curve is

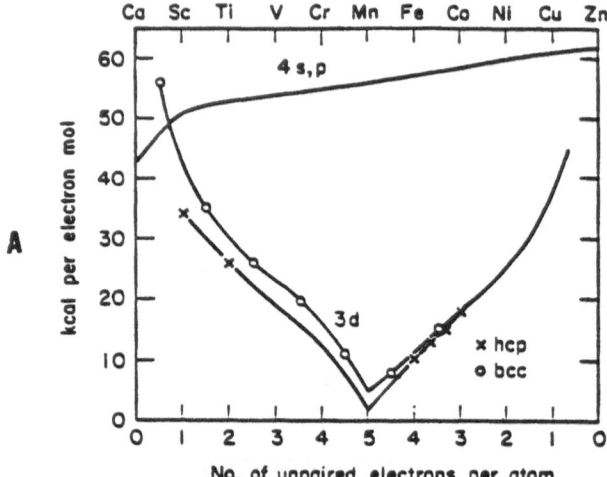

Figure A.1. Valence-state bonding enthalpy (~ energy) in kilocalories per mole of electrons. Upper curves are for E_{sp}, bonding enthalpies of s- and p-electrons, versus the elements on upper horizontal axis. Lower curves are bonding enthalpies versus numbers of unpaired d-electrons on lower horizontal axis. (From Brewer[4] with permission.)

compensated by the energy assigned to the d-electrons by the lower curves; therefore, the exact shape of the upper curve is not considered to be greatly important for the succeeding computations. Also shown in Fig. A.1 are the bonding energies of the d-electrons, E_d, versus the number of unpaired d-electrons as indicated by the lower curves, which refer to the lower horizontal axis. Each point was obtained by calculating the value of total bonding energy E_{dsp} due to all unpaired d-, s-, and p-electrons, then subtracting E_{sp}, obtained by multiplying the number of $(s + p)$ electrons with the value from the upper curve, and then dividing $E_{bonding} - E_{sp}$ by the total number of unpaired d-electrons. The proximity of the curves for bcc and cph indicates that an approximate curve that could have been drawn slightly above the curve for cph might well have served for the fcc structure. Recomputation of the bonding energies from these figures may often show which phases have high enough sublimation energy to be the most stable. For example, the ground state for Nb is d^4s and Nb crystallizes as bcc. From Fig. A.1(B), $E_{sp} = 52 \times 1$ (one s-electron) and (total E_d) = 36×4 (four d-electrons), so that the bonding energy $E_{bonding} = 196$, and since the promotion energy is zero this is also equal to the sublimation or atomization energy E_{subl}. The cph form of Nb has never been observed, and for cph, the required configuration is d^3sp, for which the promotion energy is 48 kcal/mole. The values of the energies from Fig. A.1(B) are $E_{sp} = 52 \times 2$ and (total E_d) = 35×3; hence, $E_{bonding} = 209$, and subtraction

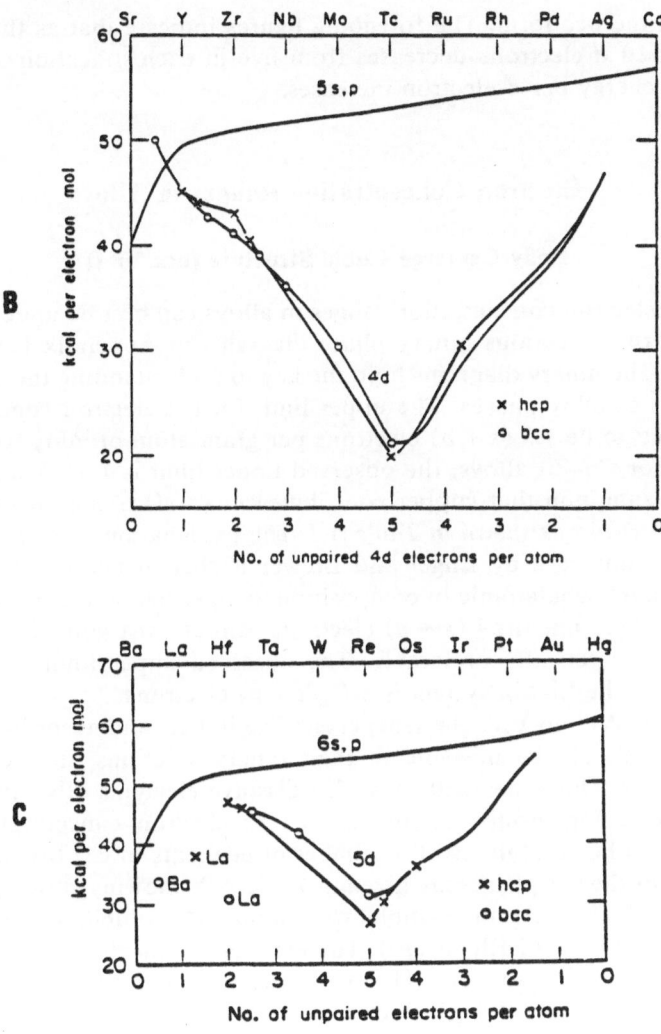

Figure A.1 (*continued*)

of 48, the promotion energy, yields $E_{\text{atomiz}} = 161$ kcal/mole, indicating that the cph form is about 35 kcal less stable than the bcc form. If we use $d^{3.3}sp^{0.7}$ configuration, the lower limit of 1.7 $(s + p)$ electrons for cph, the result would be $E_{\text{bonding}} = 52 \times 1.7 + 35 \times 3.3 = 204$, and $E_{\text{atomiz}} = 204 - 48 \times 0.7 = 170$, where 0.7 originates from the fact that 0.7 mole of d^3sp and 0.3 mole of d^4s make up $d^{3.3}sp^{0.7}$. Again, even with the use of the lower limit for 1.7 $(s + p)$ electrons, the result does not make the cph form as stable as the bcc form. The foregoing figures indicate that as the number of unpaired d-electrons decreases from five in each transition series, the bonding energy per d-electron increases.

Electron Concentration Ranges in Alloys

Body-Centered Cubic Structure (bcc, or I)

The electron concentration ranges in alloys can best be understood by referring to the various binary phase diagrams in Appendix E whenever possible. The binary diagrams hold the key to understanding the electronic structures of alloy phases. The upper limit for the electron concentration is assumed to be 1.5 $(s + p)$ electrons per gram atom of alloy for the bcc phases. For Cu–Zn alloys, the observed upper limit is 1.58. This is based on the assumption that copper contributes *one s-electron* instead of *three $(s + p)$ electrons as shown in Table A.1*. The explanation of one s-electron for Cu in this case by Engel and Brewer is that in bcc brass, which is approximately equiatomic in composition, copper has so many zinc neighbors that three unpaired $(s + p)$ electrons cannot exist and $d^{10}s$ structure becomes energetically preferable. The observed upper limit for the bcc structure in the Li–Mg system is 1.75 $(s + p)$ electrons.

Group VA and VIA elements crystallize in bcc, and according to E–B rules they should be miscible in their binary solutions, in accord with observations. These elements may also dissolve elements other than those in the preceding groups up to 1.5 $(s + p)$ electron concentration. The maximum solid solubilities of a number of elements in Cr, Mo, and W as solvents for the bcc phases are given in Table A.2, showing good agreement with the E–B rules. As an example, we compute the solubility of Rh in Mo, in atom fraction x of Rh (d^6sp^2). The electronic contribution from Mo is 6 $(d + s)$ electrons so that $6(1 - x)$ is the total contribution from Mo. Rhodium has 9 $(d + s + p)$ electrons per atom; hence, the contribution from Rh is $9x$, and the sum must correspond to $d^5sp^{0.5}$, or 6.5 electrons including 1.5 $(s + p)$ electrons; thus, $6.5 = 6(1 - x) + 9x$, and $x = 0.167$ or 16.7 at.% Rh in Mo. If the integral approximation of d^5s were used for the alloy, instead of $d^5sp^{0.5}$, x would be zero. This example shows that although d-electrons do not control the structure directly by remaining as d-electrons, they do so indirectly by contributing electrons to p by unpairing the d-electrons as from d^8s to d^6sp^2 in Rh. For Ru(d^6sp) as a solute, similar calculations from $6.5 = 6(1 - x) + 8x$ yield $x = 0.25$ as shown in Table A.2.

The values listed in Table A.2 refer to the temperatures at which the solubility is a maximum or usually at the eutectic temperature. The disagree-

Table A.2. Maximum bcc Solid Solubility Limits of Selected Elements in Cr, Mo, and W as Solvents[a]

Solute	Calculated values in Cr, Mo, and W	Experimental values		
		in Cr	in Mo	in W
Tc	50	—	50	—
Re	50	48.5	42	37
Ru	25	34	30	23
Os	25	31	19	18.5
Rh	16.7	~15	~20	~6
Ir	16.7	~12	16	~10
Pd	12.5	~5	6.5	~5
Pt	12.5	~10	15	4

[a] Values are concentrations in at.% of solute element in the first column.

ment between the calculated and experimental values becomes larger as the difference between the group numbers of the solvents and solutes increases.

The transition elements with fewer than five d-electrons act as a sink for the d-electrons of the elements having more than five d-electrons. In addition, the $(s + p)$ electron concentration may be lower than 1.5 so that the structure for bcc alloys of Hf(d^3s)-Ru(d^7s) is $d^{4.5}s$, for Hf as the solvent, corresponding to 62.5 at.% Hf as calculated from $5.5 = 4x + 8(1 - x)$. For Ru as the solvent, the structure is $d^{5.5}s$, and $6.5 = 4x + 8(1 - x)$ yields 37.5 at.% Hf. Calculation of the appropriate numbers of $(s + p)$ electrons is discussed in detail in Ref. 4. If for a known binary system the electron concentration and configuration are fitted, then for a nearby system the same concentration and configuration can be used to estimate the unknown solubility. In essence, the known maximum solubilities may dictate the electronic structure, and the unknown maximum solubilities may be based on the rules as adjusted according to the rules derived from similar binary systems if they have been established by experiment.

The Laves phases, for which the ideal compositional range is AB_2 to AB_5, generally having cubic, hexagonal, and complex structures, are formed not only on the basis of electron to atom ratio, but also on the basis of atomic radius ratio of A to B as discussed elsewhere in detail.[20,23,25] No definite $(s + p)$ electron concentration has been assigned to these phases by Brewer. Some of these, such as $LaNi_5$, are useful and interesting media for hydrogen storage as was discussed in Chapter 6.

Close-Packed Hexagonal Structure (cph, or II)

The electron concentration for cph structure is generally limited to the range of 1.7 to 2.1 $(s + p)$ electrons per gram atom of alloy (cf. Brewer[3] and Hume-Rothery[16]). The lower limit is for the solubilities of solute elements with one s-electron per atom in solvent elements with two or more $(s + p)$ electrons, whereas the higher limit is for those solutes having three or more $(s + p)$ electrons per atom in cph solvent elements. For example, Mo and W have d^5s structure, and as solutes, they should dissolve in solvent metals Tc and Re, having d^5sp structure to the extent of $d^5sp^{0.7}$, or 30 at.% as confirmed from $6.7 = 6x + 7(1 - x)$; the experimental values are 24 at.% Mo in Tc, and 20 at.% W in Re. To improve the results, Brewer suggests $d^{5.1}sp^{0.7}$ corresponding to a maximum of 20 at.% Mo for the cph structure. For Rh(d^6sp^2) as a solvent, Mo should dissolve until the solution assumes $d^6sp^{1.1}$ corresponding to 0.30 mole fraction of Mo, but only 0.18 mole fraction of Mo is dissolved in Rh when cph structure appears. For $d^6sp^{0.7}$, 0.65 mole fraction Mo is dissolved in Rh; hence, Rh can no longer be considered as the solvent. In this case, Mo is the solvent, but since Mo is d^5s, this configuration should be raised to $d^5sp^{0.7}$, and assuming that Rh(d^6sp^2) contributes one d and three $(s + p)$ electrons, then cph phase should appear at $x = 0.25$ mole fraction of Rh, computed from $6.7 = 9x + 6(1 - x)$, but the experimental value is 0.43 Rh. Here, it may be assumed that Rh cannot contribute three more electrons than Mo. On the high electron side, and therefore Rh as solvent, Hume-Rothery's configuration of $d^{6.36}sp^{1.1}$ corresponds to the experimental value of 18 at.% Mo. Again, the degree to which the rules are adjusted depends on the systems, and such rules may often appear to explain the solubilities and the electronic structures rather than to predict the solubilities.

For the mutual solubilities of the elements capable of contributing large numbers of electrons, clearly the values toward the upper limit of 2.1 must be used. Thus, for Ir(d^6sp^2) as a solute in Os(d^6sp), with x as the mole fraction of Ir, $d^6sp^{1.1}$ structure requires $9x + 8(1 - x) = 2.1 + 6$, which yields $x = 0.10$, whereas the actual value of x is 0.38. Note that here 1.7 instead of 2.1 would give a negative value for x!

Face-Centered Cubic Structure (fcc, or III)

The range of electron concentration for the fcc structure was listed as 2.5 to 3.0 electrons per atom of alloy[3,16]. For example, the E–B calculation requires that Os(d^6sp) as a solute can be dissolved in Ir(d^6sp^2) up to 50 at.%, in close agreement with the measured value of 45 at.% Os. Other solutes (in at.%) in Ir as the solvent are as follows, with the calculated E–B values in parentheses: Mo, 22 (17); W, 22 (17); Nb, 16 (12.5); Ta, 16 (12.5); Ti, 14 (10). The value for $Mo(d^5s)$ can be calculated by assuming that Ir contributes nine electrons, and Mo, six electrons, and the alloy configuration is $d^6sp^{1.5}$ so that $x = 0.17$ for Mo in Ir as calculated from $6x + 9(1 - x) = 8.5$; hence, fcc structure exists from zero to 17 at.% Mo according to the E–B postulates. The maximum measured concentration of $Nb(d^4s)$ in $Ni(d^7sp^2)$ is about 14 at.% Nb which corresponds to $d^{6.8}sp^{1.5}$. Here $d^7sp^{1.5}$ configuration for the alloy would yield a reasonably close value of 10 at.% Nb. The contribution from the d-electrons is again based on how a series of alloys for which an empirical set of electronic configurations can be assigned in order to obtain the solubility close to the experimental results.

Other Phases

Several other phases exist in addition to I, II, and III, within the electron concentration range up to 3.5. They follow, with increasing electron concentration, the structures indicated by

$$\text{Cubic } Cr_3Si(A15) \rightarrow \sigma \rightarrow P \rightarrow \mu \rightarrow R \rightarrow \chi$$

These phases appear with increasing $(d + s + p)$ electrons up to cph structure in numerous phase diagrams; they will be described and discussed briefly in the following sections.

Cubic Cr₃Si

The cubic Cr_3Si structure (Strukturbericht A15)* is such that each Si atom is surrounded by 12 Cr atoms without Si–Si bonds. In addition to Cr, the elements of IVA to VIA may be substituted for Cr, and the elements of VIIA, VIIIA to VB may be substituted for Si. These phases are ordered phases in which atomic radii should not differ by more than 15%. Exceptional A15 phases are Zn_4Sn and MoTc. In Cr_3Si-type phases, the $(s + p)$ electron concentration is 1.5 to 1.75 according to Brewer when the d configuration is chosen properly. Thus, for Cr_3Si, there are 7 $(s + p)$ electrons per 4 atoms or 1.75 $(s + p)$ electrons per atom. For Cr_3Pt, there are $6/4 = 1.5$ electrons per atom, assuming that Pt does not contribute any d atoms, but if Pt retains d^6 structure, or yields only one d-electron, then 1.75 $(s + p)$ electron concentration still prevails.

---

*These structures are discussed in standard texts.[17,18] See also Nevitt.[20] A15 and σ phases compete with bcc and cph phases when atomic sizes are favorable.

σ Phase

The σ phase occurs at the atomic radius ratios of 0.93 to 1.15 according to Nevitt.[20] It is formed by quasi-hexagonal layers of atoms, with a tetragonal unit cell. The $(s + p)$ concentration is generally 1.2 to 1.9, overlapping bcc and cph structures. The range of $(d + s + p)$ concentration is between 5.6 and 7.7 electrons per atom, with a very large number of phases occurring in the vicinity of 6.6 electrons. The assignment of nonintegral configurations, such as $d^{5.5}sp^{0.5}$ for Mn, $d^{6.5}sp^{0.5}$ for Fe, retains the range of $(s + p)$ recommended by Brewer. For example, Cr–Mn, Mn–Fe, Mn–Re, Mn–Tc, Mo–Tc, Re–W binary systems form σ phases.

χ or α-Mn Phase

The remaining phases, P, μ, R, and χ, are very closely related to the σ phase in their crystal structures. The P, μ, and R phases are labeled as the μ phase by Brewer in his multicomponent diagrams. The electron concentrations for these phases are determined from the phase diagrams where they occur. The χ phase for Mn is at 7 $(d + s + p)$ electrons per atom, and for the binary alloys, the range is 6.3 to 7.0 $(d + s + p)$ electrons per atom. Since all the transition elements forming χ phases assume the d^5 configuration, the $(s + p)$ concentration range is from 1.3 to 2.0, overlapping the σ and cph phases.

Phase-Stabilizing Elements in Fe, Mn, Ti, and Zr

The transition elements to the left of iron (VIIIA), except Mn, stabilize α- and δ-Fe (both bcc-Fe) with Fe as the solvent by enclosing the γ-Fe (fcc-Fe) inside a loop known as the γ-loop. Nontransition elements to the right of Fe forming substitutional solid solutions, e.g., Al, Si, also stabilize bcc Fe. Aluminum stabilizes α-Fe by enhancing the pairing of d-electrons in Fe by contributing the excess outer electrons in Al. This is also the case for Si in Fe.

The following elements stabilize the γ-Fe phase (fcc):

Mn   (Fe)   Co   Ni   Cu

Ru   Rh   Pd   —

Os   Ir   Pt   Au

This is explained on the basis that the elements to the right, e.g., Ni, can contribute electrons from the d-orbital to the s- and p-orbitals, and thus stabilize the fcc $(d^x sp^2)$ structure.

Similar explanations are also applicable to bcc → fcc phase stabilization in Mn-base alloys[15]; i.e., the elements to the left of Mn in the periodic table stabilize bcc-Mn, and those to the right, fcc-Mn.

The structure of Ti and Zr is cph at low temperatures and bcc at high temperatures. The transition elements to the right of these elements VA to VIIIA, and Cu, Ag, and Au, all as solutes, lower the transition temperature and favor the bcc forms of the Ti- and Zr-rich alloys because the empty d-orbitals in Ti and Zr act as sinks for the $(d + s + p)$ electrons. The presence of vacant orbitals enhances unloading of d-electrons from such metals as Ir and Pt and thus the formation of highly exothermic phases with Ti and Zr.[8] In the case of nontransition elements such as Al and Sn as solutes, the excess available s- and p-electrons stabilize the cph structures of Ti and Zr.

Effect of Pressure

Brewer postulated that increased pressure will allow the nuclei to move closer to permit the d-orbitals to overlap better and stabilize the crystal structure corresponding to the largest number of bonding d-electrons. For the metals having fewer than five d-electrons per atom, e.g., Ti in cph form (d^2sp), increasing pressure would favor d^3s corresponding to the bcc form of Ti, in agreement with experimental results. Similar results have been verified for Zr and Hf. In general, for an element with fcc and d^3sp^2 structure, increasing pressure would favor $d^3sp^2 \rightarrow d^4sp \rightarrow d^5s$, or fcc → cph → bcc. For d^4sp^2, increasing pressure would stabilize the structures corresponding to d^5sp which has the largest number of unpaired d-electrons. The elements that have bcc structure at ordinary pressures, V, Nb, Ta, Cr, Mo, and W, remain bcc up to the highest pressures investigated, because (1) for V, Nb, and Ta, d^4s structure cannot be changed to d^5, and (2) for Cr, Mo, and W, the d^5s structure already has the optimum number of d-electrons. On the other hand, for elements having more than five d-electrons per atom, e.g., d^7s, increasing pressure would favor $d^7s \rightarrow d^6sp \rightarrow d^5sp^2 \rightarrow$, or bcc → cph → fcc. In summary, increasing pressure would tend to stabilize the structures toward the maximum number of unpaired, or bonding d-electrons, i.e., toward d^m where m tends as close to five as possible.[3,15,21]

Multicomponent Diagrams

Various phases and their maximum compositional limits in ternary and multicomponent systems can be mapped by using the E–B postulates. The method appears to have its best success for the second and third transition

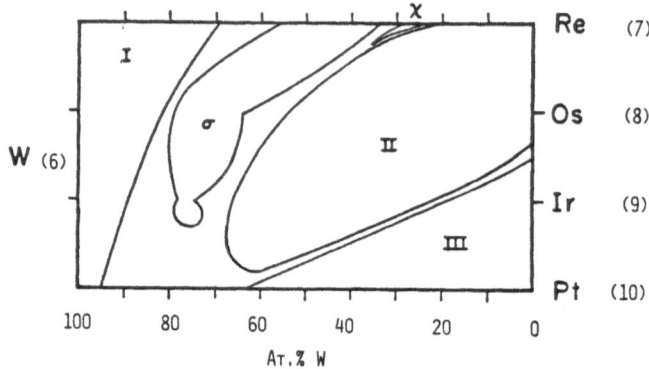

Figure A.2. Optimum phase boundaries in alloys of W with Re, Os, Ir, and Pt. Numbers in parentheses are $(d + s)$ electrons. I = bcc, II = cph, and III = fcc. (Courtesy of L. Brewer; see also Reference 3.)

series of Group IVA, VA, and VIA elements alloyed with the remaining groups of elements to the right.

One such diagram is shown in Fig. A.2 for the alloys of W, where the horizontal axis represents at.% W. The vertical axis is the total outer electron concentration, which increases from 7 for Re to 10 for Pt. A horizontal line from Os to the left represents the binary W–Os system for which the maximum compositional ranges of existence for various phases occur at the intersections of this line with the boundaries for the single phases. The areas marked I, II, and III are for bcc, cph, and fcc, respectively; σ and χ represent the phases designated by these letters.

Such diagrams are not isothermal since the maximum range of composition for one phase is generally not at the same temperature as that for another phase. In the W–Os system, the range of composition for II is from 0 to 53 at.% W. The upper horizontal line is for W–Re and the lower horizontal line is for W–Pt. The range of composition for II in W–Re is 0 to 20 at.% W but this range refers to a different temperature than the preceding range for II in W–Os. The horizontal line that can be drawn starting halfway between Os and Ir refers to an electron concentration of 8.5 at 0 at.% W and this line is for the ternary alloy of W–Os–Ir in which the atomic ratio of Os to Ir is unity. This same line is also for the ternary W–Re–Pt, and pentanary W–Re–Os–Ir–Pt system, for all of which the electron concentration is 8.5 at 0 at.% W. The boundaries in this diagram are largely obtained from the binary diagrams, and their validity for the ternary and multicomponent systems has not yet been tested. Where experimental data were absent, E–B rules were used to obtain the phase boundaries in Fig. A.2. Since each point on a phase field refers to a different temperature, it is not possible to draw the tie lines joining one point on one phase boundary to another point on another phase boundary. It is also not possible to designate the areas between the single-phase fields as the multiphase fields for the ternary and multicomponent alloys.

Concluding Remarks

The basic E–B postulate that bcc, cph, and fcc correspond respectively to one, two, and three $(s + p)$ electrons is not applicable to cph forms of Li and Na, and fcc forms of Ca and Sr. These metals have relatively simple electronic structures; hence, the inability of E–B postulates to account for these structures is fundamental. Nevertheless, the numbers of structures and transition phenomena successfully explained by these theories are quite impressive.

In addition to the electronic configuration, Brewer considered additional important factors, i.e., the size factor, and attractive and repulsive force factors [see equation (B.19)]. All these factors, considered judiciously, yield the appropriate solubility or stability limits for various structures and their energies as discussed in Brewer's publications.

There are many aspects of the E–B approach that disagree with the methods of Friedel,[13,14] Nevitt,[20] and Miedema et al.[22] It is nevertheless generally felt that the E–B theories and postulates represent the most comprehensive existing treatment of solubility ranges for various phases in binary and multicomponent systems. The theories and schemes to be developed in the future will likely take advantage of the E–B theories, along with other theories, correlations, and postulates. The immediate problem is to test the validity of the diagrams by obtaining experimental data on a number of binary and multicomponent alloys of the transition elements for which E–B theories claim the greatest success.

Additional recent publications by Brewer et al.[23–27] are recommended to the reader interested in pursuing this topic further.

References

1. For a brief summary, see N. Engel, *Powder Metall. Bull.* **7**, 8 (1954).
2. L. Brewer, in *Electronic Structure and Alloy Chemistry of Transition Elements*, edited by P. A. Beck, Interscience, New York, p. 221 (1963).
3. L. Brewer, in *High-Strength Materials*, edited by V. F. Zackay, Wiley, New York, p. 12 (1965).
4. L. Brewer, in *Phase Stability in Metals and Alloys*, edited by P. Rudman, J. Stringer, and R. Jaffee, McGraw-Hill, New York, pp. 39–61, 241–249, 344–346, 560–568 (1967).

5. L. Brewer, *Acta Metall.* **15**, 553 (1967).
6. L. Brewer, *Science* **161**, 115 (1968).
7. L. Brewer, in *Plutonium and Other Actinides*, edited by W. N. Miner, Metallurgical Society of AIME, New York, p. 650 (1970).
8. L. Brewer and P. R. Wengert, *Metall. Trans. AIME* **4**, 83 (1973).
9. L. Brewer, in *Transition Metal Alloys—A Chemist's View*, Am. Inst. Phys. Conf. Proc., 1972, edited by H. C. Wolfe, No. 10 (Part 1), p. 1 (1973).
10. L. Brewer, *J. Nucl. Mater.* **51**, 2 (1974).
11. L. Brewer, *Rev. Chim. Miner.* **11**, 616 (1974).
12. J. Friedel, in *Electronic Structure and Alloy Chemistry of Transition Elements*, edited by P. A. Beck, Interscience, New York, p. 70 (1963).
13. W. Hume-Rothery, *Acta Metall.* **13**, 1039 (1965), **15**, 567 (1967).
14. J. Friedel, W. Hume-Rothery, R. Jaffee, L. Kaufman, W. M. Lomer, T. B. Massalski, and W. B. Pearson, in *Phase Stability in Metals and Alloys*, edited by P. Rudman, J. Stringer, and R. Jaffee, McGraw-Hill, New York (1967).
14a. S. L. Altmann, C. A. Coulson, and W. Hume-Rothery, *Proc. R. Soc. London* **240A**, 145 (1957).
15. W. Hume-Rothery, *Prog. Mater. Sci.* **13**, 228 (1968).
16. W. Hume-Rothery, R. E. Smallman, and C. W. Haworth, *The Structure of Metals and Alloys*, Institute of Metals, London (1969).
17. C. S. Barrett and T. B. Massalski, *Structure of Metals*, McGraw-Hill, New York (1980).
18. C. Kittel, *Introduction to Solid State Physics*, Wiley, New York (1971).
19. L. Pauling, *The Nature of the Chemical Bond*, Cornell University Press, Ithaca, N.Y. (1960).
20. M. V. Nevitt, in *Electronic Structure and Alloy Chemistry of Transition Elements*, edited by P. A. Beck, Interscience, New York, p. 101 (1963).
21. L. Brewer, in *Structure and Bonding in Crystals*, Volume I, edited by M. O'Keefe and A. Navrotski, Academic Press, New York, p. 155 (1981).
22. A. R. Miedema, R. Boom, and F. R. deBoer, *J. Less-Common Met.* **41**, 283 (1975).
23. L. Brewer, *Systematics of the Properties of the Lanthanides*, edited by S. P. Sinha, D. Reidel, Hingham, Massachusetts (1983).
24. L. Brewer, *J. Chem. Ed.* **61**, 101 (1984).
25. L. Brewer, *J. Materials Ed.* **6** (5), 734 (1984).
26. J. K. Gibson, L. Brewer, and K. A. Gingerich, *Met. Trans. AIME* **15A**, 2075 (1984).
27. L. Brewer and D. G. Davis, *Met. Trans. AIME* **15A**, 67 (1984).

B

Estimation of Enthalpy of Alloy Formation

Introduction

Various methods based on the properties of pure elements have been developed for estimating the enthalpies of formation of a limited number of alloys. We describe one of the more recent semiempirical methods developed by Miedema et al.,[1-5] applicable to a very large number of alloys. This method is partly based on the Wigner–Seitz[6] atomic cell theory for pure metals, which will be described first since Miedema et al. assume that this theory is also applicable to the alloys.

The Wigner–Seitz cells for pure metals are constructed by drawing planes perpendicularly bisecting the lines joining the atoms in order to form polyhedra. Each polyhedron for a pure metal contains one positively charged atomic core, and the valence electrons of each atom; such cells are shown schematically in Fig. B.1(a) for two-dimensional crystals of pure A and pure B. At the cell boundary, the potential energy, and density of electrons n_{ws} are at their minimum values. The binding energy $Ze_{ii}/2$ of each atom forming Z bonds with its neighbors is equal to the energy of sublimation of the atom, and this energy is assumed to be equal to the energy of valence electrons outside the positive ion core. The total energy of a crystal of N atoms calculated by Wigner and Seitz by introducing a large number of corrections agrees well with the experimental data only for the alkali metals and disagrees for the metals containing d-electrons. The details of these calculations are not relevant for our purposes, but the concept itself is useful as will be seen shortly.

The method of Miedema et al.[1-5] for the estimation of enthalpies of formation, ΔH, for alloys assumes that the Wigner–Seitz theory for pure metals can be expanded to include the binary alloys. When two metals A

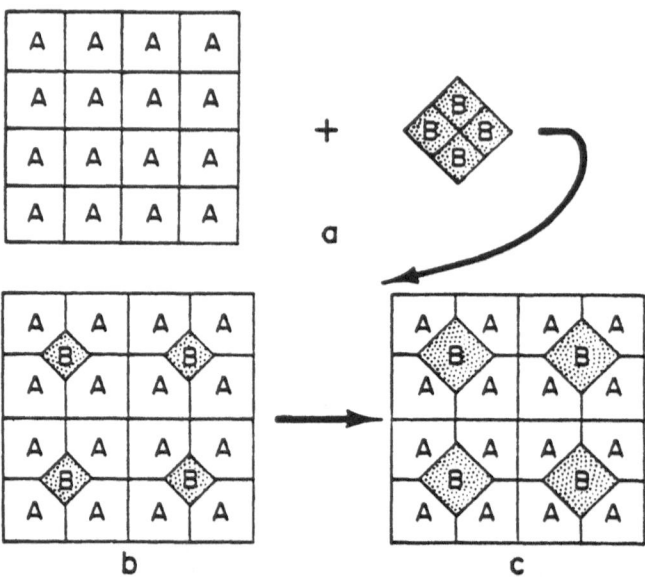

Figure B.1. Wigner–Seitz cells for two-dimensional metals and alloys. (a) Cells for pure metals A and B; they are mixed to form a mechanical mixture as shown in (b). The cell sizes (areas) are identical in (a) and (b). (c) The mechanical mixture becomes an alloy phase and phase boundaries move to smooth out the discontinuity of electron density in (b). In this figure $n_{ws}^A < n_{ws}^B$; hence, the cell boundaries in (b) move to make the B cells larger as shown in (c). Volume effects are related to the change transfer; hence, to the differences in the electronegativities.

and B form a hypothetical mechanical mixture of atoms, as shown in Fig. B.1(b), the shapes of the cells change but the volume of each cell for each type of atom remains unaltered. The cells are therefore drawn to leave no void and no change in each atomic volume after mixing. The shapes of cells in a mechanical mixture have no effect on the energies of atoms so long as their cell volumes remain the same in pure and mixed states. In this condition, ΔH is zero since the alloy is a mechanical atomic mixture. However, at the boundaries in Fig. B.1(b), a discontinuity in the Wigner–Seitz electron density n_{ws} exists because the electron densities for A and B are different, and we assume that $n_{ws}^A < n_{ws}^B$ (superscripts identify the pure metals). This discontinuity is smoothed out upon alloy formation when the cells for B expand at the expense of those for A as shown in Fig. B.1(c). The cell boundary is again where the density of electrons passes through a minimum. Miedema *et al.* assume that the resulting change in the density of electrons at the boundary of the Wigner–Seitz cell, Δn_{ws}, contributes a positive term $q(\Delta n_{ws}^{1/3})^2$ to ΔH where q is an empirical constant. This is the well-known size effect in alloying, i.e., greater differences between the atomic radii of A and B contribute greater positive values to ΔH.

The values of n_{ws} are in electrons per atomic unit cube with one atomic unit cube equal to 0.529 Å (0.1 nm = 1 Å), i.e., $n_{ws} = 1$ corresponds to about 4×10^{22} electrons/cm^3; they were found to be empirically related by $n_{ws}^2 = \phi/V$, where ϕ is the compressibility (or bulk modulus), $\phi = -V(\partial P/\partial V)$, with P in units of 10^4 bars and V the molar volume in cubic centimeters per gram atom. This relationship was originally obtained empirically by plotting $0.5 \ln(\phi/V)$ versus $\ln n_{ws}$, which yielded a straight line for a number of elements.[1] The results for n_{ws}, either calculated directly by the Wigner-Seitz method or obtained by using ϕ/V, are listed in Table B.1, where the values are empirically adjusted to obtain the best fit for the available data on the enthalpy of formation per gram atom of alloy, ΔH, for various alloys. The adjusted values[1-5] differ slightly from the actual values of n_{ws}, but for a limited number of elements they differ significantly; e.g., for Mn and U, the assigned values of n_{ws} are higher by 44 and 36%, respectively. In most cases, however, the actual and empiriclly adjusted values are considerably

Table B.1. Model Parameters for Transition and Noble Elements

	$n_{ws}^{1/3}$	ϕ	$V^{2/3}$		$n_{ws}^{1/3}$	ϕ	$V^{2/3}$
Sc	1.27	3.25	6.1	Rh	1.76	5.40	4.1
Ti	1.47	3.65	4.8	Pd	1.67	5.45	4.3
V	1.64	4.25	4.1	Ag	1.39	4.45	4.7
Cr	1.73	4.65	3.7	La	1.09	3.05	8.0
Mn	1.61	4.45	3.8	Hf	1.43	3.55	5.6
Fe	1.77	4.93	3.7	Ta	1.63	4.05	4.9
Co	1.75	5.10	3.5	W	1.81	4.80	4.5
Ni	1.75	5.20	3.5	Re	1.86	5.40	4.3
Cu	1.47	4.55	3.7	Os	1.85	5.40	4.2
Y	1.21	3.20	7.3	Ir	1.83	5.55	4.2
Zr	1.39	3.40	5.8	Pt	1.78	5.65	4.4
Nb	1.62	4.00	4.9	Au	1.57	5.15	4.7
Mo	1.77	4.65	4.4	Th	1.28	3.30	7.3
Tc	1.81	5.30	4.2	U	1.56	4.05	5.6
Ru	1.83	5.40	4.1	Pu	1.44	3.80	5.2

[a]Electronegativity, ϕ, is in volts; electron density, n_{ws}, is in electrons per (0.529 Å)^3; molar volume, V, is in cm^3 at room temperature. For values related to compressibility, ϕ, see, e.g., V. S. Fromenko and G. W. Samsonov, *Handbook of Thermionic Properties*, Plenum Press, New York (1966), and D. E. Eastman, *Phys. Rev.* **B 2**, 1 (1970); for ϕ, V, and n_{ws}, see K. A. Gschneidner, *Solid State Phys.* **16**, 275 (1964), and V. L. Moruzzi, J. F. Janak, and A. R. Williams, *Calculated Electronic Properties of Metals*, Pergamon Press, Elmsford, New York (1978).

closer to each other, but it must be emphasized that the adjustments in n_{ws} are essential for the usefulness of the proposed semiempirical method. The difference Δn_{ws} is not known exactly, but it is assumed that $\Delta n_{ws}^{1/3}$ for the alloy is equal to $[(n_{ws}^{A})^{1/3} - (n_{ws}^{B})^{1/3}]$ for the pure components A and B.

The second effect is due to the electronegativity of elements, i.e., the potential of electrons in metals, ϕ. The more electronegative elements tend to attract the electrons from the less electronegative elements upon alloying. The electronegativity scales differ depending on the types of measurement, such as the standard electrode potential method, ionization of monatomic gases, and formation of halogen bonds.[7-9] Since the electronegativity of electrons in pure component A is different from that in pure component B, the electrons tend to spend more time about the more electronegative atom after alloying. Therefore, the potential of electrons is lowered upon alloying, and consequently ΔH is lowered by $-p(\phi_A - \phi_B)^2 = -p(\Delta\phi)^2$ where p is an empirical constant. A term of this type has been used previously by Pauling.[7] Again, $\Delta\phi$ for an alloy is not known, but it is assumed that $\Delta\phi = \phi_A - \phi_B$ where ϕ_A and ϕ_B refer to pure A and pure B, respectively. The initial values of ϕ, as used by Miedema *et al.*, were the experimentally measured values in a number of compilations cited in Table B.1. However, these values have been adjusted slightly for empirical fitting of the experimental values for ΔH. The adjustments are small in most cases,[1-5] e.g., the adjusted values for Ti and Zr are approximately 15% smaller than the experimental values, but the remaining values for the transition elements differ less than 15%.

The contribution to ΔH resulting from the changes in electron concentration and electronegativity for certain classes of alloys is expressed by

$$\Delta H = f(x_A, V)g(x_A, n_{ws})[q(\Delta n_{ws}^{1/3})^2 - p(\Delta\phi)^2] \qquad (B.1)$$

where $f(x_A, V)$ is dependent on x_A and volume V, and $g(x_A, n_{ws})$ on x_A and n_{ws} as will be seen later, and p and q are empirical constants. The next task is the determination of p and q.

Empirical Coefficients

The determination of empirical coefficients p and q requires experimental data on ΔH, and in the absence of such data, binary phase diagrams from which the algebraic sign of ΔH can be estimated. Initially required information is whether ΔH is positive or negative. In the absence of experimental data for ΔH, the sign of ΔH was obtained by the following established empirical rules: (1) ΔH is negative for all the binary alloys in which intermetallic compounds or ordered phases have been observed, and (2) ΔH is positive for all the binary alloys when there are no intermetallic

Table B.2. Model Parameters for Nontransition and Nonnoble Metals[a]

	$n_{ws}^{1/3}$	ϕ	$V^{2/3}$		$n_{ws}^{1/3}$	ϕ	$V^{2/3}$
H(1)	1.5	5.20	1.42	B(3)	1.55	4.75	2.8
Li	0.98	2.85	5.5	Al	1.39	4.20	4.6
Na	0.82	2.70	8.3	Ga	1.31	4.10	5.2
K	0.65	2.25	12.8	In	1.17	3.90	6.3
Rb	0.60	2.10	14.6	Tl	1.12	3.90	6.6
Cs	0.55	1.95	16.8	C(4)	1.90	6.20	1.8
Ca	0.91	2.55	8.8	Si	1.50	4.70	4.2
Sr	0.84	2.40	10.2	Ge	1.37	4.55	4.6
Ba	0.81	2.32	11.3	Sn	1.24	4.15	6.4
Be(2)	1.60	4.20	2.9	Pb	1.15	4.10	6.9
Mg	1.17	3.45	5.8	N(5)	1.60	7.00	2.2
Zn	1.32	4.10	4.4	P	1.53	4.95	3.8
Cd	1.24	4.05	5.5	As	1.44	4.80	5.2
Hg	1.24	4.20	5.8	Sb	1.26	4.40	6.6
				Bi	1.16	4.15	7.2

[a] For units, see footnote to Table B.1. Values of $V^{2/3}$ for H, Si, Ga, Ge, C, N, As, Sb, Bi are for hypothetical, close-packed, metallic structures. The numbers in parentheses indicate the valence in the group headed by that element.

compounds and the mutual solubilities are less than 10 at.%. For all the binary alloys of the elements in Tables B.1 and B.2, except the alloys of (Class α element + Class β element) in Table B.3, equation (B.1) is valid. For $\Delta H = 0$, equation (B.1) requires that a plot of $\Delta\phi$ versus $\Delta n^{1/3}$ should leave positive ΔH values, indicated by +, on one side of the straight line passing through the origin, while negative values of ΔH, indicated by −,

Table B.3. Values of Parameter r/p[a]

					Class α							Class β			
IIA			Transition metals, IIIA–VIIIA							IB		Be 0.4	B 1.9	C 2.1	N 2.3
Ca 0.4	Sc 0.7	Ti 1.0	V 1.0	Cr 1.0	Mn 1.0	Fe 1.0	Co 1.0	Ni 1.0		Cu 0.3		Mg 0.4	Al 1.9	Si 2.1	—
Sr 0.4	Y 0.7	Zr 1.0	Nb 1.0	Mo 1.0	Tc 1.0	Ru 1.0	Rh 1.0	Pd 1.0		Ag 0.15		Zn 1.4	Ga 1.9	Ge 2.1	As 2.3
Ba 0.4	La 0.7	Hf 1.0	Ta 1.0	W 1.0	Re 1.0	Os 1.0	Ir 1.0	Pt 1.0		Au 0.3		Cd 1.4	In 1.9·	Sn 2.1	Sb 2.3
	Th 0.7	U 1.0	Pu 1.0									Hg 1.4	Tl 1.9	Pb 2.1	Bi 2.3

[a] To obtain r/p for a binary solid alloy formed by one Class α and one Class β element, multiply the numbers under the elements. For liquid alloys r/p is reduced by a factor of 0.73. For all other alloys, $r/p = 0$. Class β elements, except those in the first column, contain p-type outer electrons.

should be on the opposite side. A mapping of this type for solid alloys is shown in Fig. B.2, wherein the slope of the straight line yields

$$\left| \frac{\Delta\phi}{\Delta n_{ws}^{1/3}} \right|^2 = \frac{q}{p} = 9.4 \tag{B.2}$$

For liquid alloys of the same types of components, a similar mapping exists,[4] yielding again $q/p = 9.4$. Not all the alloys are on the correct side of the straight line in Fig. B.2, and in a similar figure for liquid alloys, but there are very few such alloys.

The corresponding analysis of the results on ΔH for (Class α + Class β) alloys shows that an additional large term, r, is required in equation (B.1):

$$\Delta H = f(x_A, V)g(x_A, n_{ws})p\left[\frac{q}{p}(\Delta n_{ws}^{1/3})^2 - (\Delta\phi)^2 - \frac{r}{p} \right] \tag{B.3}$$

The term r is justified heuristically as the hybridization or interaction energy between the electrons of Class α and Class β elements, particularly when a transition element with d-electrons interacts with the elements having outer p-electrons on the right side of the periodic table. Introduction of r makes the relationship between $\Delta\phi$ and $\Delta n_{ws}^{1/3}$ hyperbolic for $\Delta H = 0$. Therefore, a plot of $\Delta\phi$ versus $\Delta n_{ws}^{1/3}$, as shown in Fig. B.3, should yield

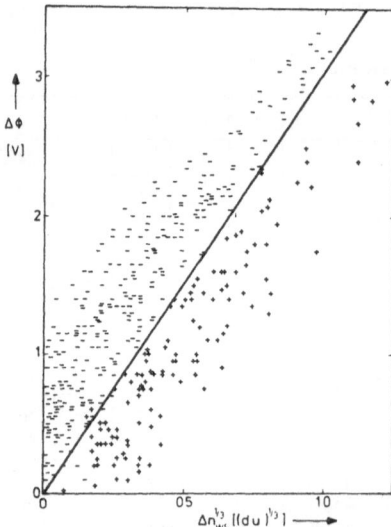

Figure B.2. Mapping of $\Delta\phi$ in volts and $\Delta n_{ws}^{1/3}$ in (density units)$^{1/3}$ for positive, +, and negative, −, values of ΔH according to equation (B.1). Straight line is for $\Delta H = 0$, with slope = $(9.4)^{0.5}$. Results are for binary solid alloys of (transition or noble element) + (transition, noble, alkali, or alkaline earth). (From Miedema and de Chatel[5] with permission.)

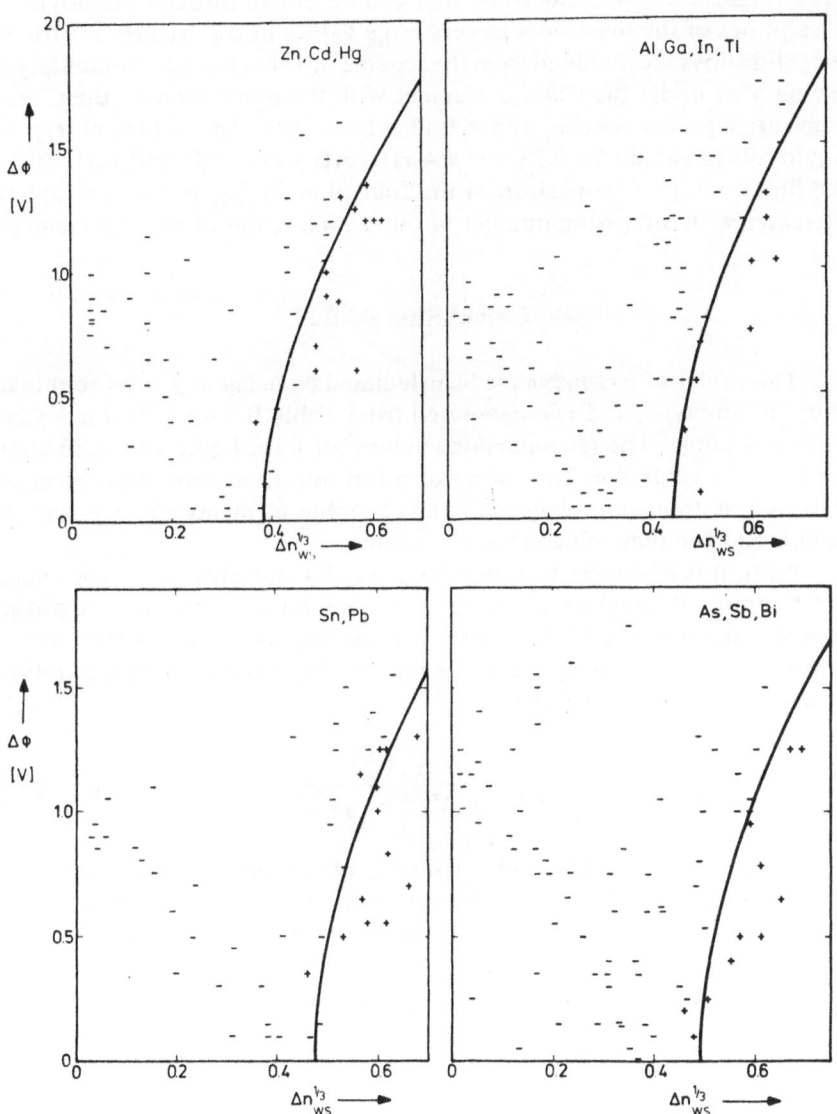

Figure B.3. Mapping of $\Delta\phi$ in volts and $\Delta n_{ws}^{1/3}$ in (density units)$^{1/3}$ for positive, +, and negative, −, values of ΔH according to equation (B.3). Solid curve separates + and − for binary solid alloys of (Class α + Class β). Class β elements are listed in each panel. (From Miedema and de Chatel[5] with permission.)

the value of r/p. Further analysis has shown that q/p is again 9.4; hence, the straight lines with this slope that can be drawn through the origin are asymptotes of the hyperbola at very large values of $\Delta\phi$. The results for r/p for solid alloys are obtained from the scheme in Table B.3, i.e., by multiplying the number under the Class α element with the number under the Class β element; e.g., for Ag–Zn, $r/p = 0.15 \times 1.4 = 0.21$. The values of r/p for liquid alloys should be $0.73 \times r/p = r(l)/p(l)$ where $r(l)$ and $p(l)$ refer to the liquid alloys. Comparison of the four plots in Fig. B.3 shows that r/p increases with increasing number of outer p-electrons of Class β elements.

Calculation of ΔH

The values of ΔH may now be calculated by using $q/p = 9.4$ for binary alloys of any type, and r/p computed from Table B.3 for (Class α + Class β) binary alloys. The recommended values[5] of p for liquid and solid alloys are given in Table B.4. The values of p fall into three sets: those elements classified as transition elements (trans.), noble elements Cu, Ag, and Au (nob.), and the remaining elements (remn.).

Next, it is necessary to formulate $f(x_A, V)$ and $g(x_A, n_{ws})$. The energy effects upon alloying are assumed to be generated at the contact surface between dissimilar cells; therefore, it is further assumed that the surface concentrations x_A^s and x_B^s are more relevant than the bulk concentrations x_A and x_B, and x_A^s is defined by

$$x_A^s = \frac{x_A V_A^{2/3}}{x_A V_A^{2/3} + x_B V_B^{2/3}} \tag{B.4}$$

where V_A and V_B are the molar volumes of the pure elements and $x_B^s = 1 - x_A^s$. Since the cell volumes change upon alloying, V_A for pure A, and V_B for pure B cannot be the same as V_A(alloy) and V_B(alloy) in the alloy,

Table B.4. Values of p for Solid and Liquid Binary Alloys[a]

Alloy type	p
(trans. or nob.) + (trans. or nob.)	14.1
(trans. or nob.) + (remn.)	12.3
(remn.) + (remn.)	10.6

[a]Transition elements (trans.) are shown in Table B.3; noble elements (nob.) are Cu, Ag, and Au; others are designated as remaining elements (remn.). Listed values, with ϕ, $n_{ws}^{1/3}$, and $V^{2/3}$ in Tables B.1 and B.2, yield ΔH in kJ/g-atom (4.184 J = 1 cal).

but the use of V_A and V_B for the pure elements in equation (B.4) is assumed to be valid for a large number of alloys. It is further assumed that $f(x_A, V)$ follows the zeroth approximation to the regular solutions with x_i^s substituted for x_i:

$$f(x_A, V) \equiv f(x_A^s, x_B^s) = x_A^s x_B^s \qquad (B.5)$$

The function $g(x_A, n_{ws}) = g$ is assumed to be

$$g = \frac{2x_A V_A^{2/3} + 2x_B V_B^{2/3}}{(n_{ws}^A)^{-1/3} + (n_{ws}^B)^{-1/3}} \qquad (B.6)$$

where the denominator is called the screening length of electrons. An approximate relationship exists between V_i and n_{ws}^i ($i = $ A or B), and for this reason g varies within a narrow range for all the alloys. However, taking g as a constant leads to a less satisfactory correlation with the experimental values of ΔH. In the original treatment,[1] the product fg was taken as equal to $x_A x_B$; therefore, the product of equations (B.5) and (B.6) is a requirement[4] largely based on empiricism, because fg provided a better fit for ΔH than $x_A x_B$. Substitution of equations (B.4)–(B.6) in equation (B.3) yields

$$\Delta H = \frac{2px_A V_A^{2/3} x_B^s [9.4(\Delta n_{ws}^{1/3})^2 - (\Delta\phi)^2 - (r/p)]}{(n_{ws}^A)^{-1/3} + (n_{ws}^B)^{-1/3}} \qquad (B.7)$$

The partial molar enthalpy of solution of A, $\Delta\bar{H}_A(x_A \to 0)$, when x_A approaches zero, can readily be derived from equation (B.7):

$$\Delta\bar{H}_A(x_A \to 0) = \frac{2pV_A^{2/3}[9.4(\Delta n_{ws}^{1/3})^2 - (\Delta\phi)^2 - (r/p)]}{(n_{ws}^A)^{-1/3} + (n_{ws}^B)^{-1/3}} \qquad (B.8)$$

This equation justifies the use of $V_A^{2/3}$ in equation (B.7), because $\Delta\bar{H}_A(x_A \to 0)/\Delta\bar{H}_B(x_B \to 0)$ is roughly proportional to $(V_A/V_B)^{2/3}$ in the majority of cases considered by Miedema et al. although there are notable exceptions. Comparison of equations (B.7) and (B.8) shows that

$$\Delta H = x_A x_B^s \Delta\bar{H}(x_A \to 0) \qquad (B.9)$$

where $x_A x_B^s$ may have any value but $\Delta\bar{H}(x_A \to 0)$ refers to $x_A \to 0$.

The surface concentration x_i^s should be expressed in terms of the volumes in alloys $V_i(\text{alloy})$ instead of $V_i(\text{pure})$ listed in Tables B.1 and B.2. The crude empirical relation yielding $V_A^{2/3}(\text{alloy})$ is

$$V_A^{2/3}(\text{alloy}) = V_A^{2/3}(\text{pure})[1 + u_A x_B(\phi_A - \phi_B)] \qquad (B.10)$$

$$u_A = \begin{cases} 0.14 \text{ for monovalent A} \\ 0.10 \text{ for divalent A} \\ 0.07 \text{ for trivalent A} \\ 0.04 \text{ for higher valent and for} \\ \qquad \text{all transition and noble A} \end{cases}$$

This equation is used only for large values of $(\phi_A - \phi_B)$ and for alkaline and alkaline earth metals and small-atom stongly electronegative elements such as B(boron), C, H, and N. For example, $V_A(\text{pure}) = 5.5$ and $V_A(\text{alloy}) = 4.98$ for A = Li, B = Al, and $x_B = 0.5$. In contrast, $V_A(\text{pure}) = 4.8$ and $V_A(\text{alloy}) = 4.68$ for A = Ti, B = Fe, $x_B = 0.5$, and $u_A = u_B = 0.04$; the resulting difference in $V_A(\text{pure})$ and $V_A(\text{alloy})$ is considered to be negligible for Fe-Ti. A few examples of calculated and experimental values of ΔH are listed in Table B.5 for solid and liquid alloy phases of equiatomic composition. For example, ΔH for *liquid* equiatomic Au-Zn alloy can be

Table B.5. Calculated and Experimental Values of ΔH of Formation for 1 g-atom of Equiatomic Solid and Liquid Alloy Phases. Results Are for Long-Range Disordered Alloys, Except as Explained in Footnote[a]

	kJ/g-atom			kJ/g-atom	
Alloy	ΔH(calc)	ΔH(exp)	Alloy	ΔH(calc)	ΔH(exp)
	Solid alloys			Liquid alloys[b]	
Al-Au[a]	−37	−38.7[a]	Al-Au	−21	−33.8
Al-Ni[a]	−47	−58.8[a]	Al-Ni	−22	−50.2
Al-Ti	−44	−36.4	Au-Zn	−16	−22.5
Co-Pt[a]	−11	−13.4[a]	Bi-Na	−20	−30.1
Cu-Zn[a]	−10.5	−12.1[a]			
Cu-Zn	−5	−7.7	Cu-Fe	+15	+8.9
Fe-Ti	−19	−20.3	Fe-Si	−18	−37.7
Fe-V	−7	−8.4	Ga-Li	−8	−23.0
Mn-Ni	−8	−14.2	Hg-K	−10	−21.3
Nb-Ni	−31	−22.6	Hg-Pb	+1	+0.2
Ni-Ti	−38	−33.3	K-Pb	−21	−19.8
Pd-Rh	+2	+10	Ni-Ti	−38	−38.5

[a] Ordered structures. Structure of Al-Au is uncertain, but assumed to be ordered in this calculation. Magnitudes of ΔH cannot always predict the existence or the absence of long-range order.
[b] Results for liquid Al-Au, Au-Zn, Cu-Fe, Hg-Pb, and K-Pb and all those for solid phases were recomputed for this book, and experimental data were obtained from R. Hultgren, P. D. Desai, D. T. Hawkins, M. Gleiser, and K. K. Kelley, *Selected Values of the Thermodynamic Properties of Binary Alloys*, ASM, Metals Park, Ohio (1973), and from O. Kubaschewski and C. B. Alcock, *Metallurgical Thermochemistry*, Fifth Edition, Pergamon Press, Elmsford, New York (1979).

calculated by using $p = 12.3$, $x_A = x_{Au} = 0.5$, $x_B^s = x_{Zn}^s = 0.4835$, $(n_{ws}^A)^{-1/3} + (n_{ws}^B)^{-1/3} = 1.395$, and $r/p = 0.3 \times 1.4 \times 0.73 = 0.3066$; the result is

$$\Delta H = (2 \times 12.3 \times 0.5 \times 4.7 \times 0.4835/1.395)$$
$$\times [9.4(1.57 - 1.32)^2 - (5.15 - 4.10)^2 - 0.3066] = -16 \text{ kJ/g-atom}$$

The experimental value is -22.5 kJ/g-atom as listed in Table B.5. Large differences between the calculated and experimental values are attributed to unusual degrees of ordering in liquid alloys.

Ordered Phases

The value of x_B^s increases for ordered structures because a much larger fraction of the surface due to B is in contact with A. A purely empirical relation expressing this fraction, denoted by f_B, is

$$f_B = x_B^s[1 + 8(x_A^s x_B^s)^2] \tag{B.11}$$

Note that for disordered solutions $f_B = x_B^s$, but equation (B.11) must replace x_B^s in equation (B.7). Substitution of equations (B.8) and (B.11) in equation (B.7) yields another form for ΔH, i.e.,

$$\Delta H = \frac{2 p x_A V_A^{2/3}(\text{alloy}) f_B [9.4(\Delta n_{ws}^{1/3})^2 - (\Delta \phi)^2 - (r/p)]}{(n_{ws}^A)^{-1/3} + (n_{ws}^B)^{-1/3}}$$
$$= x_A f_B \Delta \bar{H}_A(x_A \to 0) \tag{B.12}$$

where x_A may assume any value for ΔH; however, $\Delta \bar{H}_A(x_A \to 0)$ after the second equality refers to $x_A \to 0$. For the ordered Cu–Zn alloy of equiatomic composition at 700 K, $\Delta H(\text{exp}) = -12.1$ kJ/g-atom, whereas $\Delta H(\text{calc}) = -10.5$ kJ/g-atom from equation (B.12), in contrast with $\Delta H(\text{exp}) = -7.7$ kJ/g-atom and $\Delta H(\text{calc}) = -5$ kJ/g-atom for the disordered phase as listed in Table B.5.

The average number of solid ordered phases or compounds in a binary system correlates roughly with the magnitude of ΔH; thus, there is often one compound when ΔH is between -4 and -10; three compounds when ΔH is between -20 and -40; five compounds when ΔH is more negative than -75, all ΔH in kJ/g-atom of A + B. Here the model proposed by Miedema *et al.* cannot estimate ΔH unless experimental results indicate that the phases are either ordered or long-range disordered.

Alloys of C, Si, Ge, and N

The alloys of C, Si, Ge, and N require a special treatment,[3,10] i.e., the assumption that they exist in a hypothetical close-packed (cp) metallic structural form. The volume of this cp form for each of these elements is largely based on the volumes of closely neighboring cp elements. The enthalpy of phase change, ΔH_{ph}, for the formation of each hypothetical phase from the stable phase is as follows:

	ΔH_{ph}, kJ/g-atom
C(graphite) → C(cp metal)	100
$0.5N_2$(gas) → N(cp metal)	240
Si(diamond cubic) → Si(cp metal)	33
Ge(diamond cubic) → Ge(cp metal)	25
$0.5H_2$(gas) → H(cp metal)	100

Hydrogen is included here for comparison; it will be discussed in detail in the next section. The derivation of these values is again semiempirical so that the best-fit values can be obtained for ΔH of formation of alloys. The values for N_2(g) include the dissociation energy into atomic N(g), and then condensation into a hypothetical cp metallic phase[10]; this is also the case[11] for H_2(g). As an example we compute ΔH for TiN (A = Ti, B = N), for which $n_{ws}^{1/3} = 1.47(1.60)$, $\phi = 3.65(7.00)$, $V^{2/3} = 4.8(2.2)$, where the values outside the parentheses are for Ti, and those inside are for N(cp metal). For Ti, $V_A^{2/3}$(alloy) = 4.478, and for N, $V_B^{2/3}$(alloy) = 2.347 are obtained by using $u_A = 0.04$ in equation (B.10), and from these values, $x_A^s = 0.6561$ and $x_B^s = 0.3439$, because x_i^s (i = A or B) must be calculated by using $V_i^{2/3}$(alloy). From equation (B.11), $f_B = 0.484$, and the substitution of the foregoing values, as well as $p = 12.3$ and $r/p = 2.3$, in equation (B.7) yields $\Delta H = -273$ kJ/g-atom for 0.5Ti(c) + 0.5N(cp metal) = 0.5TiN. The value of $0.5\Delta H_{ph} = 0.5 \times 240$ for $0.25N_2$(g) → 0.5N(cp metal) added to -273 yields $\Delta H°$ (from N_2) = -153 kJ/g-atom TiN for the enthalpy of formation from gaseous N_2 and crystalline Ti in their naturally existing states, i.e., their standard states. The experimental value for $\Delta H°$ (from N_2) = $\Delta H_g°$ is -169 kJ/g-atom TiN, in good agreement with the calculated value.

The value of $\Delta G°$ at the equilibrium pressure $P(N_2)$ and a temperature T_1 for N_2(g) + nA = $A_n N_2$ can be calculated by writing $\Delta G° = RT \ln P(N_2) = \Delta H_g° - T_1 \Delta S_g°$ from which ΔS_g can be calculated. For the simple case of $P(N_2) = 1$, it is evident that $\Delta S_g° = \Delta H_g°/T_1$. The resulting value of $\Delta S_g°$, and the experimental or calculated value of $\Delta H_g°$ can be used as constants in $\Delta G° = \Delta H_g° - T \Delta S_g°$ to compute $P(N_2)$ at other temperatures within a possible range of 200 K above or below T_1, assuming

Table B.6. Calculated and Experimental Values of $\Delta H°$ for Selected Ordered Binary Phases of H, C, N, Si, and Ge[a]

Phase	$\Delta H°$(calc)	$\Delta H°$(exp)	Phase	$\Delta H°$(calc)	$\Delta H°$(exp)
$Ti_{0.33}H_{0.67}$	−43	−42			
$V_{0.33}H_{0.67}$	−11	−18			
$Ni_{0.67}H_{0.33}$	+4	−1			
$Zr_{0.33}H_{0.67}$	−63	−55	CrN	−40	−60
$Nb_{0.33}H_{0.67}$	−28	−20	Fe_2N	−17	0
$Pd_{0.67}H_{0.33}$	+2	−7	HfN	−169	−185
$La_{0.27}H_{0.73}$	−51	−62	LaN	−150	−151
$La_{0.36}H_{0.64}$	−55	−67	$Ta_{0.67}N_{0.33}$	−88	−90
$Hf_{0.37}H_{0.63}$	−55	−42	TiN	−153	−169
$Th_{0.33}H_{0.67}$	−65	−49	VN	−86	−110
UH_3	−18	−32	ZrN	−184	−184
Fe_2C	−13	+8	FeSi	−25	−39
HfC	−79	−109	MnSi	−42	−30
Mn_7C_3	−21	−13	ThSi	−100	−63
MoC	−38	−8	TiSi	−79	−65
NbC	−71	−69	ZrSi	−100	−74
TaC	−67	−72			
TiC	−75	−92	Mg_2Ge	−12	−38
WC	−29	−19	Ni_2Ge	−20	−37
ZrC	−84	−96	UGe	−54	−31

[a] $\Delta H°$ is in kJ/g-atom of alloy formed from the elements in their standard states (Refs. 3, 10, 11).

that the stoichiometry of A_nN_2 remains unchanged and that the reactant metal is pure A.

Selected values of $\Delta H°$ from the component elements in their standard states for a number of carbides, nitrides, silicides, and germanides[3,10] are given in Table B.6.

Alloys of Hydrogen

The values of ΔH for metal hydrides can be calculated by following the procedure given by Bouten and Miedema[11] and by using the listed values in Table B.2 with $r/p = 3.9$ for H, and $\Delta H_{ph} = 100$ kJ/g-atom H. The value of $V_H^{2/3} = 1.42$ is based on the closest approach of approximately 2.3 Å between two H–H atoms in binary compounds of hydrogen with metals, wherein the atomic ratios of H to metal are high. However, $V_H^{2/3}$ is again a best-fit semiempirical value for hydrogen, as well as the remaining parameters. The value of r/p for the alkaline–transition element alloys is zero, but in contrast, despite the fact that H(cp metal) should also be an alkaline metal, r/p is not zero for H but equal to 3.9 for the [H(cp metal) + transition element] alloys. For example, $r/p = 3.9 \times 0.7$ for the La–H alloys.

The calculations of $V_A^{2/3}$(alloy) and f_B for H(cp metal) are different from those for other elements because these parameters are calculated by a different procedure. The analytical form adopted for f_B is identical with equation (B.11) but equation (B.10) is rewritten as

$$V_A^{2/3}(\text{alloy}) = V_A^{2/3}(\text{pure A})[1 + u_A f_B(\phi_A - \phi_B)] \qquad (B.13)$$

The corresponding equation for $V_B^{2/3}$(alloy) is obtained by interchanging the subscripts A and B in this equation. The value of f_B is not known since $V_A^{2/3}$(alloy) is not known for computation of x_A^s, because x_A^s must be computed by using $V_A^{2/3}$(alloy) for the interstitial elements. Therefore, x_i^s is computed first by using $V_i^{2/3}$(pure) for pure A and pure B, in equation (B.4) and then the result is substituted in equation (B.11) to compute the first value of f_B. From this value of f_B, $V_A^{2/3}$(alloy) is computed from equation (B.13). This process is repeated once more to obtain closer values of f_B and $V_i^{2/3}$(alloy); however, in calculating the second value of $V_i^{2/3}$(alloy), $V_i^{2/3}$(pure i) is always used on the right side of equation (B.13). For example, for $Ti_{0.33}H_{0.67}$ (A = Ti, B = H), the parameters are: $V_A^{2/3}$(pure) = 4.8; $V_B^{2/3}$(pure) = 1.42; $(n_{ws}^A)^{-1/3} + (n_{ws}^B)^{-1/3} = 1.347$; $\Delta n_{ws}^{1/3} = -0.03$; $\phi_A - \phi_B = -1.55$; $u_A = 0.04$; and $u_B = 0.14$. The calculated first and second values, denoted as (I) and (II) after each symbol, are as follows:

$$x_A^s(\text{I}) = 0.33 \times 4.8/(0.33 \times 4.8 + 0.67 \times 1.42) = 0.6248;$$

$$x_B^s(\text{I}) = 1 - x_A^s(\text{I}) = 0.3752$$

$$f_A(\text{I}) = 0.6248[1 + 8(0.6248 \times 0.3752)^2] = 0.8995$$

$$f_B(\text{I}) = 0.3752[1 + 8(0.3752 \times 0.6248)^2] = 0.5402$$

$$V_A^{2/3}(\text{alloy, I}) = 4.8(1 - 0.04 \times 0.5402 \times 1.55) = 4.639$$

$$V_B^{2/3}(\text{alloy, I}) = 1.42(1 + 0.14 \times 0.8995 \times 1.55) = 1.697$$

$$x_A^s(\text{II}) = 0.33 \times 4.639/(0.33 \times 4.639 + 0.67 \times 1.697) = 0.5738;$$

$$x_B^S(\text{II}) = 1 - x_A^s(\text{II}) = 0.4262$$

$$f_A(\text{II}) = 0.5738[1 + 8(0.4262 \times 0.5738)^2] = 0.8483$$

$$f_B(\text{II}) = 0.4262[1 + 8(0.5738 \times 0.4262)^2] = 0.6301$$

$$V_A^{2/3}(\text{alloy, II}) = 4.8(1 - 0.04 \times 0.6301 \times 1.55) = 4.612$$

$$V_B^{2/3}(\text{alloy, II}) = 1.42(1 + 0.14 \times 0.8483 \times 1.55) = 1.681$$

The third approximation is not necessary because the differences for $V_i^{2/3}(\text{alloy, II}) - V_i^{2/3}(\text{alloy, I})$, ($i = A$ or B), are already small and the calculated third value for $V_i^{2/3}(\text{alloy, III})$ is very close to $V_i^{2/3}(\text{alloy, II})$. The result for ΔH is as follows:

$$\Delta H = 2 \times 12.3 \times 0.33 \times 4.612 \times 0.6301[9.4(-0.03)^2 - (-1.55)^2 - 3.9]/1.347$$
$$= -110 \text{ kJ/g-atom}$$

$$\Delta H°[\text{from } H_2(g)] = -110 + 0.67 \times 100 = -43 \text{ kJ/g-atom}$$

The experimental value is -42 kJ/g-atom, in very close agreement with the calculated value. The increase in the volume of Ti upon absorption of hydrogen is calculated from

$$V(\text{alloy}) = 0.33 \, \bar{V}_{Ti} + 0.67 \, \bar{V}_H = 4.73 \qquad (B.14)$$

where $V(\text{alloy})$ is the volume of the alloy, and the partial molar volume V_i is assumed to be the same as $V_i(\text{alloy, II})$. The volume of *pure* 0.33Ti is 3.47 cm^3; hence, $\Delta V = 4.73 - 3.47 = 1.26$ which is the increase in the volume of Ti after the dissolution of hydrogen, and the calculated value, $\Delta V/V = 1.26/3.47 = 0.36$ or 36%, agrees well with the experimental value, 31%.

Similar calculations for $La_{0.36}H_{0.64}$ yield $x_A^s = 0.705$; $x_B^s = 0.295$; $V_A^{2/3}(\text{alloy}) = 7.727$; $V_B^{2/3}(\text{alloy}) = 1.826$; $f_B = 0.3966$; $\Delta H = -119$ kJ/g-atom; and $\Delta H°(\text{from } H_2) = -55$ kJ/g-atom alloy, in fair agreement with the experimental value of -67 kJ/g-atom. The calculated value of $\Delta V/V$ is 8%, in poor agreement with the experimental value[11] of 19%. The results for a number of binary hydrides selected from Bouten and Miedema[11] are listed in Table B.6.

Ternary Hydrides

Intermetallic compounds of a number of transition elements form stable hydrides. The enthalpies of formation of such hydrides can be estimated by a modification of the preceding method.[11]

The enthalpy of formation of a ternary hydride $AB_n H_{v+w}$ from H_2 and AB_n can be estimated from the standard enthalpies of formation of two binary hydrides AH_v and $B_n H_w$, and the intermetallic phase or compound AB_n. Metal A and B are both transition elements, and A is assumed to form a very stable solid hydride at a sufficiently low temperature, but B forms a much less stable hydride than A. The generalized equation for estimation of $\Delta H_g^\circ(AB_n H_{v+w})$ by reaction of AB_n with $H_2(g)$, as indicated by the subscript g on ΔH, is as follows[11]:

$$0.5(v + w)H_2 + AB_n = AB_n H_{v+w}$$

$$\Delta H_g^\circ(AB_n H_{v+w}) = \Delta H_g^\circ(AH_v) + \Delta H_g^\circ(B_n H_w) \qquad (B.15)$$

$$- (1 - F)\,\Delta H(AB_n)$$

where v, w, and F depend on stoichiometric coefficient n and on the atomic radius of A, i.e., whether A is one of the first or second group of metals as shown in the first column of Table B.7. The empirical term F is a measure of bonding between A and B atoms in the ternary hydride. Thus, when n is greater than 5, and v is sufficiently high, H atoms completely surround A and the bonding between A and B is greatly weakened so that $F = 0$; in this case, equation (B.15) refers to the following reactions:

$$0.5vH_2 + A = AH_v \qquad \Delta H_g^\circ(AH_v)$$

$$0.5wH_2 + nB = B_n H_w \qquad \Delta H_g^\circ(B_n H_w)$$

$$AB_n = A + nB \qquad -\Delta H(AB_n)$$

$$AH_v + B_n H_w = AB_n H_{v+w} \qquad \Delta H_{H\text{-}H}(\text{hydride} + \text{hydride}) \approx 0$$

$$\overline{0.5(u + v)H_2 + AB_n = AB_n H_{u+v}} \qquad \Delta H_g(AB_n H_{v+w}) = \Delta H_g^\circ(AH_v)$$

$$+ \Delta H_g^\circ(B_n H_w) - \Delta H(AB_n) \quad (B.16)$$

This equality is possible only if $\Delta H_{H\text{-}H}$ for the last reaction is zero.

Table B.7. Values of Parameters for Calculation of Enthalpies of Hydrides Formed from Intermetallic Compounds AB_n

Metal A[a]	AB_n	$AB_n H_{v+w}$	v	w	F
Ti, Hf, Zr, V, Nb, Ta, Sc	AB_5	$AB_5 H_5$	2	3	0.1
	AB_3	$AB_3 H_4$	2	2	0.2
	AB_2	$AB_2 H_{3.5}$	2	1.5	0.4
	AB	ABH_2	1.5	0.5	0.6
Lanthanides Y, Th, U, Pu	AB_5	$AB_5 H_6$	2.5	3.5	0.1
	AB_3	$AB_3 H_5$	2.5	2.5	0.2
	AB_2	$AB_2 H_4$	2.5	1.5	0.4
	AB	$ABH_{2.5}$	2	0.5	0.6

[a]The two groups of metals have considerably differing atomic sizes in alloyed states.

The values of n, v, w, and F are listed in Table B.7. As an example, we calculate $\Delta H_g(AB_5H_6) = \Delta H_g(LaNi_5H_6)$ from the values listed in Tables B.7 and B.8, wherein $v = 2.5$, $w = 3.5$, $F = 0.1$, $\Delta H_g^\circ(LaH_{2.5}) = -185$ kJ/mole, and $\Delta H_g^\circ(Ni_5H_{3.5}) = +10 \times 3.5 = 35$ kJ/mole. Calculations by using equations (B.10), (B.11), and (B.12) yield $\Delta H(LaNi_5) = -70.5$ kJ/mole, and substitution of these three values of ΔH in equation (B.14) with $F = 0.1$ gives $\Delta H_g(LaNi_5H_6) = -87$ kJ/mole, in good agreement with the experimental value,[13] $\Delta H_g(LaNi_5H_6) = -94$ kJ/mole. The compound $LaNi_5$ absorbs and desorbs hydrogen reversibly at ambient pressures and temperatures and it contains twice as much hydrogen per cm^3 of alloy upon saturation as pure cryogenic liquid hydrogen.[12-14] This property makes $LaNi_5$ an attractive medium for hydrogen storage.

Previous calculations[12,13] were based on setting $v = w$, and using equation (B.16) for a large number of hydrides, but the results from equation (B.15) should be preferred.

The pressure of $H_2(g)$ over the binary and ternary alloys can be calculated by using ΔS° for the reactions in which 1 mole of $H_2(g)$ is the

Table B.8. Calculated Values of Standard Enthalpies of Formation ΔH_g° Used for Calculation of $\Delta H_g(AB_n H_{v+w})$

ΔH_g° (kJ/mole)				ΔH_g° (kJ/g-atom H)	
A	$AH_{2.5}$	AH_2	$AH_{1.5}$	B	$B_n H_w (n > w)$
Sc	−192	−185	−158	V	−37
Y	−205	−188	−153	Nb	−59
La	−185	−173	−135	Ta	−54
Ti	−125	−136	−120	Cr	−7
Zr	−197	−188	−160	Mo	−3
Hf	−165	−161	−138	W	+8
Th	−218	−196	−158	Mn	−25
U	−95	−96	−80	Tc	+18
Pu	−123	−123	−103	Re	+26
V	−12	−34	−42	Fe	+7
Nb	−74	−83	−78	Ru	+21
Ta	−65	−75	−72	Os	+24
				Co	+9
				Rh	+12
				Ir	+21
				Ni	+10
				Pd	+5
				Pt	+13

reactant, e.g.,

$$H_2(g) + (1/3)LaNi_5 = (1/3)LaNi_5H_6 \qquad (B.17)$$

For $P_{H_2}=1$, $\Delta G°=RT \ln P_{H_2}=0$, and $(1/3)[\Delta H_g(LaNi_5H_6)]=\Delta H_g°(B.17)=T \Delta S_g°(B.17)$ for this reaction as indicated by equation (B.17). Examination of the available data[12] for a number of hydrides indicates that $\Delta S_g° \approx -125 \pm 25$ J/mole H_2-K for reaction (B.17) and other similar reactions, each with a single metal at ambient temperatures, and this value is very close to the negative value of the standard entropy, $S°$, of $H_2(g)$, indicating that the standard entropies of $LaNi_5$ and $LaNi_5H_6$ are not greatly different from each other. This value of $\Delta S_g°(B.17)$ corresponds to $T \Delta S_g°(B.16)=\Delta H_g°(B.16)\approx-37.5\pm7.5$ kJ/mole H_2. Therefore, for the binary and ternary hydrides $P_{H_2}>1$ atm for $\Delta H_g°(B.17)>-37.5$ kJ/mole H_2, and $P_{H_2}<1$ atm for $\Delta H_g°(B.17)<-37.5$ kJ/mole H_2, i.e., the more exothermic hydrides are more stable as expected. In general, the plateau pressure or the equilibrium pressure P_{H_2}, which is the pressure at which two solid phases $LaNi_5$ and $LaNi_5H_6$ coexist with H_2, does not exactly refer to reaction (B.17), because the $LaNi_5$ phase is not pure, i.e., it dissolves a small amount of hydrogen, and the number of H atoms in $LaNi_5H_6$ decreases with increasing temperature. Nevertheless, these effects for many hydrides are not too great, and from a knowledge of $\Delta H_g°(B.17)$ and $\Delta S_g°(B.17)=-125$ J/mole H_2-K, it is possible to estimate the equilibrium pressure P_{H_2} for various binary and ternary hydrides. A more exact calculation is possible when P_{H_2} at one temperature, T_1, and $\Delta H_g°(B.17)$ for the reaction are known from which $\Delta S_g°(B.17)$ can be calculated from $\Delta G°=RT \ln P_{H_2}=\Delta H_g°(B.17)-T_1 \Delta S_g°(B.17)$ and then the resulting values of $\Delta H_g°(B.17)$ and $\Delta S_g°(B.17)$ can be used as constants to calculate P_{H_2} at higher and lower temperatures in a possible range of roughly ±200 K.

Ternary and Multicomponent Alloys

The enthalpies of formation of ternary alloys other than those with H, C, and N may be computed by using equation (B.7) wherein ΔH may be reidentified as ΔH_{AB}. The resulting equation for ΔH_{ABC} for the ternary alloy is given by

$$\Delta H_{ABC} = \Delta H_{AB} + \Delta H_{AC} + \Delta H_{BC} \qquad (B.18)$$

This is expected because ΔH_{AB} in equation (B.7) is obtained from pairwise interaction of atoms through their electrons, similar to the pairwise interaction of atoms in the zeroth approximation to regular solutions, except that instead of the ordinary atom fraction, the surface atom fraction x_i^s is used in equation (B.7), and x_i^s refers to the ternary composition [see equation

(B.7)]. It is evident that x_i^s must be replaced by f_i^j if the ternary alloys are ordered. Extension of equation (B.18) to multicomponent alloys is then self-evident.

Concluding Remarks

The enthalpies of formation of the binary alloys, ΔH, from equation (B.7) with $r/p = 0$ are for the nonordered alloys, as well as for the liquid alloys. The effect of crystal structures and temperature on ΔH cannot be accounted for since this would unduly tax the method. The parameters used in this method have been adjusted and readjusted in some cases so that the results prior to Ref. 3 do not always agree with the more recent improved values. It is possible to refine the method by readjusting these parameters, particularly r/p, to obtain refined fits to specific groups of alloys, e.g., binary alloys of Ti with the transition elements, or Ti with alkaline earth elements.

The overall success of this method is very good considering the fact that it covers nearly all the possible alloys, although some important and interesting alloys of O, S, Se, and Te are not included; however, the procedure can be extended to these elements when experimental results for typical alloys of these elements become available. This method is 95% successful in predicting the sign of ΔH according to Phillips.[15] A comparison of the available values of ΔH for 40 binary systems formed by Al, Ti, Hf, Cr, Fe, Co, Ni, Nb, Mo, and W with the calculated values by the Miedema method shows that the differences generally do not exceed 8 kJ/g-atom, the largest difference being 17 kJ/g-atom within a range of +47 to −62 kJ/g-atom for ΔH according to Kaufman.[16] The disagreement is large for silicides and germanides, and for relatively small atoms alloyed with large atoms, leading to large coordination numbers.[17-19]

The second term, $p(\Delta\phi)^2$, in equation (B.1) is quite similar to the term $-96.487b(\Psi_A - \Psi_B)^2$ used by Pauling[7] with his electronegativities, Ψ_i, and the effective number of bonds, b, which is not well-defined for alloys. In fact, there is an approximate linear relationship between the electronegativity scales ϕ and Ψ, and the latter is closely related to other electronegativity scales as discussed in detail by Bennett and Watson.[8] On the average, ϕ is larger by a factor of 2.5 than Ψ, as can be seen by comparing Ψ in the periodic table of Appendix G with those for ϕ in Tables B.1 and B.2.

A form of the regular solution equation proposed by Hildebrand and Scott[20] contributes a positive term to the enthalpy of alloy formation, i.e.,

$$\Delta H(\text{kJ/g-atom of alloy}) = (x_A V_A + x_B V_B)\left[\left(\frac{\Delta E_A}{V_A}\right)^{0.5} - \left(\frac{\Delta E_B}{V_B}\right)^{0.5}\right]^2 x_A^v x_B^v$$

$$(B.19)$$

where $x_A^v = x_A V_A/(x_A V_A + x_B V_B)$ is the volume fraction in terms of the atomic volumes of pure metals V_A and V_B and their atomic fractions x_A and x_B. The term $\Delta E_A/V_A$ is the energy of vaporization or sublimation of the pure metal in kJ/g-atom divided by V_A in cm^3, and the square root of this ratio is denoted by δ_A, which is called the solubility parameter. For example, at 1000 K, $\Delta E_{Al} = 305.2$ kJ/g-atom, $V_{Al} = 11.3$ cm^3, $\Delta E_{Al}/V_{Al} = 27.0$ kJ/cm$^3 = \delta_A^2$, and the solubility parameter is $\delta_A = 5.20$. In the simple case of $V_A = V_B$, equation (B.19) assumes the familiar form $\Delta H = NW_{AB}x_A x_B$ given by equation (4.5). Boom et al.[4] have shown that there is a linear relationship between $\log(\Delta E_i/V_i)$ and $\log(n_{ws}^i)$ ($i =$ metals A or B); therefore, the positive term in equation (B.1) is related to equation (B.19). The mathematical handling of equation (B.19) is simplified if the coefficient $(x_A V_A + x_B V_B)$ is approximated by $(V_A + V_B)/2$. In addition, the Pauling term may be added to the coefficient of $x_A^v x_B^v$ to write

$$\Delta H(\text{kJ/g-atom}) = \left[-96.487 b(\Psi_A - \Psi_B)^2 + \frac{V_A + V_B}{2}(\delta_A - \delta_B)^2 \right] x_A^v x_B^v$$

(B.20)

as proposed by Mott.[21] A two-term equation for ΔH has also been proposed by Shimoji and Niwa.[22] Further, Predel and Sandig[23] have shown that the positive contribution to ΔH in liquid alloys is due to the differences in the atomic volumes which is related to the second term in equation (B.1), because V_i and n_{ws}^i are also related. However, some metals with very small size differences form alloys that have large positive enthalpies of formation. Numerous other correlations have been attempted to account for attractive (or negative), and repulsive (or positive) terms in equations containing two terms for ΔH. Earlier methods have been presented in an excellent monograph by Prigogine et al.[24]

The compositional variables in equation (B.20) are the volume fractions instead of atomic or mole fractions. Comparison of equations (B.20) and (4.9)–(4.14) shows that

$$2RT \leqslant -96.487 b(\Psi_A - \Psi_B)^2 + \frac{V_A + V_B}{2}(\delta_A - \delta_B)^2 \qquad \text{(B.21)}$$

where the equal sign determines the critical point T_c and the inequality is for the existence of immiscibility in the binary system (cf. also Fig. 4.1). Mott[21] has found that for an overwhelming majority of binary liquid alloys, equation (B.21) is valid; therefore, it can be used to predict the presence or absence of immiscibility in liquid alloys for which phase diagrams have not yet been determined. The values of b for this purpose range from 1 to 6 as tabulated and discussed in detail by Mott.

A few more comments based on the more recent publications are appropriate. Minor changes in ϕ and n_{ws} have been recommended by Niessen *et al.*[25] for Ti, Zr, Nb, La, Hf, Re, U, Cu, and Ag. For example, the largest change is 4% in ϕ for Ti, U, and Re, and 8% in $n_{ws}^{1/3}$ for La, and 4% for Ti. It is significant that the values of n_{ws} are very closely related to the values of the interatomic electron densities from the self-consistent band structure calculations.[26] (See Ref. 31 with comments.)

A more recent attempt has been made[27] to improve the success of the Miedema method for the binary terminal solid solutions of transition metals having $4d$ electrons. For this purpose, an elastic term, and a structural term are added to ΔH of liquid alloy formation. Further, the Miedema method has been extended to the phosphorus alloys by Niessen,[28] and it is possible to devise similar methods for the alloys of oxygen and sulfur.

Watson and Bennett[29] have developed a method for estimation of ΔH for the equiatomic binary alloys of transition elements. According to Gachon, Charles, and Hertz,[30] this method, on the average, agrees well with the Miedema method.

References

1. A. R. Miedema, *J. Less-Common Met.* **32**, 117 (1973).
2. A. R. Miedema, R. Boom, and F. R. deBoer, *J. Less-Common Met.* **41**, 283 (1975).
3. A. R. Miedema, *J. Less-Common Met.* **46**, 67 (1976).
4. R. Boom, F. R. deBoer, and A. R. Miedema, *J. Less-Common Met.* **45**, 237 (1976), **46**, 271 (1976).
5. A. R. Miedema and P. F. de Chatel, in *Theory of Alloy Phase Formation*, edited by L. H. Bennett, Metall. Soc. AIME, p. 344 (1980).
6. E. Wigner and F. Seitz, *Phys. Rev.* **43**, 804 (1946).
7. L. Pauling, *The Nature of the Chemical Bond*, Second Edition, Cornell University Press, Ithaca, N.Y. (1960).
8. J. A. Alonso and L. A. Girifalco, in *Theory of Alloy Phase Formation*, edited by L. H. Bennett, Metall. Soc. AIME, p. 484 (1980); see also L. H. Bennett and R. E. Watson, in *Theory of Alloy Phase Formation*, p. 390.
9. C. H. Hodges, *J. Phys. F* **7**, L247 (1977); see also Ref. 5, p. 503.
10. P. C. P. Bouten and A. R. Miedema, *J. Less-Common Met.* **65**, 217 (1979).
11. P. C. P. Bouten and A. R. Miedema, *J. Less-Common Met.* **71**, 147 (1980).
12. H. H. van Mal, K. H. J. Buschow, and A. R. Miedema, *J. Less-Common Met.* **35**, 65 (1974).
13. A. R. Miedema, K. H. J. Buschow, and H. H. van Mal, *J. Less-Common Met.* **49**, 463, 473 (1976). See also K. H. J. Buschow, H. H. van Mal, and A. R. Miedema, *J. Less-Common Met.* **42**, 163 (1975).
14. J. H. N. van Vucht, F. A. Kuijpers, and H. C. A. M. Bruning, *Philips Res. Rep.* **25**, 133 (1970).
15. J. C. Phillips, in *Theory of Alloy Phase Formation*, edited by L. H. Bennett, Metall. Soc. AIME, p. 330 (1980).
16. L. Kaufman, *Calphad* **1**, 300 (1977).
17. K. C. Mills, *Thermodynamic Data for Inorganic Sulphides, Selenides, and Tellurides*, Butterworths, London (1974).

18. O. Kubaschewski, *High Temp. High Pressures* **4**, 1 (1972).

19. L. B. Pankratz, J. M. Stuve, and N. A. Gokcen, *Thermodynamic Data for Mineral Technology*, Bureau of Mines Bull. 677 (1984).

20. J. H. Hildebrand and R. L. Scott, *Regular Solutions*, Prentice-Hall, Englewood Cliffs, New Jersey (1962); see also *The Solubility of Nonelectrolytes*, Third Edition, Reinhold, New York (1950).

21. B. W. Mott, *J. Mater. Sci.* **3**, 424 (1968).

22. M. Shimoji and K. Niwa, *Acta Metall.* **5**, 496 (1957).

23. B. Predel and H. Sandig, *Z. Metallkd.* **60**, 208 (1969).

24. I. Prigogine, A. Bellemans, and V. Mathot, *The Molecular Theory of Solutions*, North-Holland, Amsterdam (1957).

25. A. K. Niessen, F. R. de Boer, R. Boom, P. F. de Chatel, W. C. M. Mattens, and A. R. Miedema, *Calphad* **7**, 51 (1983).

26. F. R. de Boer, R. Boom, and A. R. Miedema, *Physica* **101B**, 294 (1980).

27. A. R. Miedema and A. K. Niessen, *Calphad* **7**, 27 (1983).

28. A. K. Niessen, *High Temp. High Pressures* **14**, 649 (1982); see also *J. Less-Common Met.* **82**, 75 (1981).

29. R. E. Watson and L. H. Bennett, *Calphad* **5**, 25 (1981).

30. J. C. Gachon, J. Charles, and J. Hertz, *Calphad* **9**, 29 (1983).

31. F. v. d. Woude and A. R. Miedema, *Solid State Commun.* **39**, 1097 (1980); *Physica* **B100**, 145 (1980); A. R. Miedema, private communication. According to these references, the most recent view is that the volume effects are also related to the differences in the electronegativities.

C

Correlation of Thermodynamic Properties in Dilute Solutions

Several empirical attempts have been made to correlate the excess partial molar enthalpies and entropies of solution for metals, and the standard enthalpies and entropies of reactions of gases with their dilute solutions in metals. We shall limit this appendix to the correlation of (1) the excess partial molar enthalpy, $\Delta \bar{H}_i = \bar{H}_i^e$, with the excess partial molar entropy, $\bar{S}_i^e$, in dilute solutions of solute metals i in solvent metals and (2) $\Delta H^{\bullet}$ with $\Delta S^{\bullet}$ in equilibria of gaseous oxygen with dissolved oxygen in metals as discussed in two recent publications.[1,2] Thermodynamic behavior of a solute metal A in a solvent metal B dictates the processes for effective removal of A from B in order to purify B. Further, dilute solutions of hydrogen, carbon, nitrogen, oxygen, and sulfur as solutes generally play important technological roles in purification of metals. Therefore, in the absence of data, it is useful to estimate thermodynamic properties of such dilute solutions.

Correlation of $\bar{H}_i^e$ and $\bar{S}_i^e$ in Binary Metal–Metal Systems

A reasonable correlation of $\bar{H}_i^e$ and $\bar{S}_i^e$ in dilute binary solutions of metals i in solvent metals requires experimental data for sufficient numbers of systems so that the resulting correlation can be used to estimate $\bar{S}_i^e$ from the available data for either $\bar{H}_i^e$ or $\bar{G}_i^e$. A preliminary attempt of this type has recently been made by Kubaschewski.[1] The data selected for this purpose by Kubaschewski from various sources[3-12] are listed in Table C.1 and plotted in Fig. C.1. The scale in Fig. C.1(b) is expanded for clarity in plotting the values of $|\bar{H}_i^e|$ smaller than those in (a). The system deviating strongly from

Table C.1. Values of $\bar{H}_i^e$ and $\bar{S}_i^e$ for Dilute Binary Solutions of Solute i in Metals^a

Solvent	Solute	State	T, K	$\bar{H}_i^e$, kJ/g-atom	$\bar{S}_i^e$, J/g-atom-K	$\bar{H}_i^e/\bar{S}_i^e$, in K × 10^{-3}	Ref.
Ag	Au	s	800	−20.3	−5.7	3.57	7
Ag	Cd	s	673	−26.8	−6.25	4.3	7
Ag	Ge	l	1250	−12.5	−2.7	4.6	7
Ag	Ni	l	—	+76.	+19.7	3.85	12
Ag	Pb	l	1273	+10.5	+8.9	1.2	7
Al	Fe	l	1873	−128.5	−38.7	3.3	7
Al	Sn	l	973	+24.7	+9.4	2.63	7
Al	Zn	s	653	(+15.7)	(+6.0)	2.6	7
Al	Zn	l	1000	+10.6	+4.2	2.53	7
Au	Ag	s	800	−16.9	−5.8	2.93	7
Au	Ag	l	1350	−20.6	−9.05	2.3	7
Au	Cd	s	700	−65.	−34.	1.9	7
Au	Cu	s	800	−11.6	+2.55	—	7
Au	Cu	l	—	−15.6	+5.5	—	7
Au	Fe	s	1123	+25.5	+21.8	1.17	7
Au	Fe	l	1473	+65.	+53.5	1.2	7
Au	Ni	s	1150	+21.5	+5.1	4.2	7
Au	Pd	s	—	−46.7	−12.6	3.7	8
Au	Pt	s	—	−2.65	−14.6	—	8
Bi	Ag	l	1000	(+13.4)	(+3.0)	4.47	7
Bi	Al	l	1173	+19.9	+4.23	4.7	7
Bi	Cd	l	773	+2.85	+3.65	0.8	7
Bi	Cu	l	1200	+23.5	+9.8	2.4	7
Bi	In	l	900	−5.76	−0.57	—	7
Bi	Na	l	773	−32.	+35.4	—	7
Bi	Pb	l	700	−3.6	+1.34	—	7
Bi	Sn	l	600	——— Nearly ideal ———			7
Bi	Tl	l	723	−22.2	−0.12	—	7
Bi	Zn	l	873	+13.6	+6.9	—	7
Cd	Bi	l	773	(+8.5)	+9.95	—	7
Cd	Mg	s	543	−13.2	+3.35	—	7
Cd	Mg	l	930	−20.6	−4.2	4.9	7
Cd	Na	l	673	−37.	−32.5	1.15	7
Cd	Pb	l	773	+16.	+6.9	2.3	7
Cd	Sb	l	773	+5.15	+21.6	—	7
Cd	Tl	l	750	+12.8	+5.8	2.2	7
Cd	Zn	l	800	+8.8	+1.0	—	7
Co	Cr	s	1473	+2.8	+1.75	1.6	8
Co	Ni	s	—	——— Nearly ideal ———			8
Co	Si	l	1873	−147.	−25.5	5.75	6
Cr	α-Fe	s	—	+21.5	+8.1	2.65	7

Table C.1 (continued)

Solvent	Solute	State	T, K	$\bar{H}_i^e$, kJ/g-atom	$\bar{S}_i^e$, J/g-atom-K	$\bar{H}_i^e/\bar{S}_i^e$, in K × 10^{-3}	Ref.
Cu	Ag	l	1423	+16.3	+1.35	—	7
Cu	Al	l		−36.3	+24.5	—	7
Cu	Bi	l	1478	+30.2	+12.1	2.5	3, 5, 10
Cu	Ni	s	773	+6.2	−3.2	—	8
Cu	Pb	l	1473	+36.2	+10.7	3.38	7
Cu	Sn	l	1400	−33.5	+16.8	—	7
Cu	Tl	l	1573	+28.2	+1.3	—	7
Cu	Zn	s	773	−23.0	+5.35	—	7
Fe	Al	l	1873	−64.	−10.6	—	7
γ-Fe	C(gr)	s	1426	+44.2	+17.7	2.5	7
Fe	Cr	α-Fe	1550	+25.	+10.2	2.45	7
Fe	Cr	l	2000	+23.5	+10.2	2.3	7
Fe	Cu	l	1823	+47.6	(+7.5)	—	7
Fe	Mn	γ-Fe	1450	−14.1	−13.3	1.05	7
Fe	Ni	γ-Fe	1200	−15.1	−5.45	2.8	7
Fe	Ni	l	1873	−10.0	−2.15	4.65	8
Fe	Si	l	1873	−132.	−17.6	7.5	6
Ga	Al	l	1023	+2.2	+0.9	2.45	7
Ga	Cd	l	700	+12.9	+1.41	—	7
Ga	Zn	l	750	+5.45	+3.55	1.53	7
Ge	Ag	l	1250	+16.4	+8.8	1.85	7
Hg	Bi	l	594	+5.75	−2.47	—	7
Hg	Cd	l	600	−7.87	−2.35	3.35	7
Hg	Cs	l	298	(−133.)	(−90.)	(1.48)	7
Hg	In	l	433	−9.02	−1.5	6.0	7
Hg	K	l	600	−104.	−40.6	2.56	7
Hg	Na	l	648	−83.0	−29.3	2.83	7
Hg	Pb	l	600	+5.63	−6.07	—	7
Hg	Zn	l	608	+3.0	−1.1	—	7
In	Al	l	1173	+21.0	+2.28	9.2	7
In	Bi	l	900	(−6.65)	(+1.62)	—	7
In	Cd	l	800	+4.7	+1.18	3.98	7
In	Cu	l	1073	(+4.25)	(+1.57)	(2.7)	7
In	Na	l	713	(−17.6)	(−23.5)	—	7
In	Sb	l	900	−12.3	+4.5	—	7
In	Tl	l	723	+2.9	−1.6	—	7
In	Zn	l	700	+11.0	+4.3	2.56	7
Mg	Al	l	—	(−9.9)	(−3.87)	(2.56)	7
Mg	Cd	s	543	−16.9	−3.2	5.28	7
Mg	Cd	l	923	−23.4	−8.06	2.9	7
Mg	In	l	—	−34.5	−9.6	3.6	7

(continued)

Table C.1 (*continued*)

Solvent	Solute	State	T, K	$\bar{H}_i^e$, kJ/g-atom	$\bar{S}_i^e$, J/g-atom-K	$\bar{H}_i^e/\bar{S}_i^e$, in K × 10^{-3}	Ref.
Mo	C(gr)	s	2000	+143.	+52.5	2.73	11
Mo	Cr	s	1471	+20.4	+3.1	6.57	7
Na	K	l	384	+3.65	+0.9	4.05	7
Ni	Cr	s	—	−17.6	−2.83	6.2	8
Ni	Co	s	—	—————— Nearly ideal ——————			8
Ni	Cu	s	—	+16.5	+4.7	3.5	8
Ni	Fe	γ-Fe	1200	−24.3	−5.0	4.85	8
Ni	Fe	l	—	−32.3	−9.25	3.5	8
Ni	Mn	s	1050	(−60.)	(−7.2)	—	7
Ni	Si	l	1873	−188.	−23.	8.	6
Pb	Ag	l	1273	+11.7	+3.27	3.58	7
Pb	Al	l		+26.4	−3.8	—	7
Pb	Bi	l	700	−3.52	+1.1	—	7
Pb	Cd	l	773	+9.35	+1.95	4.8	7
Pb	Cu	l	1473	+27.5	+5.6	4.9	7
Pb	Fe	l	—	+103.	+22.	4.7	12
Pb	In	l	676	+4.0	+3.44	1.16	7
Pb	K	l	848	−51.6	−18.8	2.75	7
Pb	Mg	l	973	<−21.	−3.9	(5.)	7
Pb	Na	l	700	−38.5	−7.2	5.35	7
Pb	Tl	s	523	−4.9	−6.95	0.7	7
Pb	Tl	l		−3.45	−2.55	1.35	7
Pb	Zn	l	923	+20.0	+4.54	4.4	7
Pd	Cu	s	—	(−42.)	(−27.3)	—	7
Pd	γ-Fe	s	1273	(−75.)	(+6.85)	—	7
Pd	Ni	s	1273	(−11.4)	(+4.8)	—	7
Pt	Au	s	—	(+24.8)	(−3.8)	—	9
Pt	Cu	s	1350	−28.8	+3.9	—	7
Pt	Pd	s	—	−12.7	−7.1	—	9
Sb	Ag	l	1250	(+8.4)	(+11.)	—	7
Sb	Cd	l	773	−3.0	+5.65	—	7
Sb	Sn	l	905	−4.2	+2.75	—	7
Sn	Ag	l	900	+4.4	+2.35	1.87	7
Sn	Al	l	1000	+14.6	+6.65	2.2	7
Sn	Cd	l	773	+6.6	+4.5	1.47	7
Sn	Cr	l	—	(+68.5)	(+37.)	(1.85)	12
Sn	Hg	l	423	+1.85	−3.25	—	7
Sn	In	l		−0.85	−3.0	—	7
Sn	Sb	l	905	−4.5	+2.43	—	7
Sn	Tl	l	723	+4.02	−1.38	—	7

Table C.1 (*continued*)

Solvent	Solute	State	T, K	$\bar{H}_i^e$, kJ/g-atom	$\bar{S}_i^e$, J/g-atom-K	$\bar{H}_i^e/\bar{S}_i^e$, in K × 10^{-3}	Ref.
Sn	Zn	l	750	+8.7	+6.0	1.45	7
Tl	Ag	l	975	+15.0	+4.0	3.75	7
Tl	K	l	798	−42.	−11.5	3.65	7
Tl	Mg	l	923	(−11.7)	(+10.)	—	7
Tl	Na	l	673	−33.5	−2.5	—	7
Zn	Al	s	653	+25.9	+12.3	2.1	7
Zn	Al	l	1000	+11.1	+2.95	3.77	7

[a]Solute and solvent in pure states have the same state of aggregation. Greatly uncertain values are in parentheses. States of aggregation are denoted by l for liquid and s for solid.

the proposed correlation have been labeled as Au–Fe, Bi–Na, Cd–Au, and so on. The values of $\bar{H}_i^e$ and $\bar{S}_i^e$ calculated from Margules-type equations for concentrated solutions and extrapolated to infinite dilution have been disregarded. All the data are for the systems in which both the solute and the solvent have the same state of aggregation. The arithmetic average of $\bar{H}_i^e/\bar{S}_i^e = \kappa$, listed in the seventh column of Table C.1, was found to be $\kappa = 3400$ K and the line showing the empirical relation

$$\bar{H}_i^e = \kappa \bar{S}_i^e = 3400\bar{S}_i^e; \quad \text{(J/g-atom)} \tag{C.1}$$

is plotted in Fig. C.1. Substitution of this equation in $\bar{G}_i^e$ yields

$$\bar{G}_i^e = \bar{H}_i^e\left(1 - \frac{T}{\kappa}\right) = \bar{S}_i^e(\kappa - T); \quad \text{(J/g-atom)} \tag{C.2}$$

Therefore, if $\bar{G}_i^e$ is known, then both $\bar{H}_i^e$ and $\bar{S}_i^e$ can be estimated, or if $\bar{H}_i^e$ is known, then both $\bar{S}_i^e$ and $\bar{G}_i^e$ can be estimated from equation (C.2).

Equation (C.1) is approximate, and attempts to refine it by using other methods in Appendixes A and B have not yet been made, although it might well be an interesting subject to be pursued in the future.

Estimation of $\Delta H^\bullet$ and $\Delta S^\bullet$ for Dilute Solutions of Oxygen in Metals

The equilibria between diatomic gaseous oxygen and dissolved oxygen have been investigated in sufficient degrees of accuracy for a number of

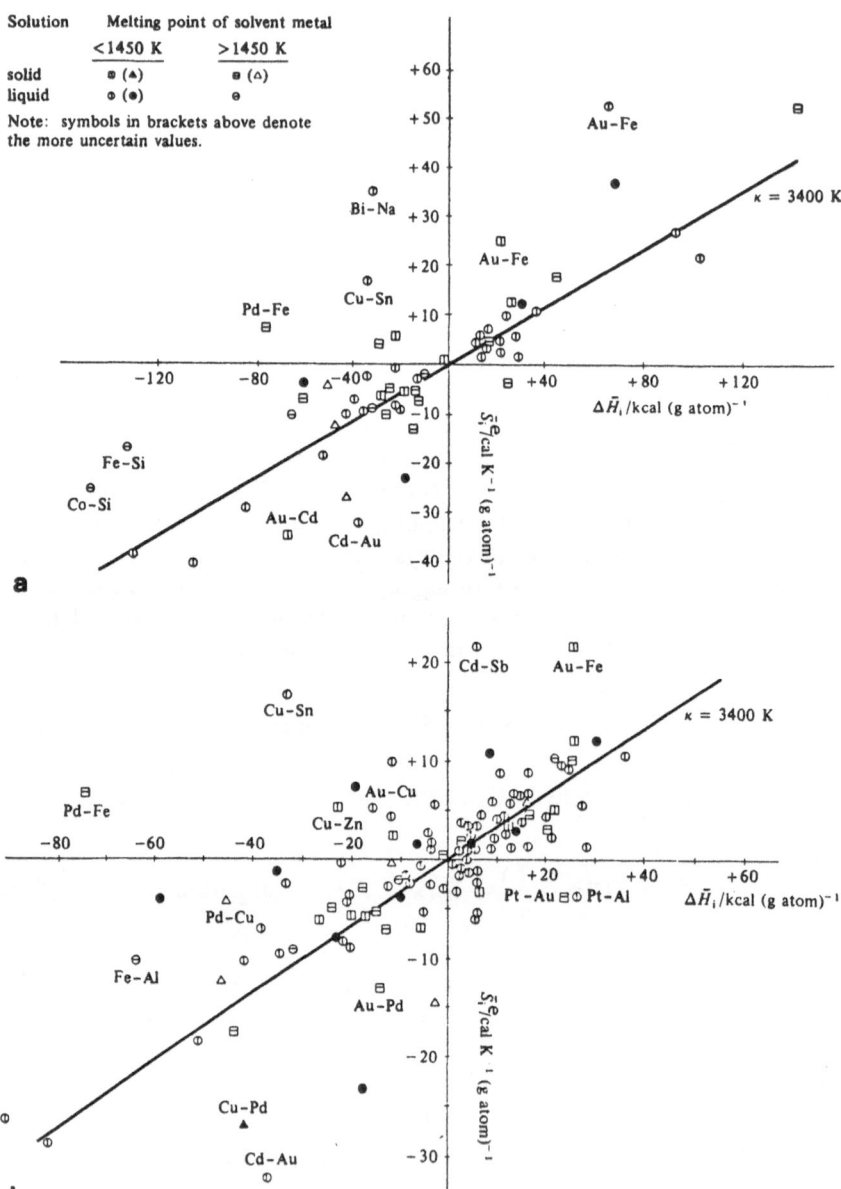

Figure C.1. Correlation of $\bar{H}_i^e$ and $\bar{S}_i^e$ for infinitely dilute binary solutions of i in solvent metals. ($\Delta\bar{H}_i = \bar{H}_{i}^e$.) (From Kubaschewski[1] with permission.)

metals^{2,13,14} such as Ag, Co, Cu, Fe, Ni, Pb, and Sn yielding $\Delta H^{\bullet}$ and $\Delta S^{\bullet}$ of adequate accuracy to devise a method of estimation.

The usual method of equilibration is either with $(H_2 + H_2O)$ or $(CO + CO_2)$ mixtures (see Chapter 6) to obtain

$$0.5O_2(g) = [O]; \qquad \Delta G_0^{\bullet} = -RT \ln \frac{x_0}{\sqrt{P_{O_2}}} = \Delta H_0^{\bullet} - T\Delta S_0^{\bullet} \qquad (C.3)$$

where the concentration of oxygen is expressed in atomic fraction x_0 (numerous publications use weight percent or atomic percent). The values of $\Delta H_0^{\bullet}$ and $\Delta S_0^{\bullet}$ are assumed to be constant within the temperature ranges of investigation. The results for the aforementioned metals are listed in Table C.2 as linear functions of temperature. In general, the disagreement among various sets of data for $\Delta H_0^{\bullet}$ and $\Delta S_0^{\bullet}$ is larger than that for $\Delta G_0^{\bullet}$ because $\Delta G_0^{\bullet}$ is the directly determined property.

Attempts have been made$^{16-18}$ to correlate $\Delta H^{\bullet}$ with ΔHf_{298}° of the metal oxide expressed as the standard enthalpy of formation per *gram atom* of oxygen. If a selected metal forms several oxides, then the oxide yielding the largest value of ΔHf_{298}° is used for this purpose. The empirical correlations proposed by Fitzner^2 for $\Delta H_0^{\bullet}$ and $\Delta S_0^{\bullet}$ for oxygen dissolved in metals with full d-orbitals are

$$\Delta H_0^{\bullet} = -1.2964 \times 10^{-5}(\Delta Hf_{298}^{\circ})^2 - 5.0 \times 10^{-11}(\Delta Hf_{298}^{\circ})^3; \quad \text{(full } d^{10}) \quad (C.4)$$

$$\Delta S_0^{\bullet} = -9.151 - 4.141 \times 10^{-9}(\Delta H_0^{\bullet})^2 - 2.2 \times 10^{-14}(\Delta H_0^{\bullet})^3; \quad \text{(full } d^{10}) \quad (C.5)$$

Table C.2. Standard Gibbs Energy Change for $0.5O_2(g) = [O]$ in Selected Metals^a

System	$\Delta G_0^{\bullet}$(exp), cal/g-atom [O]	Range for T	$-\Delta H_{298}^{\circ}$, cal/g-atom [O]^b	cal/g-atom [O]		
				$\Delta G_0^{\bullet}$(estm)	$\Delta G_0^{\bullet}$(exp)^c	$\Delta G_0^{\bullet}$(estm)^c
Ag-O	$-3,375 + 10.405\,T$	1246–1433	7,420	$-693 + 9.153\,T$	$+11,192$	$+12,121$
Co-O	$-14,730 + 4.241\,T^d$	1823–1973	56,870	$-19,929 + 4.622\,T$	$-6,672$	$-11,147$
Cu-O	$-20,710 + 10.871\,T$	1373–1773	40,800	$-18,185 + 10.388\,T$	$-5,491$	$-3,642$
Fe-O	$-28,890 + 6.177\,T^d$	1823–1973	65,000	$-26,584 + 5.664\,T$	$-17,154$	$-15,822$
Ni-O	$-11,920 + 4.289\,T^d$	1773–1973	57,300	$-20,254 + 4.667\,T$	$-3,771$	$-11,387$
Pb-O	$-28,283 + 12.053\,T$	773–1473	52,340	$-28,345 + 11.977\,T$	$-11,409$	$-11,577$
Sn-O	$-43,610 + 15.281\,T$	1173–1473	69,410	$-45,737 + 15.709\,T$	$-22,217$	$-23,744$

^{a}Concentration of oxygen is in atom fraction in pure metal.
^{b}From Pankratz *et al.*^{18}
^{c}For Ag-O, Cu-O, Pb-O, and Sn-O, numerical values are for 1400 K; remaining values are for 1900 K.
^{d}Experimental results are from Tankins *et al.*^{15} for elements having unfilled d-orbitals; others, for elements with filled d-orbitals, are from a summary by Chiang and Chang^{14} where extensive references to the original investigations are given.

where -9.151 is for the change in the standard state from atomic percent to atomic fraction, $\Delta H f^\circ_{298}$ and $\Delta H^\bullet_0$ are in cal/g-atom oxygen, and $\Delta S^\bullet_0$ is in cal/g-atom-K. The equations for the transition elements in which the d-orbitals are partially filled are

$$\Delta H^\bullet_0 = -5.252 \times 10^{-6}(\Delta H f^\circ_{298})^2 + 1.6 \times 10^{-11}(\Delta H f^\circ_{298})^3; \quad \text{(unfilled } d) \text{ (C.6)}$$

$$\Delta S^\bullet_0 = -3.151 - 4.141 \times 10^{-9}(\Delta H^\bullet_0)^2 - 2.2 \times 10^{-14}(\Delta H^\bullet_0)^3; \quad \text{(unfilled } d) \text{ (C.7)}$$

The difference between equations (C.5) and (C.7) is 6.0 cal/g-atom-K, which must be considered as an empirical term because theoretical justifications for this value are tenuous. The numerical values obtained from equations (C.4)–(C.7) are listed in Table C.2, indicating that the experimental and estimated values of $\Delta G^\bullet_0(\exp)$ and $\Delta G^\bullet_0(\text{estm})$ are in good agreement, except for the Ni-O system. The estimated equations for 18 transition metals are listed by Fitzner,[2] whose entropy term must be changed as follows to bring his results in line with the notation used in this appendix: $\Delta S^E(\text{Fitzner}) -9.151 = \Delta S^\bullet_0(\text{this appendix})$. The estimated results for $\Delta G^\bullet_0$ of oxygen in most of the transition metals are probably within 15 kcal/g-atom[O] of the experimental value for $\Delta G^\bullet_0$. However, the available data for some of these metals are scanty and greatly scattering. Therefore, new experimental data are needed for oxygen in the transition metals to test the foregoing empirical equations and, if necessary, to modify them to attain greater degrees of success for prediction of thermodynamic properties of dissolved oxygen. The method may also be extended to other interstitial solute elements such as H, C, N, and S.

References

1. O. Kubaschewski, *High Temp. High Pressures* **13**, 435 (1981).
2. K. Fitzner, *Thermochim. Acta* **52**, 103 (1982).
3. T. Azakami and A. Yazawa, *Can. Metall. Q.* **15**(2), 111 (1976).
4. C. Bernard and I. Ansara, Report LTPCM-1974-TM02, Ecole Normale Superieure d'Electrochimie et d'Electrometallurgie, St.-Martin d'Hères, France (1974).
5. J. Bode, J. Gerlach, and F. Pawlek, *Erzmetall* **24**, 480 (1971).
6. T. G. Chart, *High Temp. High Pressures* **5**, 241 (1973).
7. R. Hultgren, P. D. Desai, D. T. Hawkins, M. Gleiser, and K. K. Kelley, *Selected Values of the Thermodynamic Properties of Binary Alloys*, ASM, Metals Park, Ohio (1973).
8. O. Kubaschewski and C. B. Alcock, *Metallurgical Thermochemistry*, Fifth Edition, Pergamon Press, Elmsford, New York (1979).
9. O. Kubaschewski and J. F. Counsell, *Monatsh. Chem.* **102**, 1724 (1971).
10. B. Predel and A. Emam, *Z. Metallkd.* **64**, 496 (1973)

11. L. L. Seigle, C. L. Chang, and T. P. Sharma, *Metall. Trans. AIME* **10A,** 1223 (1979).

12. D. A. Stevenson and J. Wulff, *Trans. Metall. Soc. AIME* **221,** 271 (1961).

13. See, e.g., E. Fromm and E. Gebhard, *Gase und Kohlenstoff in Metallen,* Springer-Verlag, Berlin (1976).

14. T. Chiang and Y. A. Chang, *Metall. Trans. AIME* **7B,** 453 (1976).

15. E. S. Tankins, N. A. Gokcen, and G. R. Belton, *Trans. Metall. Soc. AIME* **230,** 820 (1964).

16. F. D. Richardson, *J. Iron Steel Inst. London* **166,** 144 (1950).

17. K. Fitzner, K. T. Jacob, and C. B. Alcock, *Metall. Trans. AIME* **8B,** 669 (1977).

18. L. B. Pankratz, J. M. Stuve, and N. A. Gokcen, *Thermodynamic Data for Mineral Technology,* Bureau of Mines Bull. 677 (1984).

D

Selected Properties of the Elements

Element	Atomic weight	Condensed phase, 298.15 K			Phase transitions				Ideal gas, 298.15 K			
		C_p°	S°	T_m	ΔH_m°	T_v	ΔH_v°	C_p°	S°	$\Delta H f^\circ$	$\Delta G f^\circ$
Ac	227.0278	6.5?	—	1324	—	3480?	—	4.975	—	—	—
Ag	107.8682	6.070	10.17	1235.08	2.70	2440	59.9	4.968	41.321	68.090	58.802
Al	26.98154	5.82	16.776	933.602	2.580	2798	70.1	5.112	39.302	78.8	69.102
Am	243?	6.178	13.023	1449	3.440	—	—	4.968	46.473	67.90	57.927
Ar	39.948	—	—	83.798	2.81	87.29	1.558	4.968	36.983	0	0
As	74.9216	5.892	8.534	876	6.62?	876	—	4.968	41.611	72.12	62.258
At	210?	—	—	575?	—	—	—	—	—	—	—
Au	196.9665	6.075	11.330	1337.58	2.957	3130	80.0	4.968	43.116	87.5	78.023
B	10.81	2.65	1.410	2365?	12.0	4275	117.0	4.971	36.649	132.80	122.293
Ba	137.33	6.713	14.918	1002	1.852	2171	33.8	4.968	40.663	43.0	35.324
Be	9.01218	3.93	2.27	1562	2.919	2745	69.9	4.968	32.545	77.40	68.374
Bi	208.9804	6.10	13.56	544.592	2.700	1837	36.200	4.968	44.669	49.5	40.225
Bk	247?	—	—	1256	—	—	—	—	—	—	—
Br	79.904	18.09	36.379	265.90	2.527	332.6	7.065	8.616	58.64	7.388	0.751
C	12.011	2.036	1.372	4130	—	—	—	4.981	37.76	171.29	160.441
Ca	40.08	6.05	9.94	1113	2.04	1757	36.626	4.968	36.992	42.5	34.434
Cd	112.41	6.20	12.38	594.258	1.48	1040	23.809	4.968	40.066	26.73	18.475
Ce	140.12	6.44	17.2	1071	1.305	3700	99.0	5.515	45.807	101.0	92.471
Cf	251?	—	—	1213	—	—	—	—	—	—	—
Cl	35.453	—	—	172.18	1.531	239.10	4.878	8.111	53.29	0	0
Cm	247?	6.617	—	1613	3.500	—	—	6.717	47.158	92.60	83.668
Co	58.9332	5.93	7.18	1768	3.7	3200	90.0	5.503	42.879	101.5	90.856
Cr	51.996	5.58	5.65	2133	4.047	2945	82.3	4.968	41.635	95.0	84.271

Atomic weights are from *Pure Appl. Chem.* **55**, 1101 (1983). Melting points are from *Bull. Alloy Phase Diagrams* 2(1), 146 (1981) where references to original sources are given. Remaining data are from Pankratz et al., *Thermodynamic Data for Mineral Technology*, Bureau of Mines Bull. 677 (1984), and from R. Hultgren, P. D. Desai, D. T. Hawkins, M. Gleiser, K. K. Kelley, and D. D. Wagman, *Selected Values of the Thermodynamic Properties of the Elements*, ASM, Metals Park, Ohio (1973), corrected for 1968 temperature scale. Melting points of C and Fe are author's values. C_p° and S° are in cal/mole-K; ΔH_m° is in kcal/mole-K; ΔH_v°, $\Delta H f^\circ$ and $\Delta G f^\circ$ refer to formation of ideal stable gas from stable condensed phase. For H, N, O, and halogens, the gas phase is diatomic; all others are monatomic.

Element											
Cs	132.9054	7.695	20.37	31.54	0.50	952	16.198	4.968	41.942	18.32	11.888
Cu	63.546	5.841	7.924	1358.62	3.12	2839	71.9	4.968	39.743	80.60	71.113
Dy	162.50	—	—	1685	—	—	—	—	—	—	—
Er	167.26	—	—	1802	—	—	—	—	—	—	—
Es	252?	—	—	1093	—	—	—	—	—	—	—
Eu	151.96	5.970	6.52	1095	—	—	—	—	—	—	—
F	18.99840	—	—	53.48	0.122	85.02	1.562	7.481	48.443	0	0
Fe	55.847	—	—	1809	3.30	3135	83.6	6.136	43.112	99.3	88.390
Fm	257?	—	—	1800?	—	—	—	—	—	—	—
Fr	223?	—	—	300?	—	—	—	—	—	—	—
Ga	69.72	6.250	9.758	302.9241	1.336	2478	61.8	6.058	40.375	66.2	57.072
Gd	157.25	8.860	16.24	1586	2.403	3540	85.9	6.584	46.416	95.0	86.003
Ge	72.59	5.58	7.43	1211.5	8.83	3107	79.1	7.345	40.104	89.5	79.758
H	1.00794	—	—	13.81	0.028	2039	0.2158	6.892	31.207	0	0
He	4.00260	—	—	3.5	0.005	4.22	0.020	4.968	30.125	0	0
Hf	178.49	6.150	10.41	2504	5.68	5500?	158?	4.972	44.643	148.0	137.793
Hg	200.59	6.687	18.14	234.314	0.549	629.81	14.151	4.968	41.792	14.67	7.618
Ho	164.9304	—	—	1747	—	—	—	—	—	—	—
I	126.9045	13.011	27.758	386.7	3.709	458.4	10.021	8.814	62.277	14.919	4.627
In	114.82	6.389	13.82	429.784	7.80	2346	55.3	4.979	41.508	58.15	49.895
Ir	192.22	5.996	8.481	2720	6.247	4400?	134.7?	4.968	46.241	159.0	147.742
K	39.0983	7.050	15.457	336.34	0.558	1043.7	19.038	4.968	38.297	21.31	14.500
Kr	83.80	—	—	115.765	0.391	119.75	2.158	4.968	39.191	0	0
La	138.9055	6.48	13.6	1191	1.481	3730	98.9	5.438	43.564	103.0	94.066
Li	6.941	5.887	6.954	453.7	0.717	1638	35.160	4.968	33.143	38.410	30.602
Lr	260?	—	—	1900?	—	—	—	—	—	—	—
Lu	174.967	6.40	12.18	1936	4.457	—	—	4.986	44.142	102.2	92.671
Md	258?	—	—	1100?	—	—	—	—	—	—	—
Mg	24.305	5.950	7.810	922	2.139	1363	30.250	4.968	35.501	35.00	26.744
Mn	54.9380	6.280	7.650	1519	2.882	2335	54.0	4.968	41.493	67.10	57.010
Mo	95.94	5.750	6.850	2896	7.777	4919?	140.8?	4.968	43.461	157.50	146.584

(continued)

Element	Atomic weight	Condensed phase, 298.15 K		Phase transitions				Ideal gas, 298.15 K			
		C_p°	S°	T_m	ΔH_m°	T_v	ΔH_v°	C_p°	S°	ΔH_f°	ΔG_f°
N	14.0067	—	—	63.1458	0.172	77.36	1.335	6.961	45.770	0	0
Na	22.98977	6.730	12.298	371.0	0.622	1177	23.285	4.968	36.714	25.755	18.475
Nb	92.9064	5.880	8.70	2742	6.302	5200?	163.0	7.208	44.490	175.2	164.529
Nd	144.24	—	—	1294	—	—	—	—	—	—	—
Ne	20.179	—	—	24.563	0.08	27.07	0.422	4.968	34.947	0	0
Ni	58.69	6.210	7.140	1728	4.100	31.87	88.5	5.583	43.519	102.80	91.954
No	259?	—	—	1100?	—	—	—	—	—	—	—
Np	237.0482	—	—	910	—	—	—	—	—	—	—
O	15.9994	—	—	54.361	0.106	90.19	1630	7.021	49.005	0	0
Os	190.2	5.905	7.8	3306	7.6	5285	178.32	4.968	46.00	189.0	177.611
P(wh)	30.97376	5.698	9.820	317.29	0.157	550	2.908	4.968	38.980	75.620	66.926
Pa	231.0359	6.601	12.400	1848	2.950	—	—	5.476	47.308	145.0	134.592
Pb	207.2	6.370	15.490	600.652	1.147	2023	42.5	4.968	41.890	46.750	38.879
Pd	106.42	6.188	9.013	1828	3.970	3237	85.4	4.968	39.902	90.4	81.190
Pm	145?	—	—	1315	—	—	—	—	—	—	—
Po	209?	—	—	527	—	—	—	—	—	—	—
Pr	140.9077	6.560	17.670	1204	1.646	3785	70.9	5.105	45.339	85.0	76.750
Pt	195.08	6.180	9.950	2042.1	4.696	4100	121.85	6.102	45.960	135.1	124.364
Pu	244?	7.850	13.420	913	0.675	3503	82.14	4.984	42.317	82.500	73.884
Ra	226.0254	—	—	973	—	—	—	—	—	—	—
Rb	85.4678	7.424	18.35	312.63	0.524	974.5	17.23	4.968	40.626	19.33	12.688
Re	186.207	6.035	8.730	3459	7.9	5869	170.85	4.968	45.131	184.00	173.147
Rh	102.9055	5.958	7.542	2236	6.356	3970	117.89	5.022	44.387	133.10	122.115
Rn	222?	—	—	202	—	—	—	—	—	—	—
Ru	101.07	5.730	6.839	2607	9.200	4423	142.34	5.144	44.550	153.6	142.356
S	32.06	5.425	7.661	388.37	0.413	—	—	5.658	40.086	66.20	56.533
Sb	121.75	6.030	10.880	903.905	4.750	—	—	4.968	43.059	62.70	53.106
Sc	44.9559	—	—	1814	—	—	—	—	—	—	—

Se	78.96	5.987	10.144	494.3	1.472	—	—	4.990	42.210	56.25	46.689
Si	28.0855	4.780	4.50	1687	12.082	—	—	5.318	40.121	107.70	97.080
Sm	150.36	—	—	1347	—	—	—	—	—	—	—
Sn(wh)	118.69	6.450	12.236	505.118	1.680	2876	70.7	5.081	40.244	71.99	63.639
Sr	87.62	6.393	12.500	1042	1.960	1654.1	32.730	4.968	39.323	39.20	31.203
Ta	180.9479	6.060	9.920	3293	7.560	5731	177.61	4.985	44.242	186.90	176.667
Tb	158.9254	6.91	17.52	1629	2.580	3496	79.1	5.895	48.552	92.90	83.648
Tc	98?	—	8.0?	2477	—	—	—	4.970	43.248	164.5?	153.38?
Te	127.60	6.140	11.880	722.72	4.180	5061	122.96	4.970	43.640	50.60	41.131
Th	232.0381	6.532	12.760	2031	3.30	5061	122.96	4.969	45.425	-142.73	132.991
Ti	47.88	5.980	7.320	1943	3.30	3562	100.6	5.839	43.066	113.00	102.342
Tl	204.383	6.290	15.340	577	0.990	1746	39.215	4.968	43.226	43.25	34.936
Tm	168.9342	6.460	17.690	1818	4.025	2220	45.6	4.968	45.412	55.50	47.235
U	238.0289	6.612	12.000	1407	2.185	4407	110.92	5.663	47.724	127.00	116.349
V	50.9415	5.950	6.915	2202	5.461	3694	106.8	6.217	43.544	123.20	112.279
W	183.85	5.800	7.800	3695	8.5	5828	196.92	5.092	41.549	203.40	193.338
Xe	131.29	—	—	161.3918	0.55	165.03	3.02	4.968	40.530	0	0
Y	88.9059	6.34	10.62	1795	2.724	—	—	6.181	42.869	100.7	91.085
Yb	173.04	6.39	14.30	1092	1.830	1467	30.800	4.968	41.352	36.35	28.284
Zn	65.38	6.070	9.950	692.73	1.750	1180	27.565	4.968	38.451	31.17	22.672
Zr	91.22	6.186	9.320	2128	5.000	4682	139.11	6.367	43.315	148.30	138.164

E

Selected Binary Phase Diagrams and Binary Thermodynamic Properties

The binary phase diagrams selected[1,2] in this appendix are for convenient reference in pursuing the topics covered in this book. The objective is to present a representative number of various types of diagrams with various types of phase equilibria.

Selected thermodynamic data from Hultgren *et al.*[2] are for the excess gram atomic (or molar) enthalpy and Gibbs energy of formation from which the corresponding partial molar properties and activities can be calculated. The procedure yields values as accurate as those listed by Hultgren *et al.* at the composition intervals of 0.1 if a power series of the type given by equation (1.63) with five terms is used for H^e and for S^e, and the results are then combined in $G^e = H^e - TS^e$ to obtain five terms dictated by the data for G^e. We illustrate the procedure by using only two parameters:

$$H^e = A_2 x_2(1 - x_2) + A_3 x_2(1 - x_2^2) \qquad (E.1)$$

Substitution of two selected values of H^e at the corresponding compositions yields the values of A_2 and A_3. The equation for S^e is similar, i.e.,

$$S^e = C_2 x_2(1 - x_2) + C_3 x_2(1 - x_2^2) \qquad (E.2)$$

Substitution of equations (E.1) and (E.2) in $G^e = H^e - TS^e$ gives

$$G^e = (A_2 - C_2 T)x_2(1 - x_2) + (A_3 - C_3 T)x_2(1 - x_2^2) \qquad (E.3)$$

Next, two values of G^e at the corresponding compositions yield the coefficients in equation (E.3) and since T is known, then C_2 and C_3 can be calculated. The use of equations (1.64), (1.66), and (1.68) yields

$$\bar{H}_1^e = A_2 x_2^2 + 2A_3 x_2^3 \tag{E.4}$$

$$\bar{H}_2^e = (A_2 + 3A_3)x_1^2 - 2A_3 x_1^3 \tag{E.5}$$

The same procedure also yields the equations for $\bar{G}_1^e$ and $\bar{G}_2^e$, as well as $\bar{S}_1^e$ and $\bar{S}_2^e$.

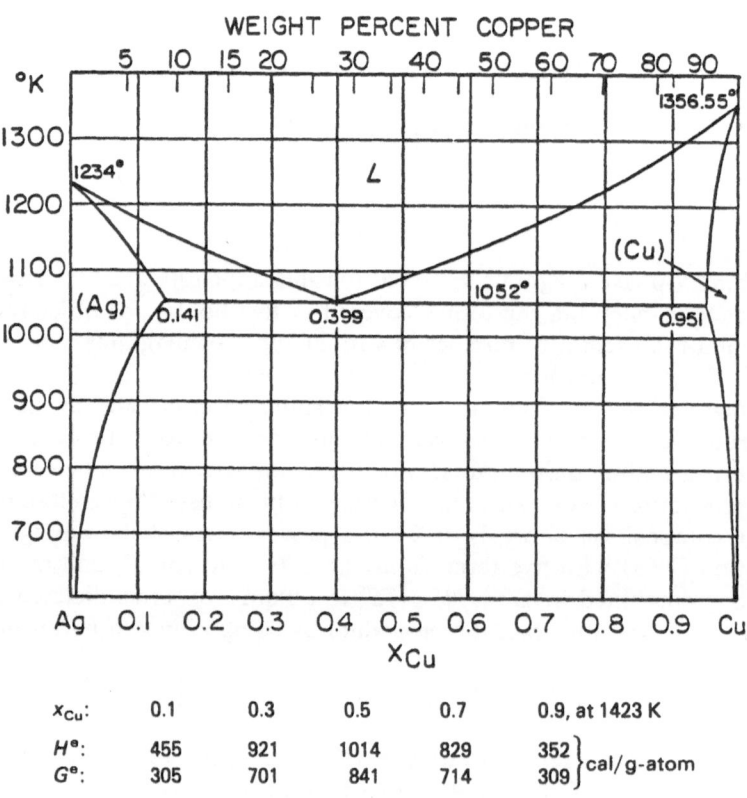

Figure E.1. Ag–Cu system.

x_{Cu}:	0.1	0.3	0.5	0.7	0.9, at 1423 K	
H^e:	455	921	1014	829	352	cal/g-atom
G^e:	305	701	841	714	309	

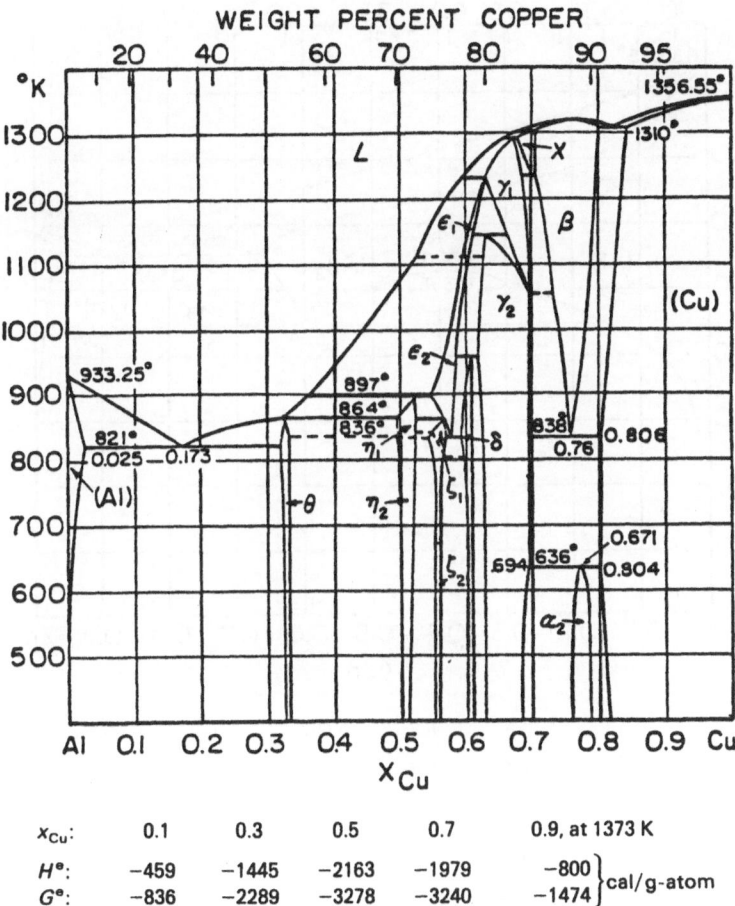

x_{Cu}:	0.1	0.3	0.5	0.7	0.9, at 1373 K	
$H°$:	−459	−1445	−2163	−1979	−800	} cal/g-atom
$G°$:	−836	−2289	−3278	−3240	−1474	

Figure E.2. Al–Cu system.

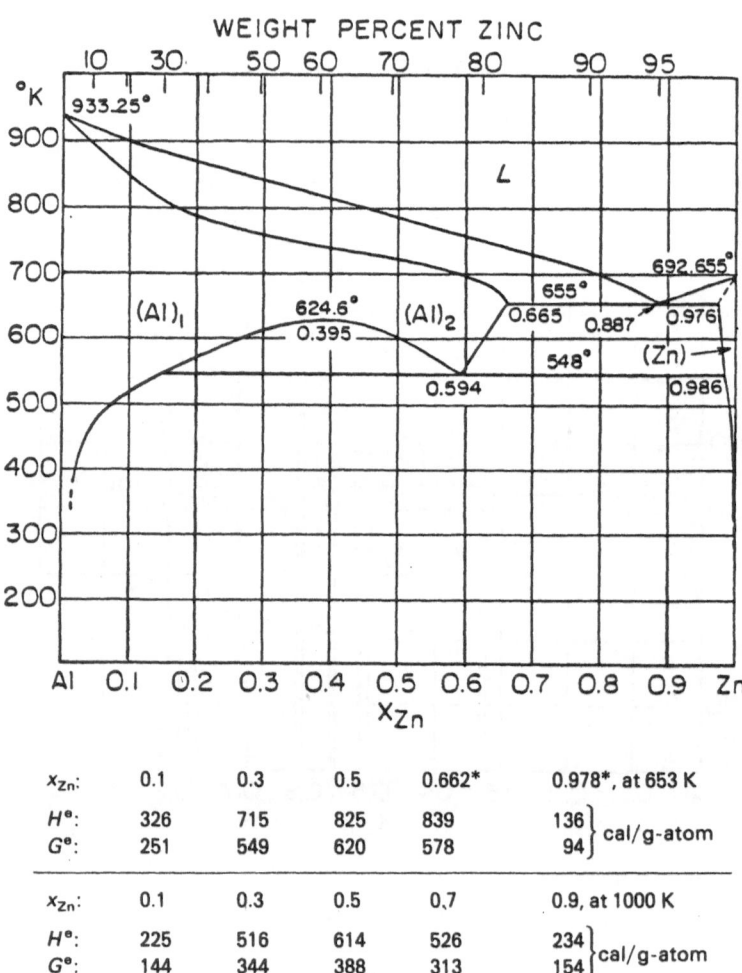

x_{Zn}:	0.1	0.3	0.5	0.662*	0.978*, at 653 K	
H^e:	326	715	825	839	136	cal/g-atom
G^e:	251	549	620	578	94	

x_{Zn}:	0.1	0.3	0.5	0.7	0.9, at 1000 K	
H^e:	225	516	614	526	234	cal/g-atom
G^e:	144	344	388	313	154	

Figure E.3. Al–Zn system. Asterisks indicate phase boundary.

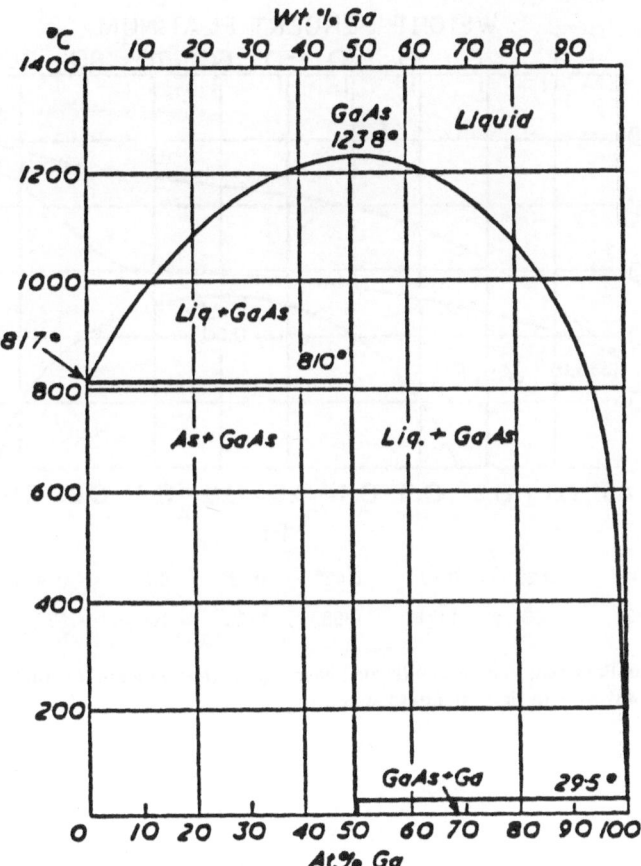

Figure E.4. As–Ga system.

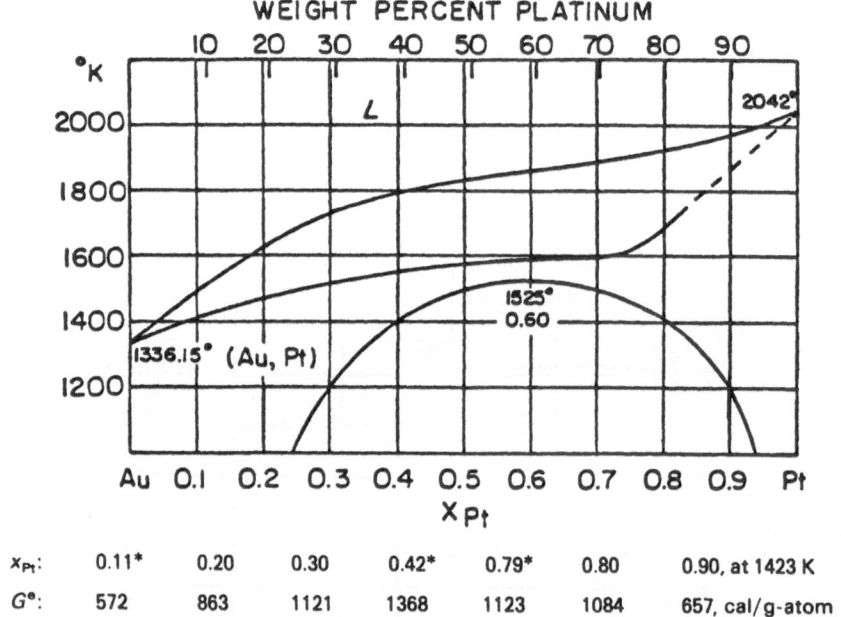

x_{Pt}:	0.11*	0.20	0.30	0.42*	0.79*	0.80	0.90, at 1423 K
G°:	572	863	1121	1368	1123	1084	657, cal/g-atom

Figure E.5. Au–Pt system. Asterisks indicate phase boundaries; data are for single solid phase region with solid Au and Pt as standard states.

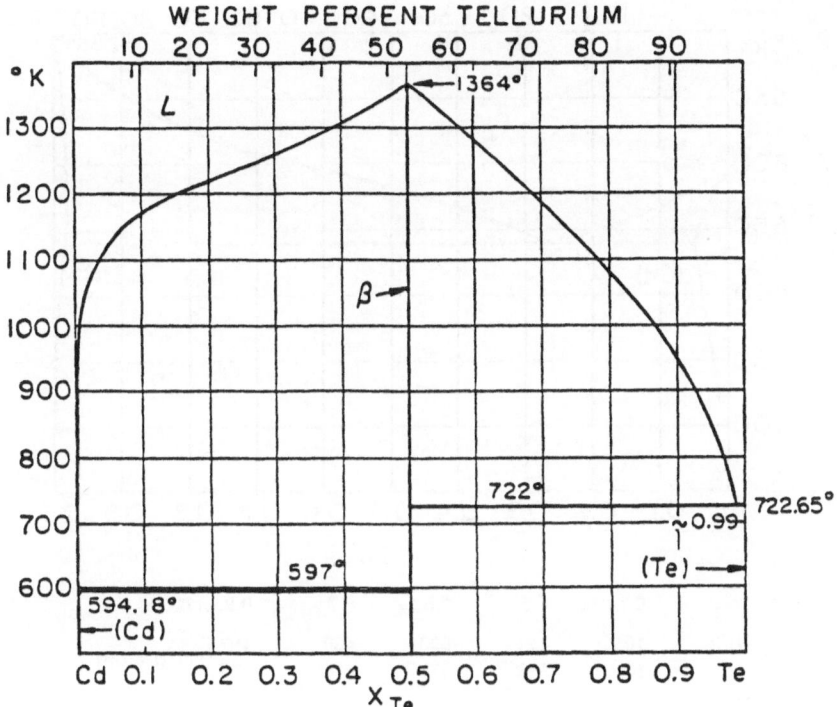

Figure E.6. Cd–Te system.

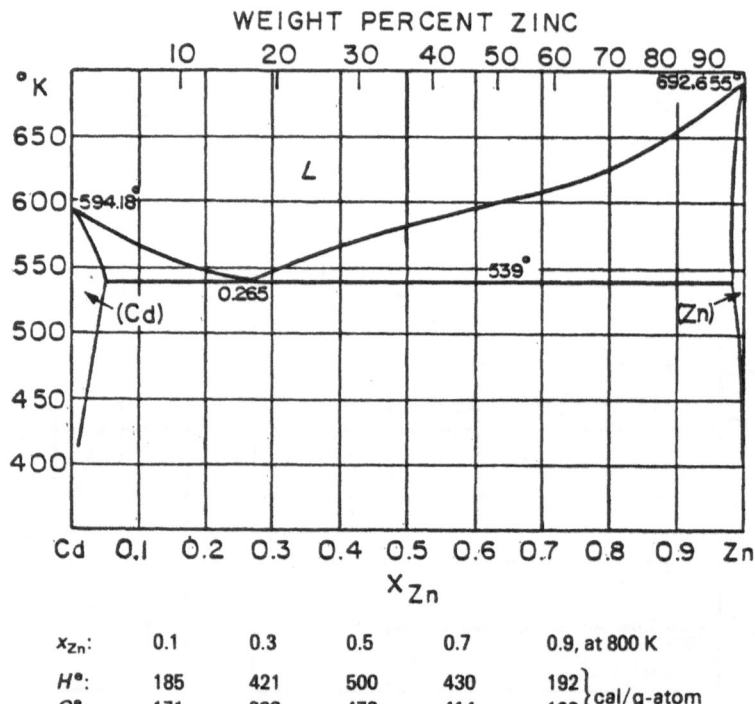

Figure E.7. Cd–Zn system.

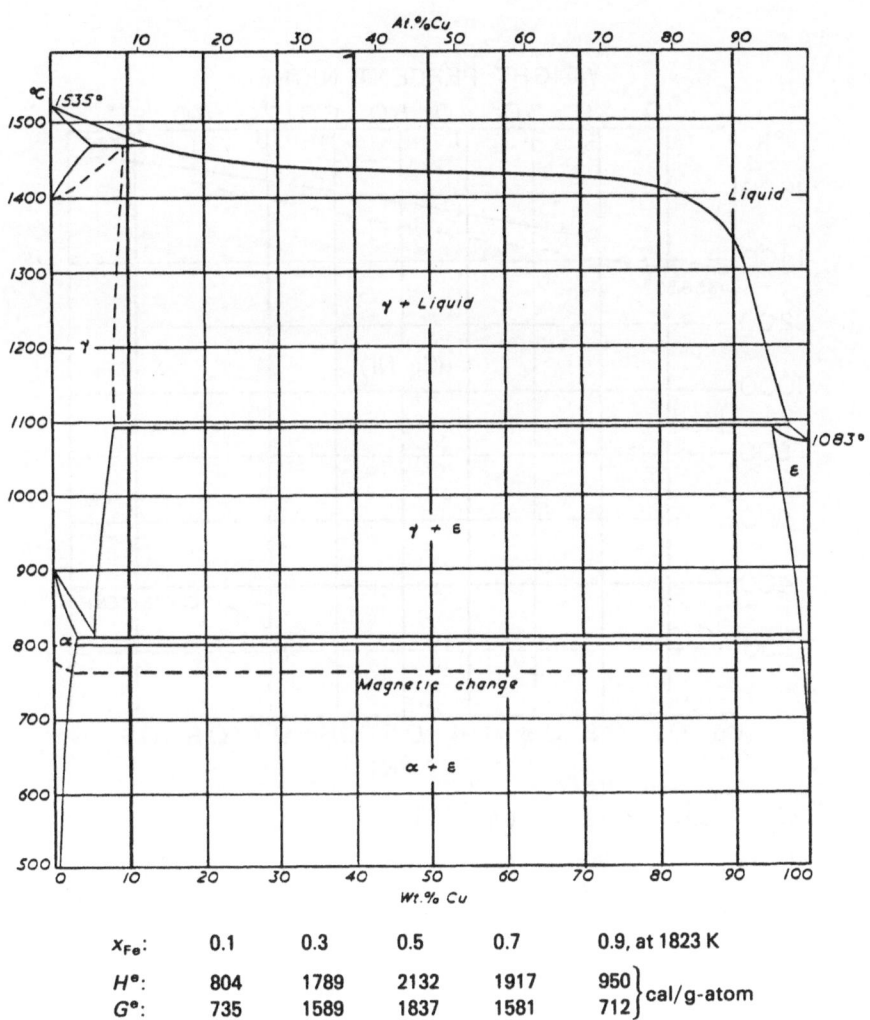

x_{Fe}:	0.1	0.3	0.5	0.7	0.9, at 1823 K
H°:	804	1789	2132	1917	950 } cal/g-atom
G°:	735	1589	1837	1581	712

Figure E.8. Cu–Fe system.

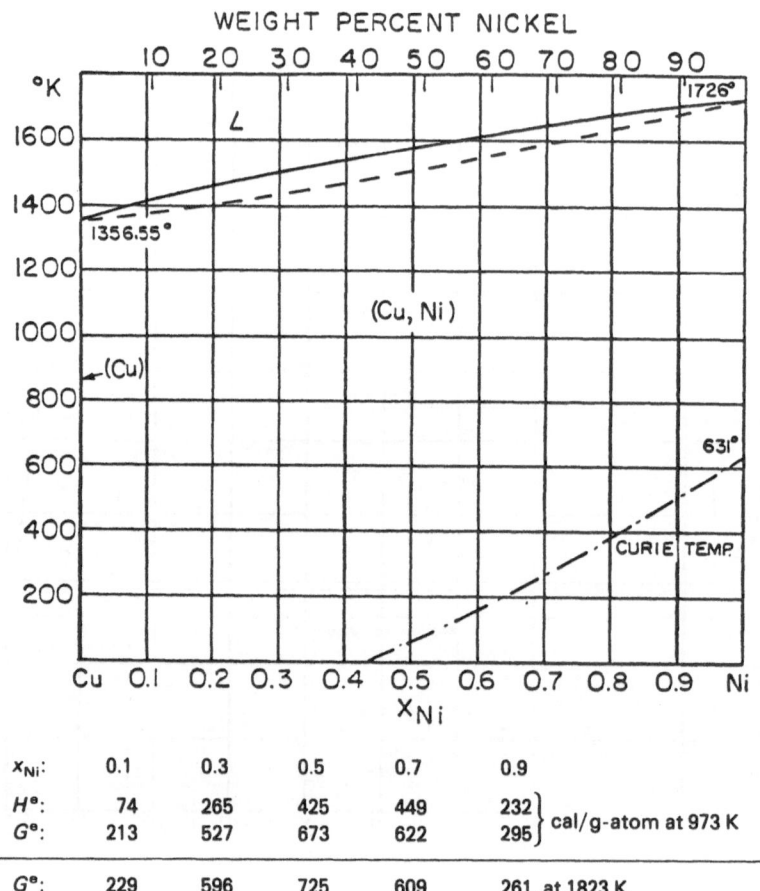

Figure E.9. Cu–Ni system.

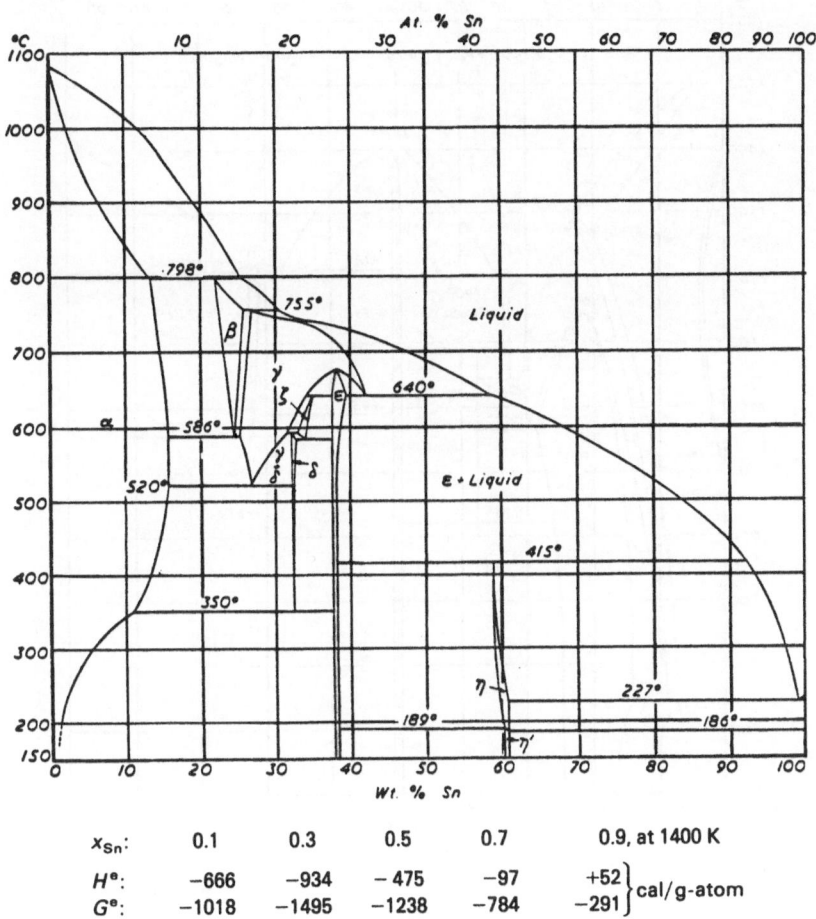

x_{Sn}:	0.1	0.3	0.5	0.7	0.9, at 1400 K	
$H°$:	−666	−934	− 475	−97	+52	cal/g-atom
$G°$:	−1018	−1495	−1238	−784	−291	

Figure E.10. Cu–Sn system.

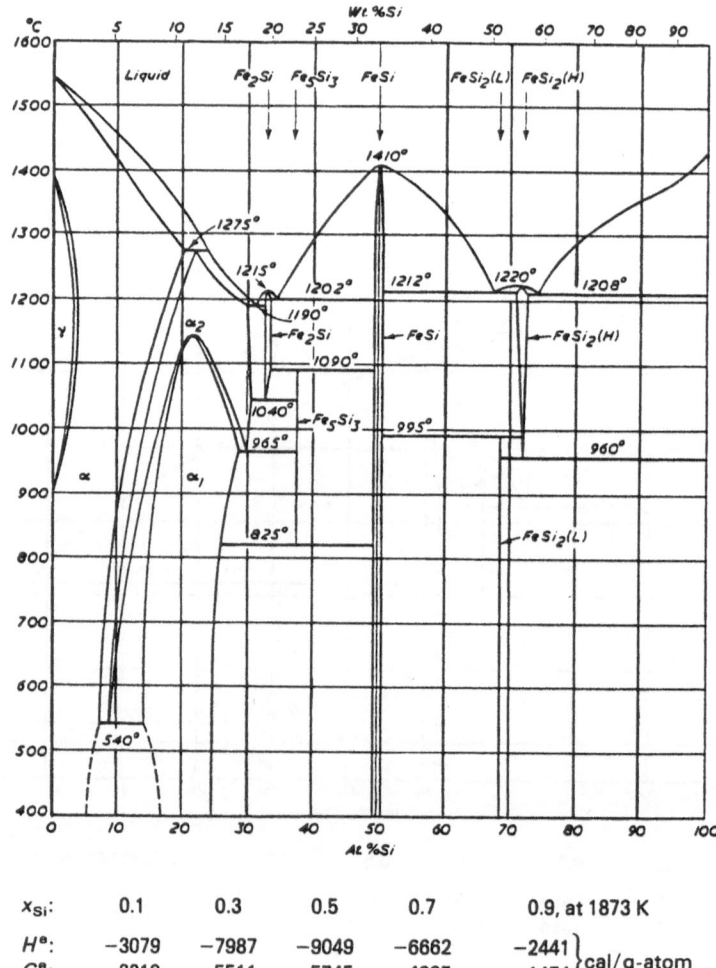

x_{Si}:	0.1	0.3	0.5	0.7	0.9, at 1873 K	
H°:	−3079	−7987	−9049	−6662	−2441	cal/g-atom
G°:	−2319	−5511	−5745	−4005	−1474	

Figure E.11. Fe–Si system.

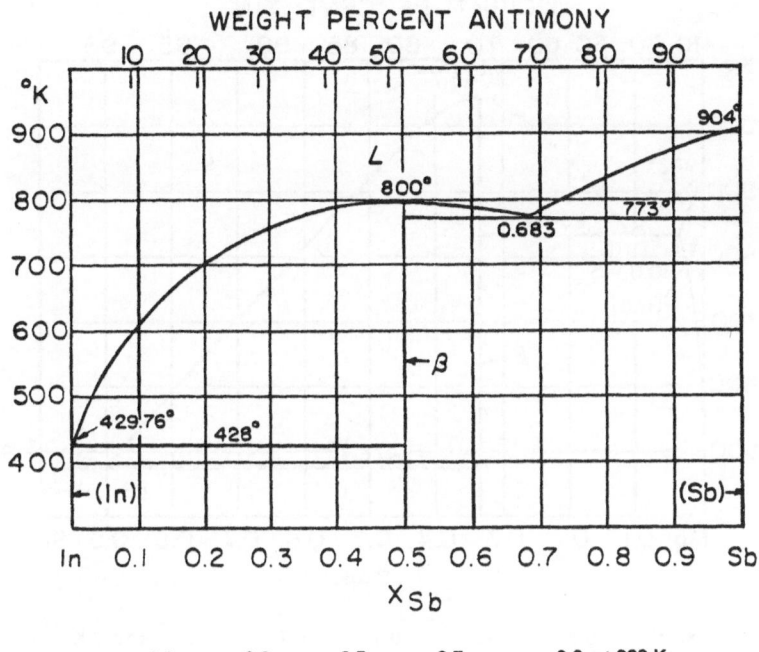

x_{Sb}:	0.1	0.3	0.5	0.7	0.9, at 900 K	
H°:	−283	−718	−769	−569	−211	cal/g-atom
G°:	−364	−848	−928	−694	−257	

Figure E.12. In–Sb system.

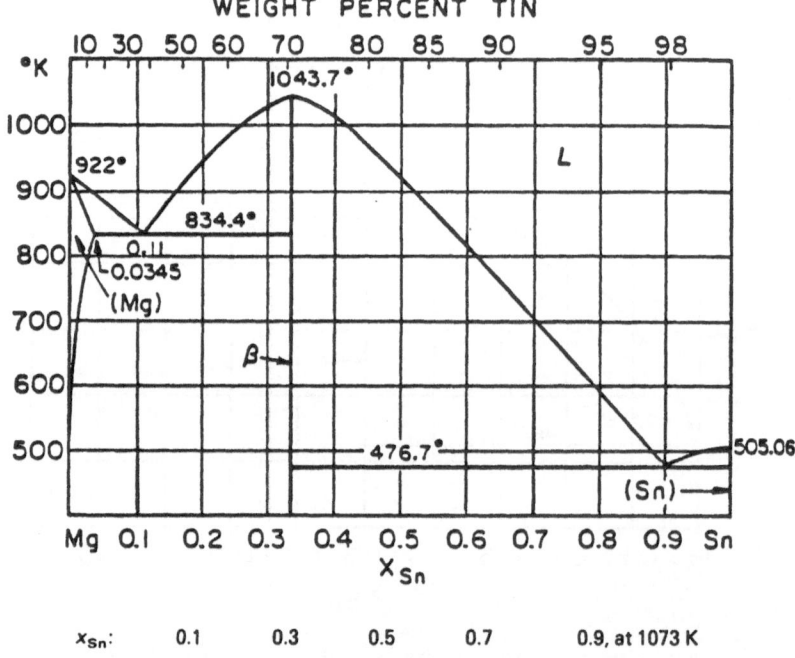

WEIGHT PERCENT TIN

x_{Sn}:	0.1	0.3	0.5	0.7	0.9, at 1073 K
$G^\bullet$:	−1607	−3291	−3385	−2410	−852, cal/g-atom

Figure E.13. Mg–Sn system.

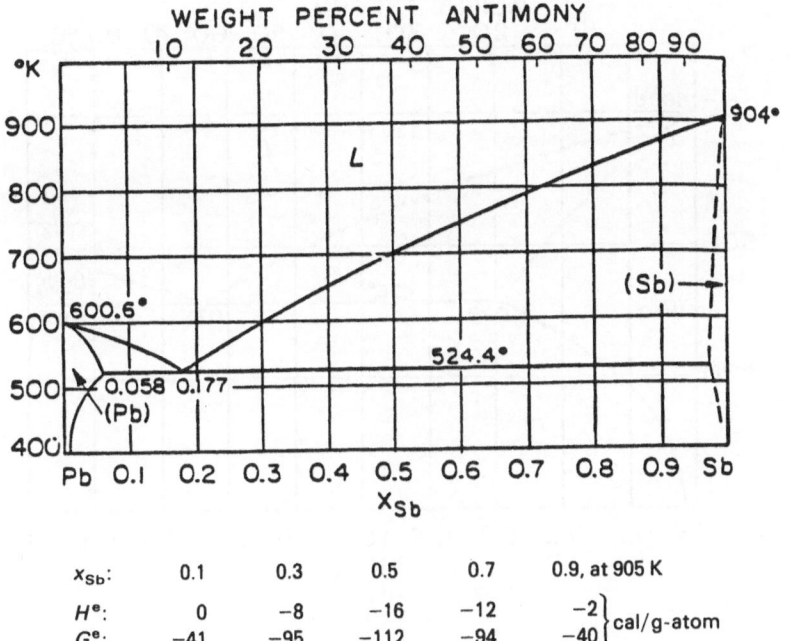

x_{Sb}:	0.1	0.3	0.5	0.7	0.9, at 905 K	
H^e:	0	−8	−16	−12	−2	cal/g-atom
G^e:	−41	−95	−112	−94	−40	

Figure E.14. Pb–Sb system.

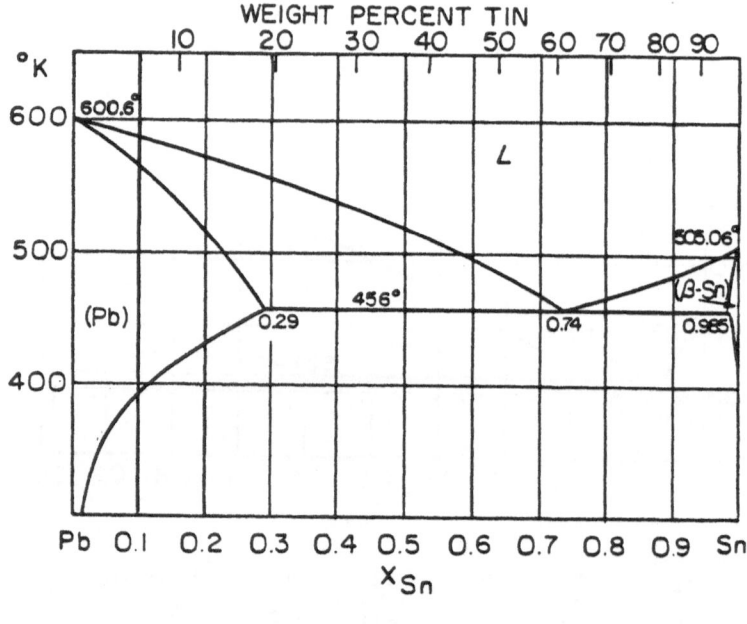

x_{Sn}:	0.1	0.3	0.5	0.7	0.9, at 1050 K
$H°$:	130	285	327	274	120 $\}$ cal/g-atom
$G°$:	323	604	583	416	156

Figure E.15. Pb–Sn system.

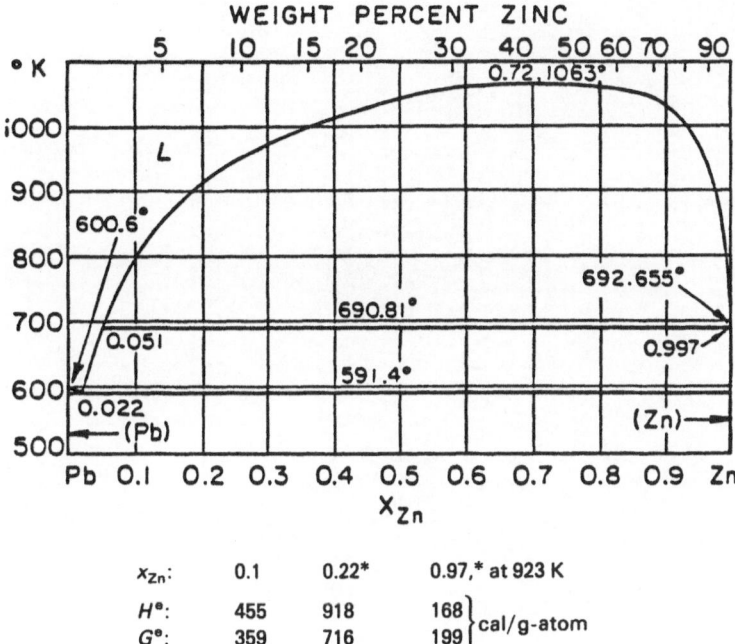

Figure E.16. Pb–Zn system. Asterisks indicate phase boundary.

x_{Zn}:	0.1	0.22*	0.97,* at 923 K
H°:	455	918	168 ⎱ cal/g-atom
G°:	359	716	199 ⎰

References

1. C. J. Smithells and E. A. Brandes, *Metals Reference Book*, Sixth Edition, Butterworths, London (1983). The diagrams for As–Ga, Cu–Fe, Cu–Sn, and Fe–Si are reproduced from this source with permission.
2. R. Hultgren, P. D. Desai, D. T. Hawkins, M. Gleiser, and K. K. Kelley, *Selected Values of the Thermodynamic Properties of Binary Alloys*, ASM, Metals Park, Ohio (1973). The remaining 12 diagrams in this appendix are reproduced from this source with permission.

F

List of Symbols

1. Latin Symbols

(Symbols not defined here are empirical constants usually used on only one or two pages.)

A	Helmholtz energy ($A = E - TS$)
A	amperes; used only after a number or a prefix
$A_{12}, A_{13}, \ldots$	empirical coefficient, equations (1.70)-(1.72)
A, B, ...	components A, B, ...
A_q	empirical coefficient, equations (1.64), (1.65), (1.67), (1.68)
AGN	average group number in alloys. For Ti–V equiatomic alloy, AGN = 4.5, since Ti = 4 and V = 5
a_i	activity of component "i" based on pure "i" as the standard, equation (1.31); a for interstitials in Chapter 6
$\dot{a}_i$	activity of "i" on Henrian scale, equation (1.36)
B	metal B
B_t	empirical coefficient in equation (1.66), cf. equations (1.67) and (1.68)
b	empirical correction term, equation (4.28)
bcc	body-centered cubic structure
C	solar intensity in number of suns, equation (7.71); used in this context only in Chapter 7
C, C_v, C_p	heat capacity; C_v and C_p, heat capacities at constant volume and pressure, respectively; equations (1.4) and (1.6)
C	coulomb; used only after a number or a prefix
°C	temperature in Celsius. (Although use of the degree sign is not currently recommended, it is used in this book to avoid confusion with other C.)

c	empirical correction term, equation (4.27)
c	number of interstitial sites per solvent metal atom in Chapter 6; $c = 1$ for bcc, $c = 3$ for close-packed crystals
c	subscript for number of components in Chapter 2
(c)	crystalline state or phase
$\cancel{c}$	bulk modulus on page 257
cph	close-packed hexagonal structure
D	actual number of configurations
D_c	corrected distribution of particles, equations (4.27) and (4.29)
D_{rm}	random distribution of particles; D_a, D_b, and D_r are defined on page 132
d	total differential
d, s, p	atomic orbitals (shells) in Appendix A
E	molar energy of a system
E_c	energy level of conduction band in semiconductors, Chapter 7
E_f	Fermi energy, Chapter 7
E_g	band-gap energy, $E_g = E_c - E_v$, Chapter 7
E_v	energy level of valence band, $E_v = 0$, Chapter 7
e_{ij}	bond energy for each bond between i and j atoms
e_0	electron on valence band edge
e^-, e^+	electron in conduction band and hole in valence band, respectively
eV	electron volts
F	fugacity defined by equation (1.16)
F_i^*	fugacity over pure condensed "i"
f	activity coefficient defined by equation (6.2)
f_i	activity coefficient of "i" on Henrian scale, equation (1.36)
$f!$	gamma function, i.e., factorials of fractional numbers, equation (4.26)
$f(x_A, V)$	function defined by equation (B.1)
fcc	face-centered cubic structure
fr, frz	subscript for freezing
G	any intensive property; molar Gibbs energy ($G = H - TS$)
$\cancel{G}$	any extensive thermodynamic property; extensive Gibbs energy
$\bar{G}_i$	partial molar Gibbs energy of component "i," equation (1.22)
$G^{\bullet}$	standard molar Gibbs energy on Henrian scale, equations (1.34) and (1.35)
G^0	standard molar Gibbs energy, based on pure component "i," i.e., Raoultian scale
G_0°	standard Gibbs energy of valence band electron
g	distribution of bonds, equation (4.65); g' is given by equation (4.63)

g	defined by equation (5.39)
g_i	degeneracy of ith level
$g(x_A, n_{ws})$	function defined in equation (B.1)
H	molar enthalpy ($H = E + PV$)
H^e	excess molar enthalpy, identical with ΔH of formation of 1 gram atom of alloy from its pure components, e.g., $xA(\text{pure}) + (1 - x) B(\text{pure}) = A - B$ (phase with x atom fraction of A)
$\bar{H}_i$	partial molar enthalpy of "i", equal to $\bar{H}^e$
h	Planck's constant
I	moment of inertia, equation (3.61)
I	current density, Chapter 7
I_1	line current in a solar cell circuit, Chapter 7
I_{rs}	reverse saturation current for a diode, Chapter 7
I_{sc}	short-circuit current in solar cells, Chapter 7
J	rotational quantum number in Chapter 3 only
J	electron flux density, Chapter 7
K	temperature in kelvins, follows numbers and T
K_p	equilibrium constant
k	Boltzmann's constant, i.e., gas constant per particle, R/N_0
L	liquid in phase diagrams
L	vibrational quantum number only in Chapter 3
(l)	liquid after atomic and compound symbols, and thermodynamic properties
ln	natural logarithm, base e
log	common logarithm, base 10
M	molecular weight or mass
M	Maedelung constant
m	mass of a particle
m	meter, used only after a number or a prefix
m	subscript for melting, i.e., fusion
$\langle m \rangle$	geometrical mean mass of carriers equal to $(m_- m_+)^{0.5}$
m_-, m_+	effective masses of electrons and holes, respectively, Chapter 7
m_i, m_j	atomic masses of i and j
m_r	reduced mass, $m_r = m_i m_j/(m_i + m_j)$
N	number of particles in a system; Avogadro's number in Chapter 4
N_-, N_+	densities of quantum states for electrons and holes, respectively, Chapter 7
N_i	occupation number of particles for state i in Chapter 3; number of atoms of component i
N_0	Avogadro's number; sometimes used as N

n	nano; used only as a prefix after a number
n	quantity defined by equation (4.38); $n > 0$; n_g is defined by equation (4.73)
n	number of interstitial atoms in Chapter 6
n_i	number of interstitial atoms in Chapter 7
n_{ws}	Wigner–Seitz electron density in electrons per cm^3, Appendix B
n_-, n_+	numbers of electrons and holes per cm^3 of a semiconductor
P	partial pressure of $H_2(g)$ in Chapter 6 only
P, P_i	pressure in atm (1 atm = 101, 325 Pa); P_i, partial pressure of i
P_i^*	vapor pressure of pure "i"
p	empirical parameter in equation (B.1) in Chapter 7 only
p_-, p_+	momentum of electrons, holes
Q	configurational partition function in Chapter 4, equation (4.32)
Q	partition function for canonical ensemble, Chapter 3
q	dissipative energy, i.e., heat, equation (1.3)
q	partition function for microcanonical ensemble, equation (3.23)
q	empirical parameter in equation (B.1)
q	empirical coefficient in Appendix B
q_-, q_+	partition functions in Chapter 7
q_{th}	thermal energy exchange, equation (3.30)
R	gas constant
R_D	dynamic impedance, equation (7.61)
R_{mp}	maximum power resistance for a solar cell circuit, Chapter 7
r	empirical parameter in equation (B.3)
r, r_0	distance between atoms, equation (3.69) and Fig. 3.2
rot	subscript for rotation, Chapter 3
S	solid in phase diagrams
S	molar entropy defined by equation (1.7)
$\bar{S}_i$	partial molar enthalpy
(s)	solid state
subl	subscript for sublimation
T	temperature in kelvins
t	time in seconds in Chapter 3; temperature in Celcius elsewhere
u	speed (velocity)
u_A	empirical parameter in equation (B.10)
V	molar volume
V	volume of semiconductor in cm^3, Chapter 7
V	molar volume in cm^3 in Appendix B
V_{mp}	maximum voltage in a solar cell, Chapter 7
V_{oc}	open-circuit voltage in a solar cell, Chapter 7
v	subscript for constant volume, e.g., C_v°
v, vap	subscript for vaporization

(v)   vapor state
vib   subscript for vibration in Chapter 3
W, W', W'' various types of work in Chapter 1
W_{ij} exchange energy, equation (4.6)
x_i, x_A, x_B mole fraction or atomic fraction of i, A, B
x_A^s surface atomic fraction of A atoms in Appendix B
Y, Y^* net numbers of molecules with unlike neighbors, equations
                    (4.16) and (4.17)
Y° highest value of Y
y concentration of interstitial atoms as fraction of interstitial sites,
                    i.e., $y = n/cN$, equation (6.1); used without subscript
y_i mole fraction of "i" in the second phase when x_i is used for
                    the first phase; also, y_A and y_B for A and B
y_i function defined by equation (6.85)
Z compressibility factor, PV/RT; used only in Chapter 1
Z coordination number, i.e., number of nearest neighbors of an
                    atom in solid and liquid states

2. Greek Symbols

α phase designation
α defined by equations (3.24) and (3.40), i.e., $\alpha = \bar{G}/kT$
α NW/RT in equation (4.12)
α_s short-range order parameter defined by equation (4.59)
β phase designation
β $1/kT$ in Chapter 3 only
β defined by equation (4.41) and used only in Chapters 4–6
γ_i activity coefficient, equation (1.32); γ_i° when $x_i \to 0$
$\overset{\bullet}{\gamma}$ activity coefficient in equation (6.2)
Δ difference; final state minus initial state
∂ partial differential
δ_A solubility parameter $\delta_A^2 = \Delta E_A/V_A$ where ΔE is enthalpy of
                    vaporization or sublimation
ε_i energy level, only in Chapter 3; energy of an interstitial atom
                    in Chapter 6
$\varepsilon_i^{(j)}$ Wagner interaction parameter, equation (1.91)
η efficiency, percent, equation (7.71)
κ empirical parameter equal to 3400 K, equation (C.1)
Λ defined as e/kT in Chapter 7
λ phase designation

ν	frequency of vibration, per second
Ξ	partition function for grand canonical ensemble
$\prod$	product of terms
$\prod_{mp}$	maximum power in Chapter 7
$\rho(E)$	density of states per unit energy at E, equation (7.8)
$\sum$	summation
σ	sigma phase in Appendix A only
Υ	degrees of freedom, equation (2.17)
ϕ	electronegativity in volts; Appendix A only
ϕ_-, ϕ_+	Fermi distribution function, equations (7.1) and (7.2)
Ψ	Pauling's electronegativity

G

Periodic Table

(Mendeleev Table of the Elements)

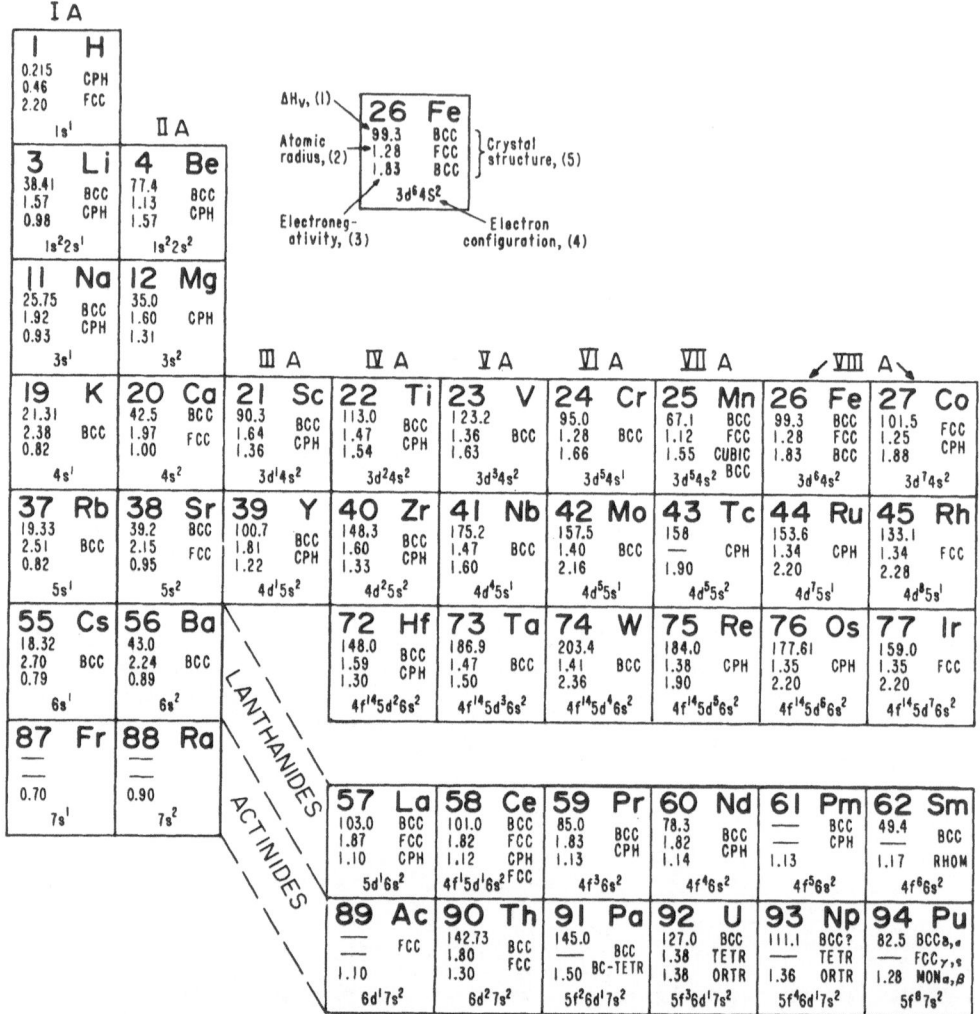

NOTES: (1) ΔH_v (in kcal/mole) is for vaporization or sublimation to monatomic gas at 298.15 K, except for H_2, N_2, O_2, F_2, Cl_2, Br_2, and I_2 to diatomic gases at their boiling points; from Refs. 6.42 and C.18. For diatomic gases at low temperatures, see Rossini *et al.*, *Selected Values of Chemical Thermodynamic Properties*, Circular 500, NBS (1952). (2) Goldschmidt atomic radius (in Å) at lowest temperature; from Ref. E.1. (3) Pauling's electronegativity from Ref. B.7, and *Periodic Table of the Elements* by Sargent–Welch Scientific Co. (1979) as sactioned by NSRDS. Most values are reliable to only one decimal place. See also *McGraw–Hill Encyclopedia of Science and Technology* (1977) where (4) electron configuration is also given. (5) Stable crystal structure in nearest geometrical shape at ambient conditions; structure at highest temperature on top and that at lowest temperature on bottom. From: Hultgren *et al.*, *Selected Values of the*

Group headers: VIII · IIIB · IVB · VB · VIB · VIIB · VIIIA · IB · IIB

Z	Element	Val 1	Val 2	Val 3	Structure(s)	Configuration
2	He	0.02	—	—	BCC / FCC / CPH	$1s^2$
5	B	132.8 RHOM	0.97 TETR	2.04 RHOM		$1s^2 2s^2 p^1$
6	C	171.29	0.77 HEX	2.55		$1s^2 2s^2 p^2$
7	N	1.33	0.71 HEX	3.04 CUBIC		$1s^2 2s^2 p^3$
8	O	1.63 CUBIC	0.60 HEX	3.44 MON		$1s^2 2s^2 p^4$
9	F	1.51	MON 2	3.98 MON 1		$1s^2 2s^2 p^5$
10	Ne	0.41	1.60	—	FCC	$1s^2 2s^2 p^6$
13	Al	78.8	1.43	1.61	FCC	$3s^2 p^1$
14	Si	107.7	1.17	1.90	DIAM	$3s^2 p^2$
15	P	75.62 ORTR	1.09	2.19 CUBIC		$3s^2 p^3$
16	S	66.2 MON	1.04 ORTR	2.58		$3s^2 p^4$
17	Cl	4.88	1.07 ORTR	3.16		$3s^2 p^5$
18	Ar	1.54 CPH	1.92 FCC	—		$3s^2 p^6$
28	Ni	102.8	1.25 FCC	1.91		$3d^8 4s^2$
29	Cu	80.5	1.28 FCC	1.90		$3d^{10} 4s^1$
30	Zn	31.17	1.37 CPH	1.65		$3d^{10} 4s^2$
31	Ga	66.2	1.35 ORTR	1.81		$3d^{10} 4s^2 p^1$
32	Ge	89.5	1.39 DIAM	2.01		$3d^{10} 4s^2 p^2$
33	As	72.12	1.25 RHOM	2.18		$3d^{10} 4s^2 p^3$
34	Se	56.25 MON	1.16 MON	2.55 HEX		$3d^{10} 4s^2 p^4$
35	Br	7.07	1.19 ORTR	2.96		$3d^{10} 4s^2 p^5$
36	Kr	2.16	1.97 FCC	—		$3d^{10} 4s^2 p^6$
46	Pd	90.4	1.37 FCC	2.20		$4d^{10}$
47	Ag	68.09	1.44 FCC	1.93		$4d^{10} 5s^1$
48	Cd	26.73	1.52 CPH	1.69		$4d^{10} 5s^2$
49	In	58.15	1.57 FC-TETR	1.78		$4d^{10} 5s^2 p^1$
50	Sn	71.99 BC-TETR	1.58 TETR	1.96 DIAM		$4d^{10} 5s^2 p^2$
51	Sb	62.7	1.61 RHOM	2.05		$4d^{10} 5s^2 p^3$
52	Te	50.6	1.43 TRI	2.10		$4d^{10} 5s^2 p^4$
53	I	10.02	1.36 ORTR	2.66		$4d^{10} 5s^2 p^5$
54	Xe	3.02	2.18 FCC	—		$4d^{10} 5s^2 p^6$
78	Pt	135.1	1.38 FCC	2.28		$4f^{14} 5d^8 6s^1$
79	Au	87.5	1.44 FCC	2.54		$4f^{14} 5d^{10} 6s^1$
80	Hg	14.67 BC-TETR	1.55 RHOM	2.00		$4f^{14} 5d^{10} 6s^2$
81	Tl	43.25	1.71 BCC	2.04 CPH		$4f^{14} 5d^{10} 6s^2 p^1$
82	Pb	46.75	1.75 FCC	2.33		$4f^{14} 5d^{10} 6s^2 p^2$
83	Bi	49.5	1.82 RHOM	2.02		$4f^{14} 5d^{10} 6s^2 p^3$
84	Po	—	1.40 CUBIC	2.00		$4f^{14} 5d^{10} 6s^2 p^4$
85	At	—	2.20	—		$4f^{14} 5d^{10} 6s^2 p^5$
86	Rn	3.92	—	—	FCC	$4f^{14} 5d^{10} 6s^2 p^6$

Lanthanides / Actinides

Z	Element	Val 1	Val 2	Val 3	Structure(s)	Configuration
63	Eu	44.9	2.04	1.20	BCC	$4f^7 6s^2$
64	Gd	95.0	1.80	1.20	BCC / CPH	$4f^7 5d^1 6s^2$
65	Tb	92.9	1.77	1.20	BCC / CPH	$4f^9 6s^2$
66	Dy	69.4	1.77	1.22	BCC / CPH	$4f^{10} 6s^2$
67	Ho	71.9	1.76	1.23	BCC / CPH	$4f^{11} 6s^2$
68	Er	75.8	1.75	1.24	CPH	$4f^{12} 6s^2$
69	Tm	55.5	1.74	1.25	BCC? / CPH	$4f^{13} 6s^2$
70	Yb	36.35	1.93	1.10	BCC / FCC / CPH	$4f^{14} 6s^2$
71	Lu	102.2	1.73	1.27	BCC / CPH	$4f^{14} 5d^1 6s^2$
95	Am	67.9	1.30	—	FCC? / CPH	$5f^7 7s^2$
96	Cm	92.6	1.30	—	FCC? / CPH	$5f^7 6d^1 7s^2$
97	Bk	—	1.30	—	FCC? / CPH?	$5f^8 7s^2$
98	Cf	—	1.30	—	FCC? / CPH?	$5f^{10} 7s^2$
99	Es	—	1.30	—		$5f^{11} 7s^2$
100	Fm	—	1.30	—		$5f^{12} 7s^2$
101	Md	—	1.30	—		$5f^{13} 7s^2$
102	No	—	1.30	—		$5f^{14} 7s^2$
103	Lr	—	—	—		$5f^{14} 6d^1 7s^2$

Thermodynamic Properties of the Elements, ASM, Metals Park, Ohio (1973); H. W. King, *Bull. Alloy Phase Diagrams* 3(2), 276 (1982); F. L. Oetting, M. H. Rand, and R. J. Ackermann, *The Chemical Thermodynamics of Actinide Elements and Compounds*, Part 1, *The Actinide Elements*, IAEC, Vienna (1976); K. A. Gschneidner and L. R. Eyring, editors, *Handbook on the Physics and Chemistry of Rare Earths*, Volume 1, *Metals*, North-Holland, Amsterdam (1978).

BCC, body-centered cubic: CPH, close-packed hexagonal; DIAM, diamond; FCC, face-centered cubic; HEX, hexagonal; MON, monoclinic; ORTR, orthorhombic; RHOM, rhombic; TETR, tetragonal; TRI, trigonal.

A recent controversial convention recommends designating the groups consecutively from 1 to 18.

Index